1990

SO-AJE-797

Ordinary differential equation

3 0301 00075246 5

ORDINARY
DIFFERENTIAL
EQUATIONS

LIBRARY

LIBRARY
College of St. Francis
JOLIET, ILLINOIS

ORDINARY DIFFERENTIAL EQUATIONS

FOURTH EDITION

Garrett Birkhoff

Harvard University

Gian-Carlo Rota

Massachusetts Institute of Technology

LIBRARY
College of St. Francis
JOLIET, ILLINOIS

WILEY

JOHN WILEY & SONS

New York · Chichester · Brisbane · Toronto · Singapore

Copyright © 1959, 1960, 1962, 1969, 1978, and 1989 by John Wiley & Sons, Inc.

All rights reserved. Published simultaneously in Canada.

Reproduction or translation of any part of
this work beyond that permitted by Sections
107 and 108 of the 1976 United States Copyright
Act without the permission of the copyright
owner is unlawful. Requests for permission
or further information should be addressed to
the Permissions Department, John Wiley & Sons.

Library of Congress Cataloging in Publication Data:

Birkhoff, Garrett, 1911–
  Ordinary differential equations.

Bibliography: p. 392
Includes index.
1. Differential equations. I. Rota, Gian-Carlo,
1932–   . II. Title.
QA372.B58   1989   515.3'52   88-14231
ISBN 0-471-86003-4

Printed in the United States of America

10 9 8 7 6 5 4 3 2 1

575.35
B618
4 ed

PREFACE

The theory of differential equations is distinguished for the wealth of its ideas and methods. Although this richness makes the subject attractive as a field of research, the inevitably hasty presentation of its many methods in elementary courses leaves many students confused. One of the chief aims of the present text is to provide a smooth transition from memorized formulas to the critical understanding of basic theorems and their proofs.

We have tried to present a balanced account of the most important key ideas of the subject in their simplest context, often that of second-order equations. We have deliberately avoided the systematic elaboration of these key ideas, feeling that this is often best done by the students themselves. After they have grasped the underlying methods, they can often best develop mastery by generalizing them (say, to higher-order equations or to systems) by their own efforts.

Our exposition presupposes primarily the calculus and some experience with the formal manipulation of elementary differential equations. Beyond this requirement, only an acquaintance with vectors, matrices, and elementary complex functions is assumed throughout most of the book.

In this fourth edition, the first eight chapters have again been carefully revised. Thus simple numerical methods, which provide convincing empirical evidence for the well-posedness of initial value problems, are already introduced in the first chapter. Without compromising our emphasis on advanced ideas and proofs, we have supplied detailed reviews of elementary facts for convenient reference. Valuable criticisms and suggestions by Calvin Wilcox have helped to eliminate many obscurities and troublesome errors.

The book falls broadly into three parts. Chapters 1 through 4 constitute a review of material to which, presumably, the student has already been exposed in elementary courses. The review serves two purposes: first, to fill the inevitable gaps in the student's mastery of the elements of the subject, and, second, to give a rigorous presentation of the material, which is motivated by simple examples. This part covers elementary methods of integration of first-order, second-order linear, and nth-order linear constant-coefficient, differential equations. Besides reviewing elementary methods, Chapter 3 introduces the concepts of transfer function and the Nyquist diagram with their relation to Green's functions. Although widely used in communications engineering for many years, these concepts are ignored in most textbooks on differential equations. Finally, Chapter

137,703

4 provides rigorous discussions of solution by power series and the method of majorants.

Chapters 5 through 8 deal with systems of nonlinear differential equations. Chapter 5 discusses plane autonomous systems, including the classification of nondegenerate critical points, and introduces the important notion of stability and Liapunov's method, which is then applied to some of the simpler types of nonlinear oscillations. Chapter 6 includes theorems of existence, uniqueness, and continuity, both in the small and in the large, and introduces the perturbation equations.

Chapter 7 gives rigorous error bounds for the methods introduced in Chapter 1, analyzing their rates of convergence. Chapter 8 then motivates and analyzes more sophisticated methods having higher orders of accuracy.

Finally, Chapters 9 through 11 are devoted to the study of second-order linear differential equations. Chapter 9 develops the theory of regular singular points in the complex domain, with applications to some important special functions. In this discussion, we assume familiarity with the concepts of pole and branch point. Chapter 10 is devoted to Sturm-Liouville theory and related asymptotic formulas, for both finite and infinite intervals. Chapter 11 establishes the completeness of the eigenfunctions of regular Sturm-Liouville systems, assuming knowledge of only the basic properties of Euclidean vector spaces (inner product spaces).

Throughout our book, the properties of various important special functions—notably Bessel functions, hypergeometric functions, and the more common orthogonal polynomials—are derived from their defining differential equations and boundary conditions. In this way we illustrate the theory of ordinary differential equations and show its power.

This textbook also contains several hundred exercises of varying difficulty, which form an important part of the course. The most difficult exercises are starred.

It is a pleasure to thank John Barrett, Fred Brauer, Thomas Brown, Nathaniel Chafee, Lamberto Cesari, Abol Ghaffari, Andrew Gleason, Erwin Kreyszig, Carl Langenhop, Norman Levinson, Robert Lynch, Lawrence Markus, Frank Stewart, Feodor Theilheimer, J. L. Walsh, and Henry Wente for their comments, criticisms, and help in eliminating errors.

Garrett Birkhoff
Gian-Carlo Rota

Cambridge, Massachusetts

CONTENTS

FIRST-ORDER DIFFERENTIAL EQUATIONS

1 INTRODUCTION

A *differential equation* is an equation between specified derivatives of an unknown function, its values, and known quantities and functions. Many physical laws are most simply and naturally formulated as differential equations (or DEs, as we will write for short). For this reason, DEs have been studied by the greatest mathematicians and mathematical physicists since the time of Newton.

Ordinary differential equations are DEs whose unknowns are functions of a single variable; they arise most commonly in the study of dynamical systems and electrical networks. They are much easier to treat than *partial* differential equations, whose unknown functions depend on two or more independent variables.

Ordinary DEs are classified according to their order. The *order* of a DE is defined as the largest positive integer, n, for which an nth derivative occurs in the equation. Thus, an equation of the form

$$\phi(x,y,y') = 0$$

is said to be of the *first order*.

This chapter will deal with first-order DEs of the special form

(1) $$M(x,y) + N(x,y)y' = 0$$

A DE of the form (1) is often said to be of the *first degree*. This is because, considered as a polynomial in the derivative of highest order, y', it is of the first degree.

One might think that it would therefore be called "linear," but this name is reserved (within the class of first-order DEs) for DEs of the much more special form $a(x)y' + b(x)y + c(x) = 0$, which are *linear* in y and its derivatives. Such "linear" DEs will be taken up in §3, and we shall call first-order DEs of the more general form (1) *quasilinear*.

A primary aim of the study of differential equations is to find their *solutions*—that is, functions $y = f(x)$ which satisfy them. In this chapter, we will deal with the following special case of the problem of "solving" given DEs.

DEFINITION. A *solution* of (1) is a function $f(x)$ such that $M(x,f(x)) + N(x,f(x))f'(x) = 0$ for all x in the interval where $f(x)$ is defined.

The problem of solving (1) for given functions $M(x,y)$ and $N(x,y)$ is thus to determine all real functions $y = f(x)$ which satisfy (1), that is, all its solutions.

Example 1.   Consider the first-order quasilinear DE

$$(2) \qquad\qquad\qquad x + yy' = 0$$

The solutions of (2) can be found by considering the formula $d(x^2 + y^2)/dx = 2(x + yy')$. Clearly, $y = f(x)$ is a solution of (2) if and only if $x^2 + y^2 = C$ is a constant.

The equation $x^2 + y^2 = C$ defines y *implicitly* as a two-valued function of x, for any positive constant C. Solving for y, we get for each positive constant C *two* solutions, the (single-valued)† functions $y = \pm\sqrt{C - x^2}$. The *graphs* of these solutions, the so-called *solution curves*, form two families of semicircles. These fill the upper half-plane $y > 0$ and the lower half-plane $y < 0$, respectively, in that there is one and only one such semicircle through each point in each half-plane.

Caution.   Note that the functions $y = \pm\sqrt{C - x^2}$ are defined only in the interval $-\sqrt{C} \leq x \leq \sqrt{C}$, and that since y' does not exist (is "infinite") when $x = \pm\sqrt{C}$, these functions are solutions of (1) only on $-\sqrt{C} < x < \sqrt{C}$. Therefore, although the pairs of semicircles in Figure 1.1 *appear* to join together to form the full circle $x^2 + y^2 = C$, the latter is *not* a "solution curve" of (1). In fact, no solution curve of (2) can cross the x-axis (except possibly at the origin), because on the x-axis $y = 0$ the DE (2) implies $x = 0$ for *any* finite y'.

The preceding difficulty also arises if one tries to solve the DE (2) for y'. Dividing through by y, one gets $y' = -x/y$, an equation which cannot be satisfied if $y = 0$. The preceding difficulty is thus avoided if one restricts attention to regions where the DE (1) is normal, in the following sense.

DEFINITION.   A *normal* first-order DE is one of the form

$$(3) \qquad\qquad\qquad y' = F(x,y)$$

In the normal form $y' = -x/y$ of the DE (2), the function $F(x,y)$ is continuous in the upper half-plane $y > 0$ and in the lower half-plane where $y < 0$; it is undefined on the x-axis.

2   FUNDAMENTAL THEOREM OF THE CALCULUS

Although the importance of the theory of (ordinary) DEs stems primarily from its many applications to geometry, science, and engineering, a clear under-

---

† In this book, the word "function" will always mean single-valued function, unless the contrary is expressly specified.

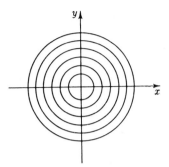

Figure 1.1   Integral curves of $x + yy' = 0$.

standing of its capabilities can only be achieved if its definitions and results are formulated precisely. Some of its most difficult results concern the *existence* and *uniqueness* of solutions. The nature of such existence and uniqueness theorems is well illustrated by the most familiar (and simplest!) class of ordinary DEs. These are the first-order DEs of the very special form

(4) $$y' = g(x)$$

Such DEs are normal; their solutions are described by the fundamental theorem of the calculus, which reads as follows.

FUNDAMENTAL THEOREM OF THE CALCULUS.   *Let the function $g(x)$ in the DE (4) be continuous in the interval $a \leq x \leq b$. Given a number c, there is one and only one solution $f(x)$ of the DE (4) in the interval such that $f(a) = c$. This solution is given by the definite integral*

(5) $$f(x) = c + \int_a^x g(t)\, dt, \qquad c = f(a)$$

This basic result serves as a model of rigorous formulation in several respects. First, it specifies the region under consideration, as a vertical strip $a \leq x \leq b$ in the xy-plane. Second, it describes in precise terms the class of functions $g(x)$ considered. And third, it asserts the *existence* and *uniqueness* of a solution, given the "initial condition" $f(a) = c$.

We recall that the definite integral

(5') $$\int_a^x g(t)\, dt = \lim_{\max \Delta t_k \to 0} \sum g(t_k)\, \Delta t_k, \qquad \Delta t_k = t_k - t_{k-1}$$

is defined for each fixed x as a limit of Riemann sums; it is not necessary to find a formal expression for the indefinite integral $\int g(x)\, dx$ to give meaning to the definite integral $\int_a^x g(t)\, dt$, provided only that $g(t)$ is continuous. Such functions as the *error function* erf $x = (2/\sqrt{\pi}) \int_0^x e^{-t^2} \, dt$ and the *sine integral function* Si $(x) = \int_0^x [(\sin t)/t] \, dt$ are indeed commonly *defined* as definite integrals; cf. Ch. 4, §1.

Quadrature. The preceding considerations enable one to solve DEs of the special form $y' = g(x)$ by inspection: for any a, one solution is the function $\int_a^x g(t) \, dt$; the others are obtained by adding an arbitrary constant C to this "particular" solution. Thus, the solutions of $y' = e^{-x^2}$ are the functions $y = \int e^{-x^2} \, dx = (\sqrt{\pi}/2) \, \text{erf } x + C$; those of $xy' = \sin x$ are the functions $y = \text{Si}(x) + C$; and so on. Note that from any one solution curve of $y' = g(x)$, the others are obtained by the vertical translations $(x,y) \mapsto (x, y + C)$.† Thus, they form a one-parameter family of curves, one for each value of the parameter C. This important geometrical fact is illustrated in Figure 1.2.

After $y' = f(x)$, the simplest type of DE is $y' = g(y)$. Any such DE is invariant under horizontal translation $(x,y) \mapsto (x + c, y)$. Hence, any horizontal line is cut by all solution curves at the same angle (such lines are called "isoclines"), and any horizontal translate $y = \phi(x + c)$ of any solution curve $y = \phi(x)$ is again a solution curve.

The DE $y' = y$ is the most familiar DE of this form. It can be solved by rewriting it as $dy/y = dx$; integrating, we get $x = \ln |y| + c$, or $y = \pm e^{x-c}$, where c is an arbitrary constant. Setting $k = \pm e^{-c}$, we get the general solution $y = ke^x$— but the solution $y = 0$ is "lost" until the last step.

Example 2. A similar procedure can be applied to any DE of the form $y' = g(y)$. Thus consider

(6) $$y' = y^2 - 1$$

Since $y^2 - 1 = (y + 1)(y - 1)$, the constant functions $y = -1$ and $y = 1$ are particular solutions of (6). Since $y^2 > 1$ if $|y| > 1$ whereas $y^2 < 1$ if $-1 < y$

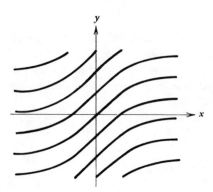

Figure 1.2 **Solution curves of $y' = e^{-x^2}$.**

---

† The symbol $\mapsto$ is to be read as "goes into".

< 1, all solutions are decreasing functions in the strip $|y| < 1$ and increasing functions outside it; see Figure 1.3.

Using the partial fraction decomposition $2/(y^2 - 1) = 1/(y - 1) - 1/(y + 1)$, one can rewrite (6) as $2\,dx = dy/(y - 1) - dy/(y + 1)$ from which we obtain, by integrating, $2(x - c) = \ln |(y - 1)/(y + 1)|$. Exponentiating both sides, we get $\pm e^{2(x-c)} = (y - 1)/(y + 1)$, which reduces after some manipulation to

(6')
$$y = \frac{1 \pm e^{2(x-c)}}{1 \mp e^{2(x-c)}} = \begin{Bmatrix} \tanh \\ \coth \end{Bmatrix} (c - x)$$

This procedure "loses" the special solutions $y = 1$ and $y = -1$, but gives all others. Note that if $y = f(x)$ is a solution of (6), then so is $1/y = 1/f(x)$, as can be directly verified from (6) (provided $y \neq 0$).

Example 3. A more complicated DE tractable by the same methods is $y' = y^3 - y$. Since $y^3 - y = y(y + 1)(y - 1)$, the constant functions $y = -1, y = 0$, and $y = 1$ are particular solutions. Since $y^3 > y$ if $-1 < y < 0$ or $1 < y$, whereas $y^3 < y$ if $y < -1$ or $0 < y < 1$, all solutions are increasing functions in the strips $-1 < y < 0$ and $y > 1$, and decreasing in the complementary strips.

To find the other solutions, we replace the DE $y' = dy/dx = y^3 - y$ by its reciprocal, $dx/dy = 1/(y^3 - y)$. We then use partial fractions to obtain the DE

(6'')
$$\frac{dx}{dy} = \frac{1}{y^3 - y} = \frac{1}{2}\left\{ \frac{1}{y + 1} + \frac{1}{y - 1} - \frac{2}{y} \right\}$$

The DE (6'') can be integrated termwise to give, after some manipulation, $x = \tfrac{1}{4}\ln |1 - y^{-2}| + C$, or $y = \pm[1 \mp \exp(2x - k)]^{-1/2}$, $k = 2c$.

Symmetry. The labor of drawing solution curves of the preceding DEs is reduced not only by their invariance under horizontal translation, but by the use of other symmetries as well. Thus, the DEs $y' = y$ and $y' = y^3 - y$ are invariant under reflection in the x-axis [i.e., under $(x,y) \mapsto (x, -y)$]; hence, so are their solution curves. Likewise, the DEs $y' = 1 + y^2$ and $y' = y^2 - 1$ (and their solution curves) are invariant under $(x,y) \mapsto (-x, -y)$—i.e., under rotation through 180° about the origin. These symmetries are visible in Figures 1.3 and 1.4.

EXERCISES A

1. (a) Show that if $f(x)$ satisfies (6), then so do $1/f(x)$ and $-f(-x)$.
   (b) Explain how these facts relate to Figure 1.2.

2. Show that every solution curve (6') of (6) is equivalent under horizontal translation and/or reflection in the x-axis to $y = (1 + e^{2x})/(1 - e^{2x})$ or to $y = (1 - e^{2x})/(1 + e^{2x})$.

3. (a) Show that if $y' = y^2 + 1$, then y is an increasing function and $x = \arctan y + c$.
   (b) Infer that no solution of $y' = y^2 + 1$ can be defined on an interval of length exceeding π.

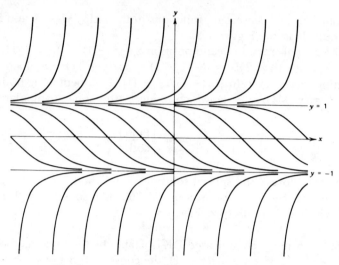

Figure 1.3   Solution curves of $y' = y^2 - 1$.

(c) Show that a nonhorizontal solution curve of $y' = y^2 \pm 1$ has a point of inflection on the x-axis and nowhere else.

4. Show that the solution curves of $y' = y^2$ are the x-axis and rectangular hyperbolas having this for one asymptote. [HINT: Rewrite $y' = y^2$ as $dy/y^2 = dx$.]

5. Sketch sample solution curves to indicate the qualitative behavior of the solutions of the following DEs: (a) $y' = 1 - y^3$, (b) $y' = \sin \pi y$, (c) $y' = \sin^2 y$.

6. Show that the solutions of $y' = g(y)$, for any continuous function g, are either all increasing functions or all decreasing functions in any strip $y_{i-1} < y < y_j$ between successive zeros of $g(y)$ [i.e., values y_j, such that $g(y_j) = 0$].

7. Show that the solutions of $y' = g(y)$ are convex up or convex down for given y according as $|g|$ is an increasing or decreasing function of y there.

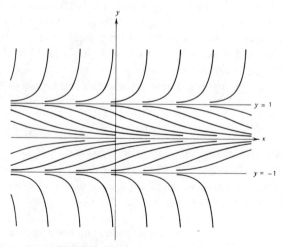

Figure 1.4   Solution curves of $y' = y^3 - y$.

*8. (a) Prove in detail that any nonconstant solution of (6) must satisfy

$$x = c + \tfrac{1}{2}\ln|(y-1)/(y+1)|$$

(b) Solve (6″) in detail, discussing the case $k = 0$ and the limiting case $k = \infty$ ($y = 0$).

*9. (a) Show that the choice $k < 0$ in (6′) gives solutions in the strip $-1 < y < 1$.

(b) Show that the choice $k = 1$ gives two solutions having the positive and negative y-axes for asymptotes, respectively.

3 FIRST-ORDER LINEAR EQUATIONS

In the next five sections, we will recall some very elementary, but extremely useful methods for solving important special families of first-order DEs. We begin with the first-order linear DE

$$(7) \qquad\qquad a(x)y' + b(x)y + c(x) = 0$$

It is called *homogeneous* if $c(x) \equiv 0$, and *inhomogeneous* otherwise.

Let the coefficient functions a, b, c be continuous. In any interval I where $a(x)$ does not vanish, the linear DE (7) can be reduced to the normal form

$$(8) \qquad\qquad y' = -p(x)y - q(x)$$

with continuous coefficient functions $p = b/a$ and $q = c/a$.

The homogeneous linear case $y' = -p(x)y$ of (8) is solved easily, if not rigorously, as follows. We separate variables, $dy/y = -p(x)\,dx$; then we integrate (by quadratures), $\ln|y| = -\int p(x)\,dx + C$. Exponentiating both sides, we obtain $|y| = Ke^{-\int p(x)dx}$, where $K = e^C$ and any indefinite integral $P(x) = \int p(x)\,dx$ may be used.

This heuristic reasoning suggests that, if $P'(x) = p(x)$, then $ye^{P(x)}$ is a constant. Though this result was derived heuristically, it is easily verified rigorously:

$$d[ye^{P(x)}]/dx = y'e^{P(x)} + p(x)ye^{P(x)} = e^{P(x)}[y' + p(x)y] = 0$$

if and (since $e^{P(x)} \neq 0$) only if y satisfies (8). This proves the following result.

THEOREM 1. *If $P(x) = \int p(x)\,dx$ is an indefinite integral of the continuous function p, then the function $ce^{-P(x)} = ce^{-\int p(x)dx}$ is a solution of the DE $y' + p(x)y = 0$ for any constant c, and all solutions of the DE are of this form.*

---

* The more difficult exercises in this book are starred.

We can treat the general case of (8) similarly. Differentiating the function $e^{P(x)}y$, where $P(x)$ is as before, we get

$$d[e^{P(x)}y]/dx = e^{P(x)}[y' + p(x)y] = -e^{P(x)}q(x)$$

It follows that, for some constant y_0, we must have $e^{P(x)}y = y_0 - \int_a^x e^{P(t)}q(t)\, dt$, whence

(8')
$$y = y_0 e^{-P(x)} - e^{-P(x)} \int_a^x e^{P(t)}q(t)\, dt$$

Conversely, formula (8') defines a solution of (8) with $y(a) = y_0$ for every y_0, by the Fundamental Theorem of the Calculus. This proves

THEOREM 2.   *If $P(x)$ is as in Theorem 1, then the general solution of the* DE (8) *is given by* (8'). *Moreover,* $y_0 = y(a)$ *if and only if* $P(x) = \int_a^x p(x)dx$.

Quadrature.   In the Fundamental Theorem of the Calculus, if the function g is nonnegative, the definite integral in (5) is the area under the curve $y = g(x)$ in the vertical strip between a and x. For this reason, the integration of (4) is called a *quadrature*. Formula (8') reduces the solution of any first-order linear DE to the performance of a sequence of quadratures. Using Tables of Indefinite Integrals,† the solutions can therefore often be expressed explicitly, in terms of "elementary" functions whose numerical values have been tabulated ("tabulated functions").

Initial Value Problem.   In general, the "initial value problem" for a first-order DE $y' = F(x,y)$ consists in finding a solution $y = g(x)$ that satisfies an initial condition $y(a) = y_0$, where a and y_0 are given constants. Theorem 2 states that the initial value problem always has one and only one solution for a *linear* DE (8), on any interval $a \leqq x \leqq b$ where $p(x)$ and $q(x)$ are defined and continuous.

Remark.   There are often easier ways to solve linear DEs than substitution in (8'). This fact is illustrated by the following example.

Example 4.   Consider the inhomogeneous linear DE

(9)
$$y' + y = x + 3$$

Trying $y = ax + b$, one easily verifies that $x + 2$ is one solution of (9). On the other hand, if $y = f(x)$ is any other solution, then $z = y - (x + 2)$ must satisfy $z' + z = (y' + y) - (x + 3) = 0$, whence $z = ce^{-x}$ by Theorem 1. It follows that the general solution of (9) is the sum $ce^{-x} + x + 2$.

---

† See the book by Dwight listed in the Bibliography. Kamke's book listed there contains an extremely useful catalog of solutions of DEs not of the form $y' = g(x)$. For a bibliography of function tables, see Fletcher, Miller, and Rosenhead.

4   SEPARABLE EQUATIONS

A differential equation that can be written in the form

(10) $$y' = g(x)h(y)$$

is said to be *separable*. Thus, the DEs $y' = y^2 - 1$ and $y' = y^3 - y$ of Examples 2 and 3 are obviously separable, with $g(x) = 1$. The DE $x + yy' = 0$ of Example 1, rewritten as $y' = (-x)(1/y)$ is separable except on the x-axis, where $1/y$ becomes infinite. As we have seen, the solutions $y = \pm\sqrt{C - x^2}$ of this DE cannot be expressed as single-valued functions of x on the x-axis, essentially for this reason.

A similar difficulty arises in general for DEs of the form

(11) $$M(x) + N(y)y' = 0$$

These can also be rewritten as

(11') $$M(x)\, dx + N(y)\, dy = 0$$

or as $y' = -M(x)/N(y)$ and are therefore also said to be "separable." Whenever $N(y)$ vanishes, it is difficult or impossible to express y as a function of x.

It is easy to solve separable DEs formally. If $\phi(x) = \int M(x)\, dx$ and $\psi(y) = \int N(y)\, dy$ are any antiderivatives ("indefinite integrals") of $M(x)$ and $N(y)$, respectively, then the *level curves*

$$\phi(x) + \psi(y) = C$$

of the function $U(x,y) = \phi(x) + \psi(y)$ are solution curves of the DEs (11) and (11'). Moreover, the Fundamental Theorem of the Calculus assures us of the *existence* of such antiderivatives. Likewise, for any indefinite integrals $G(x) = \int g(x)\, dx$ and $H(y) = \int dy/h(y)$, the level curves of

$$G(x) - H(y) = C$$

may be expected to define solutions of (10), of the form

(11'') $$y = H^{-1}[C - G(x)]$$

However, the solutions defined in this way are only *local*. They are defined by the Inverse Function Theorem,† but only in *intervals of monotonicity* of $H(y)$ where $h(y)$ and hence $H'(y) = 1/h(y)$ has constant sign. Moreover, the range of $H(y)$ may be bounded, as in the case of the DE $y' = 1 + y^2$. In this case,

---

† This theorem states that if $H(y)$ is a strictly monotonic map of $[c,d]$ onto $[a,b]$, then $H^{-1}(y)$ is single-valued and monotonic from $[a,b]$ to $[c,d]$.

$\int_{-\infty}^{\infty} dy/(1 + y^2) = \pi$. Therefore, *no* solution of the DE $y' = 1 + y^2$ can be continuously defined over an interval (a,b) of length exceeding π.

Example 5. Consider the DE $y' = (1 + y^2)e^{-x^2}$. Separating variables, we get $\int dy/(1 + y^2) = \int e^{-x^2} dx$, whose general solution is arctan $y = (\sqrt{\pi}/2)$ erf $x + C$, or $y = \tan\{(\sqrt{\pi}/2)$ erf $(x) + C\}$.

The formal transformations (10′) and (10″) can be rigorously justified whenever $g(x)$ and $h(y)$ are *continuous* functions, in any interval in which $h(y)$ does not vanish. This is because the Fundamental Theorem of the Calculus again assures us that $\phi(x) = \int g(x) \, dx$ exists and is differentiable on any interval where $g(x)$ is defined and continuous, while $\psi(y) = \int dy/h(y)$ exists and is *strictly monotonic* in any interval (y_1,y_2) between successive zeros y_1 and y_2 of $h(y)$, which we also assume to be continuous. Hence, as in Example 2, the equation

$$\psi(y) - \phi(x) = \int \frac{dy}{g(y)} - \int h(x) \, dx = c$$

gives for each c a solution of $y' = g(x)h(y)$ in the strip $y_1 < y < y_2$. Near any x with $y_1 - c < \phi(x) < y_2 - c$, this solution is defined by the inverse function theorem, by the formula $y = \psi^{-1}(\phi(x) + c)$.

Orthogonal Trajectories.   An *orthogonal trajectory* to a family of curves is a curve that cuts all the given curves at right angles. For example, consider the family of geometrically similar, coaxial ellipses $x^2 + my^2 = C$. These are integral curves of the DE $x + myy' = 0$, whose normal form $y' = -x/my$ has separable variables. The orthogonal trajectories of these ellipses have at each point a slope $y' = my/x$, which is the negative reciprocal of $-x/my$. Separating variables, we get $dy/y = m \, dx/x$, or $\ln |y| = m \ln |x|$, whence the orthogonal trajectories are given by $y = \pm|x|^m$.

More generally, the solution curves of any separable DE $y' = g(x)h(y)$ have as orthogonal trajectories the solution curves of the separable DE $y' = -1/g(x)h(y)$.

Critical Points.   Points where $\partial u/\partial x = \partial u/\partial y = 0$ are called *critical points* of the function $u(x,y)$. Note that the directions of level lines and gradient lines may be very irregular near critical points; consider those of the functions $x^2 \pm y^2$ near their critical point $(0,0)$.

As will be explained in §5, the level curves of any function $u \in \mathcal{C}^1(D)$ satisfy the DE $\partial u/\partial x + y'\partial u/\partial y = 0$ in D, except at critical points of u. Clearly, their orthogonal trajectories are the solution curves of $\partial u/\partial y = y'\partial u/\partial x$, and so are everywhere tangent to the direction of $\nabla u = \text{grad } u = (\partial u/\partial x, \partial u/\partial y)$. Curves having this property are called *gradient curves* of u. Hence the gradient curves of u are orthogonal trajectories of its level curves, except perhaps at critical points.

EXERCISES B

1. Find the solution of the DE $xy' + 3y = 0$ that satisfies the initial condition $f(1) = 1$.

2. Find equations describing all solutions of $y' = (x + y)^2$. [HINT: Set $u = x + y$.]

3. (a) Find all solutions of the DE $xy' + (1 - x)y = 0$
   (b) Same question for $xy' + (1 - x)y = 1$.

4. (a) Solve the DEs of Exercise 3 for the initial conditions $y(1) = 1$, $y(1) = 2$.
   (b) Do the same for $y(0) = 0$ and $y(0) = 1$, *or* prove that no solution exists.

5. (a) Find the general solution of the DE $y' + y = \sin 2t$.
   (b) For arbitrary (real) constants a, b, and $k \neq 0$, find a particular solution of

   (*) $y' = ay + b \sin kt$

   (c) What is the general solution of (*)?

6. (a) Find a polynomial solution of the DE

   (**) $y' + 2y = x^2 + 4x + 7$

   (b) Find a solution of the DE (*) that satisfies the initial condition $y(0) = 0$.

7. Show that if k is a nonzero constant and $q(x)$ a polynomial of degree n, then the DE $xy' + y = q(x)$ has exactly one polynomial solution of degree n.

In Exs. 8 and 9, solve the DE shown and discuss its solutions qualitatively.

8. $dr/d\theta = r^2 \sin 1/r$ (polar coordinates).

9. $dr/d\theta = 2/\log r$.

10. (a) Show that the ellipses $5x^2 + 6xy + 5y^2 = C$ are integral curves of the DE

   $$(5x + 3y) + (3x + 5y)y' = 0$$

   (b) What are its solution curves?

5   QUASILINEAR EQUATIONS; IMPLICIT SOLUTIONS

In this section and the next, we consider the general problem of solving *quasilinear* DEs (1), which we rewrite as

(12) $M(x,y)\ dx + N(x,y)\ dy = 0$

to bring out the latent symmetry between the roles of x and y. Such DEs arise naturally if we consider the *level curves* of functions. If $G(x,y)$ is any continuously differentiable function, then the DE

(12′) $\dfrac{\partial G}{\partial x}(x,y) + \dfrac{\partial G}{\partial y}(x,y)\,y' = 0$

is satisfied on any level curve $G(x,y) = C$, at all points where $\partial G/\partial y \neq 0$. This DE is of the form (1), with $M(x,y) = \partial G/\partial x$ and $N(x,y) = \partial G/\partial y$.

For this reason, any function G which is related in the foregoing way to a quasilinear DE (1) or (12), or to a nonzero *multiple* of (12) of the form

(12″) $\mu(x,y)[M(x,y)\ dx + N(x,y)\ dy] = 0, \qquad \mu \neq 0$

is called an *implicit solution* of (12). Slightly more generally, an *integral* of (1) or (12) is defined as a function $G(x,y)$ of two variables that is constant on every solution curve of (1).

For example, the equation $x^4 - 6x^2y^2 + y^4 = C$ is an implicit solution of the quasilinear DE

$$(x^3 - 3xy^2) + (y^3 - 3x^2y)y' = 0$$

or
$$y' = \frac{(x^3 - 3xy^2)}{(3x^2y - y^3)}$$

The level curves of $x^4 - 6x^2y^2 + y^4$ have vertical tangents on the x-axis and the lines $y = \pm\sqrt{3}x$. Elsewhere, the DE displayed above is of the normal form $y' = F(x,y)$.

Critical Points.   At points where $\partial\phi/\partial x = \partial\phi/\partial y = 0$, the directions of the gradient and level curves are undefined; such points are called "critical points" of ϕ. Thus, the function $x^2 + y^2$ has the origin for its only critical point, and the same is true of the function $x^4 - 6x^2y^2 + y^4$. (Can you prove it?) On the other hand, the function $\sin(x^2 + y^2)$ also has circles of critical points, occurring wherever r^2 is an odd integral multiple of $\pi/2$. Most functions have only *isolated* critical points, however, and in general we shall confine our attention to such functions.

We will now examine more carefully the connection between quasilinear DEs and level curves of functions, illustrated by the two preceding examples. To describe it accurately, we will need two more definitions. We first define a *domain*† as a nonempty *open connected set*. We call a function $\phi = \phi(x_1, \ldots, x_r)$ of *class* $\mathcal{C}^n$ in a domain D when all its derivatives $\partial\phi/\partial x_i$, $\partial^2\phi/\partial x_i\partial x_j$, ... of orders $1, \ldots, n$ exist and are continuous in D. We will write this condition in symbols as $\phi \in \mathcal{C}^n$ or $\phi \in \mathcal{C}^n(D)$. When ϕ is merely assumed to be continuous, we will write $\phi \in \mathcal{C}$ or $\phi \in \mathcal{C}(D)$.

To make the connection between level curves and quasilinear DEs rigorous, we will also need to assume the following basic theorem.

---

† See Apostol, Vol. 1, p. 252. Here and later, page references to authors refer to the books listed in the selected bibliography.

IMPLICIT FUNCTION THEOREM.† *Let $u(x,y)$ be a function of class C^n ($n \geq$ 1) in a domain containing (x_0,y_0); let u_0 denote $u(x_0,y_0)$, and let $u_y(x_0,y_0) \neq 0$. Then there exists positive numbers ϵ and η such that for each $x \in (x_0 - \epsilon, x_0 + \epsilon)$ and $C \in (u_0 - \epsilon, u_0 + \epsilon)$, the equation $u(x,y) = C$ has a unique solution $y = f(x,C)$ in the interval $(y_0 - \eta, y_0 + \eta)$. Moreover, the function f so defined is also of class C^n.*

It follows that if $u \in C^n(D)$, $n \geq 1$, the level curves of u are graphs of functions $y = f(x,c)$, also of class C^n, except where $\partial u/\partial y = 0$. In Example 1, $u = x^2 + y^2$ and there is one such curve, the x-axis $y = 0$; this divides the plane into two subdomains, the half-planes $y > 0$ and $y < 0$. Moreover, the locus (set) where $\partial u/\partial y = 0$ consists of the points where the circles $u = $ const have vertical tangents and the "critical point" $(0,0)$ where $\partial u/\partial x = \partial u/\partial y = 0$—that is, where the surface $z = u(x,y)$ has a horizontal tangent plane.

This situation is typical: for most functions $u(x,y)$, the partial derivative $\partial u/\partial y$ vanishes on isolated curves that divide the (x,y)-plane into a number of regions where $\partial u/\partial y \neq 0$ has constant sign, and hence in which the Implicit Function Theorem applies.

THEOREM 3. *In any domain where $\partial u/\partial y \neq 0$, the level curves of any function $u \in C^1$ are solution curves of the quasilinear DE*

$$(13) \qquad \phi(x,y,y') = M(x,y) + N(x,y)y' = 0$$

where $M(x,y) = \partial u/\partial x$ and $N(x,y) = \partial u/\partial y$.

Proof. By the Chain Rule, $du/dx = \partial u/\partial x + (\partial u/\partial y)y'$ along any curve $y = f(x)$. Hence, such a curve is a level curve of u if and only if

$$\frac{du}{dx} = \frac{\partial u}{\partial x} + \frac{\partial u}{\partial y}y' = 0$$

By the Implicit Function Theorem, the level curves of u, being graphs of functions $y = f(x)$ in domains where $\partial f/\partial y \neq 0$, are therefore solution curves of the quasilinear DE (13). In the normal form $y' = F(x,y)$ of this DE, therefore, $F(x,y) = -(\partial u/\partial x)/(\partial u/\partial y)$ becomes infinite precisely when $\partial u/\partial y = 0$.

To describe the relationship between the DE (13) and the function u, we need a new notion.

DEFINITION. An *integral* of a first-order quasilinear DE (1) is a function of two variables, $u(x,y)$, which is constant on every solution curve of (1).

Thus, the function $u(x,y) = x^2 + y^2$ is an integral of the DE $x + yy' = 0$

---

† Courant and John, Vol. 2, p. 218. We will reconsider the Implicit Function Theorem in greater depth in §12.

because, upon replacing the variable y by any function $\pm \sqrt{C - x^2}$, we obtain $u(x,y) = C$. This integral is most easily found by rewriting $x + y\, dy/dx = 0$ in differential form, as $x\, dx + y\, dy = 0$, and recognizing that $x\, dx + y\, dy = \frac{1}{2}d(x^2 + y^2)$ is an "exact" differential (see §6).

Level curves of an integral of a quasilinear DE are called *integral curves* of the DE; thus, the circles $x^2 + y^2 = C$ are integral curves of the DE $x + yy' = 0$, although not solution curves.

Example 6.   From the DE $yy' = x$, rewritten as $y\, dy/dx = x$, we get the equation $y\, dy - x\, dx = 0$. Since $y\, dy - x\, dx = \frac{1}{2}d(y^2 - x^2)$, we see that the integral curves of the DE are the branches of the hyperbolas $y^2 = x^2 + C$ and the asymptotes $y = \pm x$, as shown in Figure 1.5. The branches $y = \pm \sqrt{x^2 + k^2}$ are solution curves, but each level curve $y = \pm \sqrt{x^2 - k^2}$ has *four* branches separated by the x-axis (the line where the integral curves have vertical tangents).

Note that, where the level curves $y = x$ and $y = -x$ of $y^2 - x^2$ cross, the *gradient* $(\partial F/\partial x, \partial F/\partial y)$ of the integral $F(x,y) = y^2 - x^2$ vanishes: $(\partial F/\partial x, \partial F/\partial x) = (0,0)$.

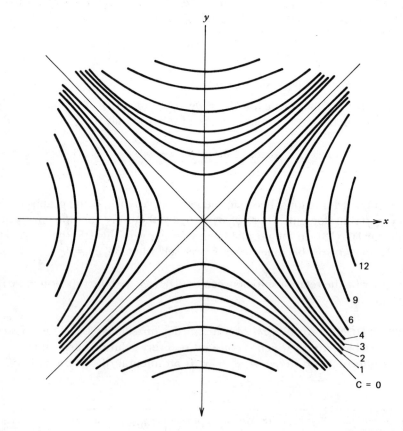

Figure 1.5   Level curves c = 0, ±1, ±2, ±3, ±4, ±6, ±9, ±12 of $y^2 - x^2$.

6 EXACT DIFFERENTIALS; INTEGRATING FACTORS

A considerably larger class of "implicit solutions" of quasinormal DEs can be found by examining more closely the condition that $M(x,y)\, dx + N(x,y)\, dy$ be an "exact differential" dU, and by looking for an "integrating factor" $\mu(x,y)$ that will convert the equation

$$(14) \qquad\qquad M(x,y)\, dx + N(x,y)\, dy = 0$$

into one involving a "total" or "exact" differential

$$\mu dU = \mu(x,y)[M(x,y)\, dx + N(x,y)\, dy] = 0$$

whose (implicit) solutions are the level curves of U.

In general, the quasinormal DE (1) or

$$(14') \qquad\qquad M(x,y) + N(x,y)y' = 0$$

is said to be *exact* when there exists a function $U(x,y)$ of which it is the 'total differential', so that $\partial U/\partial x = M(x,y)$ and $\partial U/\partial y = N(x,y)$, or equivalently

$$(14'') \qquad dU = \frac{\partial U}{\partial x}\, dx + \frac{\partial U}{\partial y}\, dy = M(x,y)\, dx + N(x,y)\, dy$$

Since $dU = 0$ on any solution curve of the DE (14), we see that solution curves of (14) must lie on level curves of U, just as in the "separable variable" case.

Since $\partial^2 U/\partial x \partial y = \partial^2 U/\partial y \partial x$, clearly a necessary condition for (14') to be an exact differential is that $\partial N/\partial x = \partial M/\partial y$. It is shown in the calculus that the converse is also true locally. More precisely, the following result is true.

THEOREM 4. *If $M(x,y)$ and $N(x,y)$ are continuously differentiable functions in a simply connected domain, then (14') is an exact differential if and only if $\partial N/\partial x = \partial M/\partial y$.*

The function $U = U(P)$ for (14) is constructed as the *line integral* $\int_0^P [M(x,y)dx + N(x,y)\, dy]$ from a fixed point 0 in the domain [perhaps $0 = (0,0)$] to a variable point $P = (x,y)$. Thus, for the DE $x + yy' = 0$ of Example 1, this procedure gives $\int_0^P (x\, dx + y\, dy) = (x^2 + y^2)/2$, showing again that the solution curves of $x + yy' = 0$ lie on the circles $x^2 + y^2 = C$ with center $(0,0)$. More generally, in the separable equation case of $g(x)\, dx + dy/h(y)$, we have $\partial[g(x)]/\partial y = 0 = \partial[1/h(y)]/\partial x$, giving $G(x) + H(y) = C$ as in §5.

Even when the differential $M\, dx + N\, dy$ is not exact, one can often find a function $\mu(x,y)$ such that the product

$$(\mu M)\, dx + (\mu N)\, dy = du$$

is an exact differential. The contour lines $u(x,y) = C$ will then again be integral curves of the DE $M(x,y) + N (x,y)y' = 0$ because $du/dx = \mu(M + Ny') = 0$; and segments of these contour lines between points of vertical tangency will be solution curves. Such a function μ is called an integrating factor.

   DEFINITION. An *integrating factor* for a differential $M(x,y) \, dx + N(x,y) \, dy$ is a nonvanishing function $\mu(x,y)$ such that the product $(\mu M) \, dx + (\mu N) \, dy$ is an exact differential.

   Thus, as we saw in §3, for any indefinite integral $P(x) = \int p(x) \, dx$ of $p(x)$, the function $\exp \{P(x)\}$ is an integrating factor for the linear DE (8). Likewise, the function $1/h(x)$ is an integrating factor for the separable DE (11).

   The differential $x \, dy - y \, dx$ furnishes another interesting example. It has an integrating factor in the right half-plane $x > 0$ of the form $\mu(x) = 1/x^2$, since $dy/x - y \, dx/x^2 = d(y/x)$; cf. Ex. C11. A more interesting integrating factor is $1/(x^2 + y^2)$. Indeed, the function

$$\theta(x,y) = \int_{(1,0)}^{(x,y)} \frac{(x \, dy - y \, dx)}{(x^2 + y^2)}$$

is the angle made with the positive x-axis by the vector (x,y). That is, it is just the polar angle θ when the point (x,y) is expressed in polar coordinates. Therefore, the integral curves of $xy' = y$ in the domain $x > 0$ are the radii $\theta = C$, where $-\pi/2 < \theta < \pi/2$; the solution curves are the same.

   Note that the differential $(x \, dy - y \, dx)/(x^2 + y^2)$ is not exact in the punctured plane, consisting of the x,y-plane with the origin deleted. For θ changes by 2π in going around the origin. This is possible, even though $\partial[x/(x^2 + y^2)]/\partial x = \partial [-y/(x^2 + y^2)]/\partial y$, because the punctured plane is not a simply connected domain.

   Still another integrating factor of $x \, dy - y \, dx$ is $1/xy$, which replaces $x \, dy - y \, dx = 0$ by $dy/y = dx/x$, or $\ln |y| = \ln |x| + C$ in the interior of each of the four quadrants into which the coordinate axes divide the (x,y)-plane. Exponentiating both sides, we get $y = kx$.

   A less simple example concerns the DE $x(x^3 - 2y^3)y' = (2x^3 - y^3)y$. Here an integrating factor is $1/x^2y^2$. If we divide the given DE by x^2y^2, we get

$$\frac{d}{dx} \left(\frac{x^2}{y} + \frac{y^2}{x} \right) = \frac{2x^3y - x^4y' - y^4 + 2xy^3y'}{x^2y^2}$$

Hence the solution curves of the DE are $(x^2/y) + (y^2/x) = C$, or $x^3 + y^3 = Cxy$.

   Parametric Solutions.   Besides "explicit" solutions $y = f(x)$ and "implicit" solutions $U(x,y) = C$, quasinormal DEs (14) can have "parametric" solutions. Here by a *parametric solution* is meant a parametric curve $x = g(t), y = h(t)$ along which the line integral $\int M(x,y) \, dx + N(x,y) \, dy$, defined as

(15) $\int [M(g(t),h(t))g'(t) + N(g(t),h(t))h'(t)] \, dt$

vanishes. Thus, the curves $x = A \cos t$, $y = A \sin t$ are parametric solutions of $x + yy' = 0$. They are also solutions of the *system* of two first-order DEs $dx/dt = -y$, $dy/dt = x$, and will be studied from this standpoint in Chapter 5.

EXERCISES C

1. Find an integral of the DE $y' = y^2/x^2$, and plot its integral curves. Locate its critical points, if any.

2. Sketch the level curves and gradient lines of the function $x^3 + 3x^2y + y^3$. What are its critical points?

3. Same question as Exercise 2 for $x^3 - 3x^2y + y^3$.

4. Find equations describing all solutions of

$$y^2 = \frac{1}{2x + y}$$

5. For what pairs of positive integers n,r is the function $|x|^n$ of class $\mathcal{C}^r$?

6. Solve the DE $xy' + y = 0$ by the method of separation of variables. Discuss its solution curves, integral curves, and critical points.

7. (a) Reduce the *Bernoulli* DE $y' + p(x)y = q(x)y^n$, $n \neq 1$, to a linear first-order DE by the substitution $u = y^{1-n}$.
   (b) Express its general solution in terms of indefinite integrals.

In Exs. 8 and 9, solve the DE exhibited, sketch its solution curves, and describe them qualitatively:

8. $y' = y/x - x^2$.   9. $y' = y/x - \dfrac{1}{\ln |x|}$.

10. Find all solutions of the DE $|x| + |y|y' = 0$. In which regions of the plane is the differential on the left side exact?

*11. Show that the reciprocal of any homogeneous quadratic function $Q(x) = Ax^2 + 2Bxy + Cy^2$ is an integrating factor of $x\,dy - y\,dx$.

*12. Show that if u and v are both integrals of the DE $M(x,y) + N(x,y)y' = 0$, then so are $u + v$, uv except where $v = 0$, $\lambda u + \mu v$ for any constants λ and μ, and $g(u)$ for any single-valued function g.

*13. (a) What are the level lines and critical points of $\sin (x + y)$?
    (b) Show that for $u = \sin (x + y)$, $(x_0,y_0) = (0,0)$, and $\delta = \epsilon = \frac{1}{4}$, $f(x,c)$ in the Implicit Function Theorem need not exist if $\eta < \frac{1}{4}$ while it may not be unique if $n > 4$.

7   LINEAR FRACTIONAL EQUATIONS

An important first-order DE is the *linear fractional equation*

(16) $$\frac{dy}{dx} = \frac{cx + dy}{ax + by}, \qquad ad \neq bc$$

which is the normal form of

(16') $(ax + by)y' - (cx + dy) = 0$

It is understood that the coefficients a, b, c, d are constants.

The integration of the DE (16) can be reduced to a quadrature by the substitution $y = vx$. This substitution replaces (16) by the DE

$$xv' + v = \frac{c + dv}{a + bv}$$

in which the variables x and v can be separated. Transposing v, we are led to the separation of variables

$$\frac{(a + bv)\,dv}{bv^2 + (a - d)v - c} + \frac{dx}{x} = 0$$

Since the integrands are rational functions, this can be integrated in terms of elementary functions. Thus, x can be expressed as a function of $v = y/x$: we have $x = kG(y/x)$, where

$$G(v) = \exp\left\{-\int\left[\frac{a + bv}{bv^2 + (a - d)v - c}\right]dv\right\}$$

More generally, any DE of the form $y' = F(y/x)$ can be treated similarly. Setting $v = y/x$ and differentiating $y = xv$, we get $xv' + v = F(v)$. This is clearly equivalent to the separable DE

$$\frac{dv}{F(v) - v} = \frac{dx}{x} = d\,(\ln x)$$

whence $x = K \exp\{\int dv/[F(v) - v]\}$.

Alternatively, we can introduce polar coordinates, setting $x = r\cos\theta$ and $y = r\sin\theta$. If $\psi = \gamma - \theta$ is the angle between the tangent direction γ and the radial direction θ, then

$$\frac{1}{r}\frac{dr}{d\theta} = \cot\psi = \frac{\cot\gamma\cot\theta + 1}{\cot\theta - \cot\gamma}$$

Since $\tan\gamma = y' = F(y/x) = F(\tan\theta)$, we have

(17) $$\frac{1}{r}\frac{dr}{d\theta} = \frac{1 + \tan\gamma\tan\theta}{\tan\gamma - \tan\theta} = \frac{1 + (\tan\theta)F(\tan\theta)}{F(\tan\theta) - \tan\theta} = Q(\theta)$$

This can evidently be integrated by a quadrature:

$$(17') \qquad\qquad r(\theta) = r(0) \exp \int_0^\theta Q(\theta)\, d\theta \qquad\qquad (17')$$

The function on the right is well-defined, by the Fundamental Theorem of the Calculus, as long as $\tan \gamma \neq \tan \theta$, that is, as long as $y' \neq y/x$.

Invariant Radii. The radii along which the denominator of $Q(\theta)$ vanishes are those where (16) is equivalent to $d\theta/dr = 0$. Hence, these radii are particular solution curves of (16); they are called *invariant radii*. They are the solutions $y = \tau x$, for constant $\tau = \tan \theta$. Therefore, they are the radii $y = \tau x$ for which $y' = \tau = (c + d\tau)/(a + b\tau)$, by (16), and so their slopes τ are the roots of the quadratic equation

$$(18) \qquad\qquad b\tau^2 + (a - d)\tau = c$$

If $b \neq 0$, Eq. (18) has zero, one, or two real roots according as its discriminant is negative, zero, or positive. This discriminant is

$$(18') \qquad \Delta = (a - d)^2 + 4bc = (a + d)^2 - 4(ad - bc)$$

In the sectors between adjacent invariant radii, $d\theta/dr$ has constant sign; this fact facilitates the sketching of solution curves. Together with the invariant radii, the solution curves (17') form a regular curve family in the *punctured plane*, consisting of the xy-plane with the origin deleted.

Similarity Property. Each solution of the linear fractional DE (16) is transformed into another solution when x and y are both multiplied by the same nonzero constant k. The reason is, that both $y' = dy/dx$ and y/x are unchanged by the transformation $(x,y) \rightarrow (kx,ky)$. In polar coordinates, if $r = f(\theta)$ is a solution of (17), then so is $r = kf(\theta)$. Since the transformation $(x,y) \rightarrow (kx,ky)$ is a similarity transformation of the xy-plane for any fixed k, it follows that the solution curves in the sector between any two adjacent invariant radii are all *geometrically similar* (and similarly placed). This fact is apparent in the drawings of Figure 1.6.

Note also that the hyperbolas in Figure 1.6a are the orthogonal trajectories of those of Figure 1.5. This is because they are integral curves of $yy' = x$ and $xy' = -y$, respectively, and x/y is the negative reciprocal of $-y/x$.

EXERCISES D

1. Sketch the integral curves of the DEs in Exs. C8 and C9 in the neighborhood of the origin of coordinates.

2. Express in closed form all solutions of the following DEs:
   (a) $y' = (x^2 - y^2)/(x^2 + y^2)$ (b) $y' = \sin (y/x)$

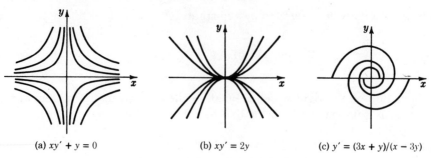

(a) $xy' + y = 0$ (b) $xy' = 2y$ (c) $y' = (3x + y)/(x - 3y)$

Figure 1.6   Integral curves.

3. (a) Show that the inhomogeneous linear fractional DE

$$(cx + dy + e)\, dx - (ax + by + f)\, dy = 0, \qquad ad \neq bc$$

   can be reduced to the form (16) by a translation of coordinates.
   (b) Using this idea, integrate $(x + y + 1)\, dx = (2x - y - 1)\, dy$.
   (c) For what sets of constants a, b, c, d, e, f, is the displayed DE exact?

4. Find all integral curves of $(x^n + y^n)y' - x^{n-1}y = 0$. [HINT: Set $u = y/x$.]

5. Prove in detail that the solutions of any homogeneous DE $y' = g(y/x)$ have the Similarity Property described in §7.

6. Show that the solution curves of $y' = G(x,y)$ cut those of $y' = F(x,y)$ at a constant angle β if and only if $G = (\tau + F)/(1 - \tau F)$, where $\tau = \tan \beta$.

7. Let A, B, C be constants, and K a parameter. Show that the coaxial conics $Ax^2 + 2Bxy + Cy^2 = K$, satisfy the DE $y' = -(Ax + By)/(Bx + Cy)$.

8. (a) Show that the differential $(ax + by)\, dy - (cx + dy)\, dx$ is exact if and only if $a + d = 0$, and that in this case the integral curves form a family of coaxial conics.
   (b) Using Exs. 6 and 7, show that if $\tan \beta = (a + d)/(c - b)$, the curves cutting the solution curves of the linear fractional DE $y' = (cx + dy)/(ax + by)$ at an angle β form a family of coaxial conics.

9. For the linear fractional DE (16) show that

$$y'' = (ad - bc)[cx^2 - (a - d)xy - by^2]/(ax + by)^3$$

   Discuss the domains of convexity and concavity of solutions.

10. Find an integrating factor for $y' + (2y/x) = a$, and integrate the DE by quadratures.

8   GRAPHICAL AND NUMERICAL INTEGRATION

The simplest way to sketch approximate solution curves of a given first-order normal DE $y' = F(x,y)$ proceeds as follows. Draw a short segment with slope $\lambda_i = F(x_i,y_i) = \tan \theta_i$ through each point (x_i,y_i) of a set of sample points sprinkled fairly densely over the domain of interest. Then draw smooth curves so as to have at every point a slope y' approximately equal to the average of the $F(x_i,y_i)$

at nearby points, weighting the nearest points most heavily (i.e., using graphical interpolation). Methods of doing this systematically are called schemes of *graphical integration*.

The preceding construction also gives a graphical representation of the direction field associated with a given normal first-order DE. This is defined as follows.

DEFINITION. A *direction field* in a region D of the plane is a function that assigns to every point (x,y) in D a *direction*. Two directions are considered the same if they differ by an integral multiple of 180°, or π radians.

With every quasinormal DE $M(x,y) + N(x,y)y' = 0$, there is associated a direction field. This associates with each point (x_k,y_k) not a *critical point* where $M = N = 0$, a short segment parallel to the vector $(N(x_k,y_k), -M(x_k,y_k))$. Such segments can be vertical whereas this is impossible for normal DEs.

It is very easy to integrate graphically the linear fractional equation (16) because solution curves have the same slope along each radius $y = vx$, $v =$ constant: each radius $y = kx$ is an isocline. We need only draw segments having the right direction fairly densely on radii spaced at intervals of, say, 30°. After tracing one approximate integral curve through the direction field by the graphical method described above, we can construct others by taking advantage of the Similarity Property stated in §7.

Numerical Integration. With modern computers, it is easy to construct accurate numerical tables of the solutions of initial value problems, where they exist, for most reasonably well-behaved functions $F(x,y)$. Solutions may exist only locally. Thus, to solve the initial value problem for $y' = 1 + y^2$ for the initial value $y(0) = 0$ on $[0,1.6]$ is impossible, since the solution $\tan x$ becomes infinite when $y = \pi/2 = 1.57086. \ldots$ We will now describe three very simple methods (or "algorithms") for computing such tables; the numerical solution of ordinary DEs will be taken up systematically in Chapters 7 and 8.

Simplest is the so-called *Euler method*, whose convergence to the exact solution (for $F \in \mathcal{C}^1$) was first proved by Cauchy around 1840 (see Chapter 7, §2). One starts with the given initial value, $y(a) = y_0 = c$, setting $X_0 = a$ and $Y_0 = y_0$, and then for a suitable step-size h computes recursively

(19) $$X_{n+1} = X_n + h, \qquad Y_{n+1} = Y_n + hF(X_n,Y_n)$$

A reasonably accurate table can usually be obtained in this way, by letting $h = .001$ (say), and printing out every tenth value of Y_n.

If greater accuracy is desired, one can reduce h to .0001, printing out $Y_0, Y_{100}, Y_{200}, Y_{300}, \ldots$, and "formatting" the results so that values are easy to look up.

Improved Euler Method. The preceding algorithm, however, is very wasteful, as Euler realized. As he observed, one can obtain much more accurate results with roughly the same computational effort by replacing (19) with the following "improved" Euler algorithm

$$(20) \qquad\qquad Z_{n+1} = Y_n + hF(X_n, Y_n)$$

$$Y_{n+1} = Y_n + \frac{h}{2}[F(X_n, Y_n) + F(X_{n+1}, Z_{n+1})]$$

With $h = .001$, this "improved" Euler method gives 5-digit accuracy in most cases, while requiring only about twice as much arithmetic per time step. Whereas with Euler's method, to use 10 times as many mesh points ordinarily gives only one more digit of accuracy, the same mesh refinement typically gives two more digits of accuracy with the improved Euler method.

As will be explained in Chapter 8, when truly accurate results are wanted, it is better to use other, more sophisticated methods that give *four* additional digits of accuracy each time h is divided by 10. In the special case of *quadrature*—that is, of DEs of the form $y' = g(x)$ (see §2)—to do this is simple. It suffices to replace (19) by *Simpson's Rule*.

$$(21) \qquad Y_{n+1} = Y_n + \frac{h}{6}\left[g(x_n) + 4g\left(\frac{x_n + h}{2}\right) + g(x_n + h)\right]$$

For example, one can compute the natural logarithm of 2,

$$y(2) = \ln 2 = \int_1^2 dx/x = .69314718\dots$$

with 8-digit accuracy by choosing $n = 25$ and using the formula

$$\ln 2 = \frac{1}{150}\sum_{k=1}^{25}\left[\frac{50}{48 + 2k} + \frac{50}{49 + 2k} + \frac{50}{50 + 2k}\right]$$

Caution.   To achieve 8-digit accuracy in summing 25 terms, one must use a computer arithmetic having at least 9-digit accuracy. Many computers have only 7-digit accuracy!

Taylor Series Method.   A third scheme of numerical integration is obtained by truncating the Taylor series formula after the term in y_n'', and writing

$$Y_{n+1} = Y(x_n + h) = Y_n + hY_n' + h^2Y_n''/2 + 0(h^3)$$

For the DE $y' = y$, since $y_n' = y_n'' = y_n$, this method gives $Y_{n+1} = (1 + h + h^2/2)Y_n$, and so it is equivalent to the improved Euler method.

For the DE $y' = 1 + y^2$, since $y'' = 2yy' = 2y(1 + y^2)$, the method gives

$$Y_{n+1} = Y_n + h(1 + Y_n^2) + h^2(Y_n + Y_n^3)$$

This differs from the result given by Euler's improved method. In general, since $d[F(x,y)]/dx = \partial F/\partial x + (\partial F/\partial y)\, dy/dx$, $Y_n'' = (F_x + FF_y)_n$. This makes the method easy to apply.

The error per step, like that of the improved Euler method, is roughly proportional to the *cube* of h. Since the number of steps is proportional to h^{-1}, the *cumulative* error of both methods is roughly proportional to h^2. Thus, one can obtain two more digits of accuracy with it by using 10 times as many mesh points.

As will be explained in Chapter 8, when truly accurate results are wanted, one should use other, more sophisticated methods that give *four* additional digits of accuracy when 10 times as many mesh points are used.

Constructing Function Tables.   Many functions are most simply defined as solutions of initial value problems. Thus e^x is the solution of $y' = y$ that satisfies the initial condition $e^0 = 1$, and $\tan x$ is the solution of $y' = 1 + y^2$ that satisfies $\tan 0 = 0$. Reciprocally, $\ln x$ is the solution of $y' = 1/y$ that satisfies $\ln 0 = 1$, while $\arctan x$ is the solution of $y' = 1/(1 + y^2)$ that satisfies $\arctan 0 = 0$.

It is instructive and enjoyable (using modern computers) to try to construct tables of numerical values of such functions, using the methods described in this section, and other methods to be discussed in Chapters 7 and 8. The accuracy of the computer output, for different methods and choices of the mesh length h, can be determined by comparison with standard tables.† One can often use simple recursion formulas instead, like

$$e^{x+h} = e^h e^x \quad \text{and} \quad \tan(x + h) = \frac{\tan x + \tan h}{1 - \tan x \tan h},$$

after evaluating $e^h = 1.01005167$, and also by its Taylor series $\tan x = x + x^3/3 + 2x^5/15 + \cdots$ $\tan(.01) = 0.0100003335\ldots$. Such comparisons will often reveal the limited accuracy of machine computations (perhaps six digits).

EXERCISES E

1. For each of the following initial value problems, make a table of the approximate numerical solution computed by the Euler method, over the interval and for the mesh lengths specified:
   (a) $y' = y$ with $y(0) = 1$, on $[0,1]$, for $h = 0.1$ and 0.02.
   (b) $y' = 1 + y^2$ with $y(0) = 0$, on $[0,1.6]$, for $h = 0.1, 0.05$, and 0.02.

2. Knowing that the exact solutions of the preceding initial value problems are e^x and $\tan x$:
   (a) Evaluate the *errors* $E_n = Y_n - y(X_n)$ for the examples of Exercise 1.
   (b) Tabulate the ratios E_i/hx, verifying when it is true that they are roughly independent of h and x.

---

† See for example Abramowitz and Stegun, which contains also a wealth of relevant material.

3. Compute approximate solutions of the initial value problems of Exercise 1 by the *improved* Euler method.

4. Find the errors of the approximate values computed in Exercise 3, and analyze the ratios Y_i/h^2x (cf. Ex. 2).

5. Use Simpson's Rule to compute a table of approximate values of the natural logarithm function $\ln x = \int_1^x dt/t$, on the interval $[1,2]$.

6. Construct a table of the function $\arctan x = \int_0^x dt/(1 + t^2)$ on the interval $[0,1]$ by Simpson's Rule, and compare the computed value of arctan 1 with $\pi/4$.

*7. In selected cases, test how well your tables agree with the identities $\arctan (\tan x) = x$ and $\ln (e^x) = x$.

*8. Let e_n be the approximate value of e obtained using Euler's method to solve $y' = y$ for the initial condition $y(0) = 1$ on $[0,1]$, on a uniform mesh with mesh length $h = 1/n$.
   (a) Show that $\ln e_n = n \ln (1 + h)$.
   (b) Infer that $\ln e_n = 1 - h/2 + h^2/3 - \cdots$.
   (c) From this, derive the formula

   (*) $$-\ln (e_n/e) = h/2 - h^2/3 + \cdots$$

   (d) From formula (*) show that, as $h \downarrow 0$, $e - e_n = (he/2)[1 - (h/6) + O(h^3)]$.

9   THE INITIAL VALUE PROBLEM

For any normal first-order differential equation $y' = F(x,y)$ and any "initial" x_0 (think of x as time), the *initial value problem* consists in finding the solution or solutions of the DE, for $x \geq x_0$, which also satisfy $f(x_0) = c$. In geometric language, this amounts to finding the solution curve or curves that issue from the point (x_0,c) to the right in the (x,y)-plane. As we have just seen, most initial value problems are easy to solve on modern computers, if one is satisfied with *approximate* solutions accurate to (say) 3–5 decimal digits.

However, there is also a basic *theoretical* problem of proving the *uniqueness* of this solution.

When $F(x,y) = g(x)$ depends on x alone, this theoretical problem is solved by the Fundamental Theorem of the Calculus (§2). Given $x_0 = a$ and $y_0 = c$, the initial value problem for the DE $y' = g(x)$ has one and only one solution, given by the definite integral (5′).

The initial value problem is said to be *well-posed* in a domain D when there is one and only one solution $y = f(x,c)$ in D of the given DE for each given $(x_0,c) \in D$, and when this solution varies continuously with c. To show that the initial value problem is well-posed, therefore, requires proving theorems of *existence* (there is a solution), *uniqueness* (there is only one solution), and *continuity* (the solution depends continuously on the initial value). The concept of a well-posed initial value problem gives a precise mathematical interpretation of the physical

Calaba h W. frances i. B. .
itorida javic.

concept of *determinism* (cf. Ch. 6, §5). As was pointed out by Hadamard, solutions which do not have the properties specified are useless physically because no physical measurement is exact.

It is fairly easy to show that the initial value problems discussed so far are well-posed. Thus, using formula (8′), one can show that the initial value problem is well-posed for the linear DE $y' + p(x)y = q(x)$ in any vertical strip $a < x < b$ where p and q are continuous. The initial value problem is also well-posed for the linear fractional DE (16) in each of the half-planes $ax + by > 0$ and $ax + by < 0$.

Actually, for the initial value problem for $y' = F(x,y)$ to be well-posed in a domain D, it is sufficient that $F \in \mathcal{C}^1$ in D. But it is not sufficient that $F \in \mathcal{C}$: though the continuity of F implies the existence of at least one solution through every point (cf. Ch. 6,§13), it does not necessarily imply uniqueness, as the following example shows.

Example 7. Consider the curve family $y = (x - C)^3$, sketched in Figure 1.7. For fixed C, we have

$$(22) \qquad y' = \frac{\partial y}{\partial x} = 3(x - C)^2 = 3y^{2/3}$$

a DE whose right side is a continuous function of position (x,y). Through every point (x_0,c) of the plane passes just one curve $y = (x - C)^3$ of the family, for which $C = x_0 - c^{1/3}$ depends continuously on (x_0,c). Hence, the initial value problem for the DE (22) always has one and only one solution of the form $y = (x - C)^3$. But there are also other solutions.

Thus, the function $y = 0$ also satisfies (22). Its graph is the envelope of the curves $y = (x - C)^3$. In addition, for any $\alpha < \beta$, the function defined by the three equations

$$(22') \qquad y = \begin{cases} (x - \alpha)^3, & x < \alpha \\ 0 & \alpha \le x \le \beta \\ (x - \beta)^3 & x > \beta \end{cases}$$

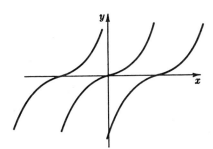

137,703

Figure 1.7   Solution curves of $y' = 3y^{2/3}$.

College of St. Francis Library
Joliet, Illinois

is a solution of (22). Hence, the first-order DE $y' = 3y^{2/3}$ has a *two-parameter* family of solutions, depending on the parameters α and β.

*10 UNIQUENESS AND CONTINUITY

The rest of this chapter will discuss existence, uniqueness, and continuity theorems for initial value problems concerning *normal* first-order DEs $y' = F(x,y)$. Readers who are primarily interested in applications are advised to skip to Chapter 2.

Example 9 shows that the mere continuity of $F(x,y)$ does not suffice to ensure the uniqueness of solutions $y = f(x)$ of $y' = F(x,y)$ with given $f(a) = c$. However, it is sufficient that $F \in \mathcal{C}^1$ (D). We shall prove this and continuity at the same time, using for much of the proof the following generalization of the standard Lipschitz condition.

DEFINITION.   A function $F(x,y)$ satisfies a *one-sided* Lipschitz condition in a domain D when, for some finite constant L

$$(23) \qquad y_2 > y_1 \quad \text{implies} \quad F(x,y_2) - F(x,y_1) \le L(y_2 - y_1)$$

identically in D. It satisfies a *Lipschitz condition*† in D when, for some nonnegative constant L (Lipschitz constant), it satisfies the inequality

$$(23') \qquad |F(x,y) - F(x,z)| \le L|y - z|$$

for all point pairs (x,y) and (x,z) in D having the same x-coordinate.

The same function F may satisfy Lipschitz conditions with different Lipschitz constants, or no Lipschitz conditions at all, as the domain D under consideration varies. For example, the function $F(x,y) = 3y^{2/3}$ of the DE in Example 9 satisfies a Lipschitz condition in any half-plane $y \ge \epsilon$, $\epsilon > 0$, with $L = 2\epsilon^{-1/3}$, but no Lipschitz condition in the half-plane $y > 0$. More generally, one can prove the following.

LEMMA 1.   *Let F be continuously differentiable in a bounded closed convex‡ domain D. Then it satisfies a Lipschitz condition there, with $L = sup_D|\partial F/\partial y|$.*

---

* In this book, starred sections may be omitted without loss of continuity.

† R. Lipschitz, *Bull. Sci. Math.* 10 (1876), p. 149; the idea of the proof is due to Cauchy (1839). See Ince, p. 76, for a historical discussion.

‡ A set of points is called *convex* when it contains, with any two points, the line segment joining them.

College of St. Francis Library
Joliet, Illinois

Proof. The domain being convex, it contains the entire vertical segment joining (x,y) with (x,z). Applying the Law of the Mean to $F(x,\eta)$ on this segment, considered as a function of η, we have

$$|F(x,y) - F(x,z)| = |y - z| \left| \frac{\partial F(x,\eta)}{\partial \eta} \right|$$

for some η between y and z. The inequality (23'), with $L = \sup_D |\partial F/\partial y|$, follows. A similar argument shows that (23) holds with $L = \max_D \partial F/\partial y$.

The case $F(x,y) = g(x)$ of ordinary integration, or "quadrature," is easily identified as the case when $L = 0$ in (23'). A Lipschitz condition is satisfied even if $g(x)$ is discontinuous.

LEMMA 2.   *Let σ be a differentiable function satisfying the differential inequality*

$$(24) \qquad \sigma'(x) \le K\sigma(x), \qquad a \le x \le b$$

where K is a constant. Then

$$(24') \qquad \sigma(x) \le \sigma(a)e^{K(x-a)}, \qquad \text{for} \qquad a \le x \le b$$

Proof. Multiply both sides of (24) by e^{-Kx} and transpose, getting

$$0 \ge e^{-Kx}[\sigma'(x) - K\sigma(x)] = \frac{d}{dx}\{\sigma(x)e^{-Kx}\}$$

The function $\sigma(x)e^{-Kx}$ thus has a negative or zero derivative and so is nonincreasing for $a \le x \le b$. Therefore, $\sigma(x)e^{-Kx} \le \sigma(a)e^{-Ka}$, q.e.d.

LEMMA 3.   *The one-sided Lipschitz condition (23) implies that*

$$[g(x) - f(x)][g'(x) - f'(x)] \le L[g(x) - f(x)]^2$$

for any two solutions $f(x)$ and $g(x)$ of $y' = F(x,y)$.

Proof. Setting $f(x) = y_1$, $g(x) = y_2$, we have

$$[g(x) - f(z)][g'(x) - f'(x)] = (y_2 - y_1)[F(x,y_2) - F(x,y_1)]$$

from the DE. If $y_2 > y_1$, then, by (23'), the right side of this equation has the upper bound $L(y_2 - y_1)^2$. Since all expressions are unaltered when y_1 and y_2 are interchanged, we see that the inequality of Lemma 3 is true in any case.

We now prove that solutions of $y' = F(x,y)$ depend continuously (and hence uniquely) on their initial values, provided that a one-sided Lipschitz condition holds.

THEOREM 5. *Let $f(x)$ and $g(x)$ be any two solutions of the first-order normal DE $y' = F(x,y)$ in a domain D where F satisfies the one-sided Lipschitz condition (23). Then*

$$(25) \qquad |f(x) - g(x)| \leq e^{L(x-a)}|f(a) - g(a)| \qquad if \qquad x > a$$

Proof. Consider the function

$$\sigma(x) = [g(x) - f(x)]^2$$

Computing the derivative by elementary formulas, we have

$$\sigma'(x) = 2[g(x) - f(x)] \cdot [g'(x) - f'(x)]$$

By Lemma 3, this implies that $\sigma'(x) \leq 2L\sigma(x)$; and by Lemma 2, this implies $\sigma(x) \leq e^{2L(x-a)}\sigma(a)$. Taking the square root of both sides of this inequality (which are nonnegative), we get (25), completing the proof.

As the special case $f(a) = g(a)$ of Theorem 5, we get uniqueness for the initial value problem: in any domain where F satisfies the one-sided Lipschitz condition (23), at most one solution of $y' = F(x,y)$ for $x \geq a$, satisfies $f(a) = c$. However, we do not get uniqueness or continuity for *decreasing* x. We now prove that we have uniqueness and continuity in both directions when the Lipschitz condition (23') holds.

THEOREM 6. *If (23') holds in Theorem 5, then*

$$(26) \qquad\qquad |f(x) - g(x)| \leq e^{L|x-a|}|f(a) - g(a)|$$

In particular, the DE $y' = F(x,y)$ has at most one solution curve passing through any point $(a,c) \in D$.

Proof. Since (23') implies (23), we know that the inequality (23) holds; from Theorem 5, this gives (26) for $x \geq a$. Since (23') also implies (23) when x goes to $-x$, we also have by Theorem 5

$$|f(x) - g(x)| \leq e^{L(a-x)}|f(a) - g(a)| = e^{L|x-a|}|f(a) - g(a)|$$

giving (26) also for $x < a$, and completing the proof.

EXERCISES F

1. In which domains do the following functions satisfy a Lipschitz condition?
   (a) $F(x,y) = 1 + x^2$ (b) $F(x,y) = 1 + y^2$
   (c) $F(x,y) = y/(1 + x^2)$ (d) $F(x,y) = x/(1 + y^2)$
2. Find all solutions of $y' = |xy|$.
3. Show that the DE $xu' - 2u + x = 0$ has a two-parameter family of solutions.
   [HINT: Join together solutions satisfying $u(0) = 0$ in each half-plane separately.]

4. Let f and g be solutions of $y' = F(x,y)$, where F is a continuous function. Show that the functions m and M, defined as $m(x) = \min (f(x), g(x))$ and $M(x) = \max (f(x), g(x))$, satisfy the same DE. [HINT: Discuss separately the cases $f(x) = g(x)$, $f(x) < g(x)$, and $f(x) > g(x)$.]

5. Let $\sigma(t)$, positive and of class $\mathcal{C}^1$ for $a \leq t \leq a + \epsilon$, satisfy the differential inequality $\sigma'(t) \leq K\sigma(t) \log \sigma(t)$. Show that $\sigma(t) \leq \sigma(a) \exp [K(t - a)]$.

6. Let $F(x,y) = y \log (1/y)$ for $0 < y < 1$, $F(y) = 0$ for $y = 0$. Show that $y' = F(x,y)$ has at most one solution satisfying $f(0) = c$, even though F does not satisfy a Lipschitz condition.

7. *(Peano uniqueness theorem).* For each fixed x, let $F(x,y)$ be a nonincreasing function of y. Show that, if $f(x)$ and $g(x)$ are two solutions of $y' = F(x,y)$, and $b > a$, then $|f(b) - g(b)| \leq |f(a) - g(a)|$. Infer a uniqueness theorem.

8. Discuss uniqueness and nonuniqueness for solutions of the DE $y' = -y^{1/3}$. [HINT: Use Ex. 7.]

9. (a) Prove a uniqueness theorem for $y' = xy$ on $-\infty < x,y < +\infty$.
   *(b) Prove the same result for $y' = y^{2/3} + 1$.

10. *(Generalized Lipschitz condition.)* Let $F \in \mathcal{C}$ satisfy

$$|F(x,y) - F(x,z)| \leq k(x)|y - z|$$

identically on the strip $0 < x < a$. Show that, if the improper integral $\int_0^a k(x)\, dx$ is finite, then $y' = F(x,y)$ has at most one solution satisfying $y(0) = 0$.

*11. Let F be continuous and satisfy

$$|F(x,y) - F(x,z)| \leq K|y - z| \log (|y - z|^{-1}), \qquad \text{for} \qquad |y - z| < 1$$

Show that the solutions of $y' = F(x,y)$ are unique.

*11 A COMPARISON THEOREM

Since most DEs cannot be solved in terms of elementary functions, it is important to be able to compare the unknown solutions of one DE with the known solutions of another. It is also often useful to compare functions satisfying the *differential inequality*

(27) $$f'(x) \leq F(x,f(x))$$

with exact solutions of the DE (3). The following theorem gives such a comparison.

THEOREM 7. *Let F satisfy a Lipschitz condition for $x \geq a$. If the function f satisfies the differential inequality (27) for $x \geq a$, and if g is a solution of $y' = F(x,y)$ satisfying the initial condition $g(a) = f(a)$, then $f(x) \leq g(x)$ for all $x \geq a$.*

Proof. Suppose that $f(x_1) > g(x_1)$ for some x_1 in the given interval, and define x_0 to be the largest x in the interval $a \leq x \leq x_1$ such that $f(x) \leq g(x)$. Then

$f(x_0) = g(x_0)$. Letting $\sigma(x) = f(x) - g(x)$, we have $\sigma(x) \geq 0$ for $x_0 \leq x \leq x_1$; and, also for $x_0 \leq x \leq x_1$,

$$\sigma'(x) = f'(x) - g'(x) \leq F(x,f(x)) - F(x,g(x)) \leq L(f(x) - g(x)) = L\sigma(x)$$

where L is the Lipschitz constant for the function F. That is, the function σ satisfies the hypothesis of Lemma 2 of §10 on $x_0 \leq x \leq x_1$, with $K = L$. Hence $\sigma(x) \leq \sigma(x_0)e^{L(x-x_0)} = 0$ and so σ, being nonnegative, vanishes identically. But this contradicts the hypothesis $f(x_1) > g(x_1)$. We conclude that $f(x) \leq g(x)$ for all x in the given interval, q.e.d.

THEOREM 8 (Comparison Theorem).   *Let f and g be solutions of the DEs*

(28) $$y' = F(x,y), \qquad z' = G(x,z)$$

respectively, where $F(x,y) \leq G(x,y)$ in the strip $a \leq x \leq b$ and F or G satisfies a Lipschitz condition. Let also $f(a) = g(a)$. Then $f(x) \leq g(x)$ for all $x \in [a,b]$.

Proof.   Let G satisfy a Lipschitz condition. Since $y' = F(x,y) \leq G(x,y)$, the functions f and g satisfy the conditions of Theorem 7 with G in place of F. Therefore, the inequality $f(x) \leq g(x)$ for $x \geq a$ follows immediately.

If F satisfies a Lipschitz condition, the functions $u = -f(x)$ and $v = -g(x)$ satisfy the DEs $u' = -F(x, -u)$ and

$$v' = -G(x, -v) \leq -F(x, -v)$$

Theorem 6, applied to the functions v, u and $H(u,v) = -F(x, -v)$ now yields the inequality $v(x) \leq u(x)$ for $x \geq a$, or $g(x) \geq f(x)$, as asserted.

The inequality $f(x) \leq g(x)$ in this Comparison Theorem can often be replaced by a strict inequality. Either f and g are identically equal for $a \leq x \leq x_1$, or else $f(x_0) < g(x_0)$ for some x_0 in the interval (a, x_1). By the Comparison Theorem, the function $\sigma_1(x) = g(x) - f(x)$ is nonnegative for $a \leq x \leq x_1$, and moreover $\sigma_1(x_0) > 0$. Much as in the preceding proof

$$\sigma_1'(x) = G(x,g(x)) - F(x,f(x)) \geq G(x,g(x)) - G(x,f(x)) \geq -L\sigma_1$$

Hence $[e^{Lx}\sigma_1(x)]' = e^{Lx}[\sigma_1' + L\sigma_1] \geq 0$; from this expression $e^{Lx}\sigma_1(x)$ is a nondecreasing function on $a \leq x \leq x_1$. Consequently, we have

$$\sigma_1(x) \geq \sigma_1(x_0)e^{-L(x-x_0)} > 0$$

which gives a strict inequality. This proves

COROLLARY 1.   *In Theorem 6, for any $x_1 > a$, either $f(x_1) < g(x_1)$, or $f(x) \equiv g(x)$ for all $x \in [a,x_1]$.*

Theorem 7 can also be sharpened in another way, as follows.

COROLLARY 2. *In Theorem 7, assume that F, as well as G, satisfies a Lipschitz condition and, instead of f(a) = g(a), that f(a) < g(a). Then f(x) < g(x) for x > a.*

Proof. The proof will be by contradiction. If we had $f(x) \geq g(x)$ for some $x > a$, there would be a first $x = x_1 > a$ where $f(x) \geq g(x)$. The two functions $y = \phi(x) = f(-x)$ and $z = \psi(x) = g(-x)$ satisfy the DEs $y' = -F(-x,y)$ and $z' = -G(-x,z)$ as well as the respective initial conditions $\phi(-x_1) = \psi(-x_1)$. Since $-F(-x,y) \geq -G(-x,y)$, we can apply Theorem 7 in the interval $[-x_1, -a]$, knowing that the function $-F(-x,y)$ satisfies a Lipschitz condition. We conclude that $\phi(-a) \geq \psi(-a)$, that is, that $f(a) \geq g(a)$, a contradiction.

*12 REGULAR AND NORMAL CURVE FAMILIES

In this chapter, we have analyzed many methods for solving first-order DEs of the related forms $y' = F(x,y), M(x,y) + N(x,y)y' = 0$, and $M(x,y) \, dx + N(x,y) \, dy = 0$, describing conditions under which their "solution curves" and/or "integral curves" constitute "one-parameter families" filling up appropriate domains of the (x,y)-plane. In this concluding section, we will try to clarify further the relationship between such first-order DEs and one-parameter curve families.

A key role is played by the Implicit Function Theorem, which shows† that the level curves $u = C$ of any function $u \in \mathcal{C}^1(D)$ have the following properties in any domain D not containing any critical point: (i) one and only one curve of the family passes through each point of D, (ii) each curve of the family has a tangent at every point, and (iii) the tangent direction is a continuous function of position. Thus, they constitute a regular curve family in the sense of the following definition.

DEFINITION. A *regular* curve family is a curve family that satisfies conditions (i) through (iii).

Thus, the circles $x^2 + y^2 = C$ ($C > 0$) form a regular curve family; they are the integral curves of $x + yy' = 0$, the DE of Example 1. Concerning the DE $y' = y^3 - y$ of Example 2, even though it is harder to integrate, we can say more: its solution curves form a normal curve family in the following sense.

DEFINITION. A regular curve family is *normal* when no curve of the family has a vertical tangent anywhere.

Almost by definition, the curves of any normal curve family are solution curves of the normal DE $y' = F(x,y)$, where $F(x,y)$ is the slope at (x,y) of the curve passing through it. Moreover, by Theorem 5', if $F \in \mathcal{C}^1$, there are no other solution curves.

The question naturally arises: do the solution curves of $y' = F(x,y)$ always form a normal curve family in any domain where $F \in \mathcal{C}^1$? They always do *locally*, but the precise formulation and proof of a theorem to this effect are very dif-

---

† Where $\partial u/\partial y = 0$ but $\partial u/\partial x \neq 0$, we can set $x = g(y)$ locally on the curve; see below.

ficult, and will be deferred to Chapter 6. There we will establish the simpler result that the initial value problem is locally *well-posed* for such DEs, after treating (in Chapter 4) the case that F is *analytic* (i.e., the sum of a convergent power series).

In the remaining paragraphs of this chapter, we will simply try to clarify further what the Implicit Function theorem does and does not assert about "level curves."

Parametrizing Curve Families.   Although the name "level curve" suggests that for each C the set of points where $F(x,y) = C$ is always a single curve, this is not so. Thus, consider the level curves of the function $F(x,y) = (x^2 + y^2)^2 - 2x^2 + 2y^2$. The level curve $F = 0$ is the lemniscate $r^2 = 2 \cos 2\theta$, and is divided by the critical point at the origin into two pieces. Inside each lobe of this lemniscate is one piece of the level curve $F = C$ for $-1 < C < 0$, while the "level curve" $F = -1$ consists of the other two critical points $(\pm 1, 0)$.

Similarly, in the infinite horizontal strip $-1 < y < 1$, every solution curve $y = \sin x + C$ of the DE $y' = \cos x$ consists of an infinite number of pieces. The same is true of the interval curves of the DE $\cos x \, dx = \sin x \, dy$, which are the level curves of $e^{-y} \sin x$. (These can also be viewed as the graphs of the functions $y = y = \ln |\sin x| + C$ and the vertical lines $y = \pm n\pi$.) In general, one cannot parametrize the level curves of $F(x,y)$ globally by the parameter C.

However, one can parametrize the level curves of any function $u \in \mathcal{C}^1$ *locally*, in some neighborhood of any point (x_0,y_0) where $\partial u/\partial y \neq 0$. For, by the Implicit Function Theorem, there exist positive ϵ and η such that for all $x \in (x_0 - \epsilon, x_0 + \epsilon)$ and $c \in (u_0 - \epsilon, u_0 + \epsilon)$, there is exactly one $y \in (y_0 - \eta, y_0 + \eta)$ such that $u(x,y) = c$. This defines a function $y(x,c)$ locally, in a rectangle of the (x,u)-plane. The parameter c parametrizes the level curves of $u(x,y)$ in the corresponding neighborhood of (x_0,y_0) in the (x,y)-plane; cf. Figure 1.8.

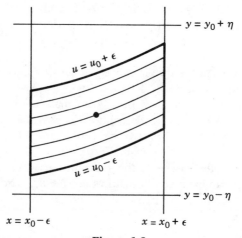

Figure 1.8

EXERCISES G

1. Let $f(u)$ be continuous and $a + bf(u) \neq 0$ for $p \leq u \leq q$. Show that the DE $y' = f(ax + by + c)$ (a,b,c are constants) has a solution passing through every point of the strip $p < ax + by + c < q$.

2. Find all solutions of the DE $y' = |x^3 y^3|$.

3. Show that, if M and N are homogeneous functions of the same degree, then (1') has the integrating factor $(xM + yN)^{-1}$ in any simply connected domain where $xM + yN$ does not vanish.

4. Show that if $g(y)$ satisfies a Lipschitz condition, the solutions of $y' = g(y)$ form a normal curve family in the (x,y)-plane. [HINT: Apply the Inverse Function Theorem to $x = \int dy/g(y) + C$.]

5. Let $g(x)$ be continuous for $0 \leq x < \infty$, $\lim_{x \to \infty} g(x) = b$ and $a > 0$. Show that, for every solution $y = f(x)$ of $y' + ay = g(x)$, we have $\lim_{x \to \infty} f(x) = b/a$.

6. Show that if $a < 0$ in Ex. 5, then there exists one and only one solution of the DE such that $\lim_{x \to \infty} f(x) = b/a$.

*7. (*Osgood's Uniqueness Theorem.*) Suppose that $\phi(u)$ is a continuous increasing function defined and positive for $u > 0$, such that $\int_\epsilon^1 du/\phi(u) \to \infty$ as $\epsilon \to 0$. If $|F(x,y) - F(x,z)| < \phi(|y - z|)$, then the solutions of the DE (3) are unique. [HINT: Use Ex. E4.]

8. Let F, G, f, g be as in Theorem 8, and $F(x,y) < G(x,y)$. Show that $f(x) < g(x)$ for $x > a$, without assuming that F or G satisfies a Lipschitz condition.

9. Show that the conditions $dx/dt = |x|^{1/2}$ and $x(0) = -1$ define a well-posed initial value problem on $[0,a)$ if $a \leq 1$, but not if $a > 1$.

10. (a) Find the critical points of the DE $x\,dy = y\,dx$.
    (b) Show that in the punctured plane (the x,y-plane with the origin deleted), the integral curves of $xy' = y$ are the lines $\theta = c$, where θ is a *periodic* angular variable only determined up to integral multiples of 2π.
    (c) What are its solution curves?
    (d) Show that the *real* variables $x/r = \cos\theta$ and $y/r = \sin\theta$ are integrals of $xy' = y$, and describe carefully their level curves.

*11. (a) Prove that there is no *real*-valued function $u \in \mathcal{C}^1$ in the punctured plane of Ex. 10 whose level curves are the integral curves of $xy' = y$.
    (b) Show that the integral curves of $y' = (x + y)/(x - y)$ are the equiangular spirals $r = ke^\theta = e^{(\theta - c)}$, $k \neq 0$.
    (b) Prove that there is no real-valued function $u \in \mathcal{C}^1$ whose level curves are these spirals.

CHAPTER 2

SECOND-ORDER LINEAR EQUATIONS

1 BASES OF SOLUTIONS

The most intensively studied class of ordinary differential equations is that of second-order linear DEs of the form

(1) $$p_0(x) \frac{d^2u}{dx^2} + p_1(x) \frac{du}{dx} + p_2(x)u = p_3(x)$$

The coefficient-functions $p_i(x)$ [$i = 0, 1, 2, 3$] are assumed continuous and real-valued on an interval I of the real axis, which may be finite or infinite. The interval I may include one or both of its endpoints, or neither of them. The central problem is to find and describe the unknown functions $u = f(x)$ on I satisfying this equation, the *solutions* of the DE. The present chapter will be devoted to second-order linear DEs and the behavior of their solutions.

Dividing (1) through by the leading coefficient $p_0(x)$, one obtains the *normal form*

(1′) $$\frac{d^2u}{dx^2} + p(x) \frac{du}{dx} + q(x)u = r(x)$$

$$p = \frac{p_1}{p_0}, \qquad q = \frac{p_2}{p_0}, \qquad r = \frac{p_3}{p_0}$$

This DE is equivalent to (1) so long as $p_0(x) \neq 0$; if $p_0(x_0) = 0$ at some point $x = x_0$, then the functions p and q are not defined at the point x_0. One therefore says that the DE (1) has a *singular point*, or *singularity* at the point x_0, when $p_0(x_0) = 0$.

For example, the Legendre DE

(*) $$\frac{d}{dx}\left[(1 - x^2) \frac{du}{dx} \right] + \lambda u = 0$$

has singular points at $x = \pm 1$. This is evident since when rewritten in the form (1) it becomes $(1 - x^2)u'' = 2xu' + \lambda u = 0$. Although it has polynomial solutions when $\lambda = n(n + 1)$, as we shall see in Ch. 4, §1, all its other nontrivial solutions have a singularity at either $x = 1$ or $x = -1$.

Likewise, the Bessel DE

(**)
$$x^2 u'' + xu' + (x^2 - n^2)u = 0$$

has a singular point at $x = 0$, and nowhere else. More commonly written in the normal form

$$u'' + \frac{1}{x} u' + \left(1 - \frac{n^2}{x^2}\right)u = 0,$$

its important *Bessel function* solution $J_0(x)$ will be discussed in Ch. 4, §8.

Linear DEs of the form (1) or (1′) are called *homogeneous* when their right-hand sides are zero, so that $p_3(x) \equiv 0$ in (1)—or, equivalently, $r(x) \equiv 0$ in (1′). The homogeneous linear DE

(2)
$$p_0(x)u'' + p_1(x)u' + p_2(x)u = 0$$

obtained by dropping the *forcing term* $p_3(x)$ from a given inhomogeneous linear DE (1) is called the *reduced* equation of (1). Evidently, the normal form of the reduced equation (2) of (1) is the reduced equation

(2′)
$$\frac{d^2 u}{dx^2} + p(x)\frac{du}{dx} + q(x)u = 0$$

of the normal form (1′) of (1).

A fundamental property of linear homogeneous DEs is the following *Super-position Principle*. Given any two solutions $f_1(x)$ and $f_2(x)$ of the linear homogeneous DE (2), and any two constants c_1 and c_2, the function

(3)
$$f(x) = c_1 f_1(x) + c_2 f_2(x)$$

is also a solution of (2). This property is characteristic of homogeneous linear equations; the function f is called a *linear combination* of the functions f_1 and f_2.

Bases of Solutions.   It is a fundamental theorem, to be proved in §5, that if $f_1(x)$ and $f_2(x)$ are two solutions of (2′), and if neither is a multiple of the other, then *every* solution of (2′) can be expressed in the form (3). A pair of functions with this property is called a *basis* of solutions.

Example 1.   The trigonometric DE is $u'' + k^2 u = 0$; its solutions include $\cos kx$ and $\sin kx$. Hence, all linear combinations $a \cos kx + b \sin kx$ of these basic solutions are likewise solutions.

Evidently, the zero function $u(x) \equiv 0$ is a *trivial* solution of any homogeneous linear DE. Letting $A = \sqrt{a^2 + b^2}$, and expressing $(a,b) = (A \cos \gamma, A \sin \gamma)$ in polar coordinates, we can also write

$$a \cos kx + b \sin kx = A \cos(kx - \gamma)$$

for any nontrivial solution of $u'' + k^2 u = 0$. The constant A in (2) is called the *amplitude* of the solution; γ its initial *phase*, and k its *wave number*; $k/2\pi$ is called its *frequency*, and $2\pi/k$ its *period*.

Constant-coefficient DEs. We next show how to construct a basis of solutions of any second-order *constant-coefficient* homogeneous linear DE

$$u'' + pu' + qu = 0, \quad (p,q \text{ constants}) \tag{5}$$

The trick is to set $u = e^{-px/2}v(x)$, so that $u' = e^{-px/2}[v' - pv/2]$ and $u'' = e^{-px/2}[v'' - pv' + p^2v/4]$, whence (5) is equivalent to

$$(5') \qquad\qquad v'' + (q - p^2/4)v = 0, \qquad v = e^{px/2}u$$

There are three cases, depending on whether the *discriminant* $\Delta = p^2 - 4q$ is positive, negative, or zero.

Case 1. If $\Delta > 0$, then (5′) reduces to $v'' = k^2v$, where $k = \sqrt{\Delta}/2$. This DE has the functions $v = e^{kx}$, e^{-kx} as a basis of solutions whence

$$(6a) \qquad\qquad u = e^{(\sqrt{\Delta}-p)x/2}, \qquad u = e^{(-\sqrt{\Delta}-p)x/2}$$

are a basis of solutions of (5). Actually, it is even simpler to make the "exponential substitution" $u = e^{\lambda x}$ in this case. Then (5) is equivalent to $(\lambda^2 + p\lambda + q)e^{\lambda x} = 0$; the roots of the quadratic equation $\lambda^2 + p\lambda + q = 0$ are the coefficients of the exponents in (6a).

Case 2. If $\Delta < 0$, then (5′) reduces to $v'' + k^2v = 0$, where $k = \sqrt{-\Delta}/2$. This DE has $\cos kx$, $\sin kx$ as a basis of solutions, whence

$$(6b) \qquad u = e^{-px/2} \cos(\sqrt{-\Delta}x/2), \qquad u = e^{-px/2} \sin(\sqrt{-\Delta}x/2)$$

form a basis of solutions of (5) when $\Delta < 0$.

Case 3. When $\Delta = 0$, (5′) reduces to $v'' = 0$, which has 1 and x as a basis of solutions. Hence the pair

$$(6c) \qquad\qquad u = e^{-px/2}, \qquad u = xe^{-px/2}$$

is a basis of solutions of (5) when $p^2 = 4q$.

2   INITIAL VALUE PROBLEMS

With differential equations arising from physical problems, one is often inter-ested in particular solutions satisfying additional initial or boundary conditions. Thus, in Example 1, one may wish to find a solution satisfying $u(0) = u_0$ and $u'(0) = u_0'$. An easy way to find a solution satisfying these initial conditions is to use Eq. (4) with $a = u_0$ and $b = u_0'/k$. In general, given a second-order linear DE such as (1) or (1'), the problem of finding a solution $u(x)$ that satisfies given initial conditions $u(a) = u_0$ and $u'(a) = u_0'$ is called the *initial value problem*.

Example 2.   Suppose we supplement the normal DE of Example 1 with the "forcing function" $r(x) = 3 \sin 2x$, and wish to find the solution of the resulting DE $u'' + u = 3 \sin 2x$ satisfying the initial conditions $u(0) = u'(0) = 0$.

To solve this initial value problem, we first construct a *particular* solution of this DE, trying $u = A \sin 2x$, where A is an unknown coefficient to be deter-mined. Substituting into the DE, we get $(-4A + A) \sin 2x = 3 \sin 2x$, or $A = -1$. Since $a \cos x + b \sin x$ satisfies $u'' + u = 0$ for any constants a and b, it follows that any function of the form

$$u = a \cos x + b \sin x - \sin 2x$$

satisfies the original DE $u'' + u = 3 \sin 2x$. Such a function will satisfy $u(0) = 0$ if and only if $a = 0$, so that

$$u'(x) = b \cos x - 2 \cos 2x$$

In particular, therefore, $u'(0) = b - 2 = 0$. Hence the function $u = 2 \sin x - \sin 2x$ solves the stated initial value problem.

Particular solutions of constant-coefficient DEs with polynomial forcing terms can be treated similarly. Thus, to solve

$$u'' + pu' + qu = cx + d \quad (p, q, c, d \text{ constants})$$

it is simplest to look first for a particular solution of the form $ax + b$. Substi-tuting into the DE, we obtain the equations $qa = c$ and $pa + qb = 0$. Unless $q = 0$, these give the particular solution

$$u = \frac{c}{q} x + \frac{qd - pc}{q^2}$$

When $q = 0$ but $p \neq 0$, we look for a quadratic solution; thus $u'' + u' = x$ has the solution

$$\frac{x^2}{2} - x$$

Finally, $u'' = cx + d$ has the cubic solution $u = cx^3/6 + dx^2/2$.

The procedure just followed can be used to solve initial value problems for many other second-order linear DEs of the form (1) and (1'). It requires four steps.

Step 1. Find a particular solution $u_p(x)$ of the DE.

Step 2. Find the general solution of the reduced equation obtained by setting $p_3(x) = 0$ in (1), or $r(x) = 0$ in (1'). It suffices to find two solutions $\phi(x)$ and $\psi(x)$ of the reduced DE, neither of which is a multiple of the other.

Step 3. Recognize $u = a\phi(x) + b\psi(x) + u_p(x)$, where a and b are constants to be determined from the initial conditions, as the general solution of the inhomogeneous DE.

Step 4. Solve for a and b the equations

$$\phi(0)a + \psi(0)b = u_0 - u_p(0)$$

$$\phi'(0)a + \psi'(b) = u_0' - u_p'(0)$$

For these equations to be uniquely solvable, the condition

$$\begin{vmatrix} \phi(0) & \psi(0) \\ \phi'(0) & \psi'(0) \end{vmatrix} = \phi(0)\psi'(0) - \psi(0)\phi(0) \neq 0$$

is clearly necessary and sufficient—the expression (4') is called the *Wronskian* of ϕ and ψ; we will discuss it in §5.

EXERCISES A

1. (a) Find the general solution of $u'' + 3u' + 2u = K$, where K is an arbitrary constant.
   (b) Same question for $u'' + 3u' = K$.

2. Solve the initial value problem for $u'' + 3u' + 2u = 0$, and the following initial conditions:
   (a) $u(0) = 1, u'(0) = 0$ (b) $u(0) = 0, u'(0) = 1$.

3. Answer the same questions for $u'' + 2u' + 2u = 0$.

4. Find a particular solution of each of the following DEs:
   (a) $u'' + 3u' + 2u = e^x$ (b) $u'' + 3u' + 2u = \sin x$
   (c) $u'' + 3u' + 2u = e^{-x}$ *(d) $u'' + 2u' + u = e^{-x}$

5. Find the general solution of each of the DEs of Ex. 4.

6. Solve the initial value problem for each of the DEs of Exercise 4, with the initial conditions $u(0) = u'(0) = 4$.

7. Find a particular solution of: (a) $u'' + 2u' + 2u = e^{-x}$,
   (b) $u'' + 2u' + 2u = \sin x$,   *(c) $u'' + 2u' + 2u = e^{-x}\sin x$.

8. Solve the initial value problem for each of the DEs of Exercise 7, and the initial conditions $u(0) = u'(0) = 0$.

9. Show that any second-order linear homogeneous DE satisfied by $x \sin x$ must have a singular point at $x = 0$.

3   QUALITATIVE BEHAVIOR; STABILITY

Note that when $\Delta < 0$ (i.e., in Case 2), all nontrivial solutions of (5) reverse sign each time that x increases by π/k. Qualitatively speaking, they are *oscillatory* in the sense of changing sign infinitely often. These facts become evident if we rewrite (6b) in the form (4), as

$$(7) \qquad\qquad u(x) = Ae^{-px/2}\cos[k(x - \phi)]$$

Contrastingly, when $\Delta \geq 0$, a nontrivial solution of (6) can vanish only when $ae^{\alpha x} = -be^{\beta x}$. This implies that $e^{(\alpha - \beta)x} = -b/a$, so that (i) a and b must have opposite signs, and (ii) $x = \ln|b/a|/(\beta - \alpha)$. Hence, a nontrivial solution can change sign at most once: it is *nonoscillatory*. Likewise, in Case 3, a nontrivial solution can vanish only where $a + bx = 0$, or $x = (-a/b)$, giving the same result. We conclude:

THEOREM 1.   *If $\Delta \geq 0$, then a nontrivial solution of (5) can vanish at most once. If $\Delta < 0$, however, it vanishes periodically with period $\pi/\sqrt{-\Delta}$.*

Stability.   Even more important than being oscillatory or nonoscillatory is the property of being *stable* or *unstable*, in the sense of the following definitions.

DEFINITION.   The homogeneous linear DE (2) is *strictly stable* when every solution tends to zero as $x \to \infty$; it is *stable* when every solution remains bounded as $x \to \infty$. When not stable, it is called *unstable*.

THEOREM 2.   *The constant-coefficient DE (5) is strictly stable when $p > 0$ and $q > 0$; it is stable when $p = 0$ but $q > 0$. It is unstable in all other cases.*

Proof.   This result can be proved very simply if complex exponents are used freely (see Chapter 3, §3). In the real domain, however, one must distinguish several possibilities, viz.:

(A) If $q < 0$, then $\Delta > 0$ and $\lambda^2 + p\lambda + q = 0$ must have two real roots of opposite sign. Instability is therefore obvious.

(B) If $p < 0$, instability is obvious from (6a)-(6b), if one keeps in mind the sign of p in each case.

(C) If $p = 0$ and $q > 0$, then we have Example 2: the DE (5) is stable but not strictly stable.

(D) If $p > 0$ and $q > 0$, there are two possibilities: (i) $\Delta \leq 0$, in which case we have strict stability by (6b) and (6c); (ii) $\Delta > 0$, in which case $\sqrt{\Delta} < p$ since $\Delta = p^2 - q < p^2$, and strict stability follows from (6a).

Second-order linear DEs with constant coefficients have so many applications that it is convenient to summarize their qualitative properties in a diagram; we have done this in Figure 2.1. (The words "focal," "nodal," and "saddle" point will be explained in §7; to have a focal point is equivalent to having oscillatory solutions.)

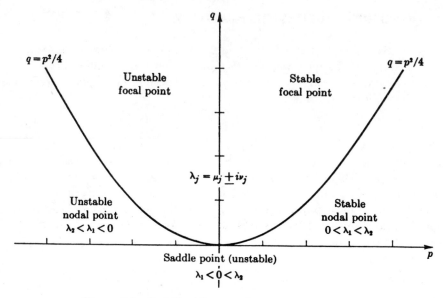

Figure 2.1   **Stability Diagram for** $\ddot{u} + p\dot{u} + qu = 0$.

4   UNIQUENESS THEOREM

We are now ready to treat rigorously the initial value problem stated in §1. The first basic concept involved is very general and applies to any normal second-order DE $u'' = F(x,u,u')$, whether linear or not.

Think of x as *time,* and of the possible pairs (u,u') as *states* of a physical system, which is governed (or modeled mathematically) by the given DE. Since u' expresses the rate of change of u at any "time" x, while $u'' = du'/dx$ gives the rate of change of u', it is natural to surmise that the present state of any such system *uniquely* determines its state at all future times. Indeed, the *theoretical* initial value problem is to prove this result as generally as possible.

In this section, we will prove it for second-order linear DEs of the form (1) having continuous coefficient-functions $p_j(x)$ and no singular points. Since $p_0(x) \neq 0$, it suffices to consider the normal form (1').

One would like to prove also that there always *exists* a solution for any initial (u_0, u_0'); this will be proved for second-order linear DEs having *analytic* coefficient-functions in Chapter 4, and (locally) for linear DEs having continuously *differentiable* coefficient-functions in Chapter 6. For the present, we will have to construct "particular" solutions and bases of solutions for homogeneous DEs $u'' + p(x)u' + q(x)u = 0$ by other methods.

Linear Operators.   We begin by discussing carefully the general concept of a "linear operator." Clearly, the operation of transforming a given function f into a new function g by the rule

$$g = p_0 f'' + p_1 f' + p_2 f$$

(for continuous p_i) is a transformation from one family of functions [in our case, the family $\mathcal{C}^2(I)$ of continuously twice-differentiable functions on a given interval I] to another family of functions [in our case, $\mathcal{C}(I)$]. Such a functional transformation is called an *operator,* and is written in operator notation

$$L[f] = p_0 f'' + p_1 f' + p_2 f$$

In our case, the operator L is *linear;* that is, it satisfies

$$L[cf + dg] = cL[f] + dL[g]$$

for any constants c and d.

As a special case (setting $c = 1, d = -1$), if u and v are any two solutions of the inhomogeneous linear DE (1), then their difference $u - v$ satisfies

$$L[u - v] = L[u] - L[v] = p_3(x) - p_3(x) = 0$$

That is, their difference is a solution of the *homogeneous* second-order linear DE (2).

The preceding simple observations, whose proofs are immediate, have the following result as a direct consequence.

LEMMA 1. *If the function $v(x)$ is any particular solution† of the inhomogeneous DE (1), then the general solution of (1) is obtained by adding to $v(x)$ the general solution of the corresponding homogeneous linear DE (2).*

For, if $u(x)$ is any other solution of (1), then $u(x) = v(x) + [u(x) - v(x)]$, where $L[u(x) - v(x)] = 0$ as before. More generally, the following lemma holds.

LEMMA 2. *If $u(x)$ is a solution of $L[u] = r(x)$, if $v(x)$ is a solution of $L[u] = s(x)$, and if c,d are constants, then $w = cu(x) + dv(x)$ is a solution of the DE $L[u] = cr(x) + ds(x)$.*

The proof is trivial, but the result describes the fundamental property of linear operators. Its use greatly simplifies the solution of inhomogeneous linear DEs.

Main Theorem. Having established these preliminary results, it is easy to prove a strong *uniqueness* theory for second-order linear DEs.

THEOREM 3. *(Uniqueness Theorem). If p and q are continuous, then at most one solution of (1′) can satisfy given initial conditions $f(a) = c_0$ and $f'(a) = c_1$.*

Proof. Let v and w be any two solutions of (1′) that satisfy these initial conditions; we shall show that their differences $u = v - w$ vanishes identically.

---

† The phrase "particular solution" is used to emphasize that *only one* solution of (1) need be found, thus reducing the problem of solving it to the case $p_3(x) = 0$.

Indeed, u satisfies (8) by Lemma 1. It also satisfies the initial conditions $u = u'$ $= 0$ when $x = a$. Now consider the nonnegative function $\sigma(x) = u^2 + u'^2$. By definition, $\sigma(a) = 0$. Differentiating, we have, since $r(x) = 0$,

$$\sigma'(x) = 2u'(u + u'') = 2u'[u - p(x)u' - q(x)u]$$

$$= -2p(x)u'^2 + 2[1 - q(x)]uu'$$

Since $(u \pm u')^2 \geq 0$, it follows that $|2uu'| \leq u^2 + u'^2$. Hence

$$2[1 - q(x)]uu' \leq (1 + |q(x)|)(u^2 + u'^2)$$

and

$$\sigma'(x) \leq [1 + |q(x)|]u^2 + [1 + |q(x)| + |2p(x)|]u'^2$$

Therefore, if $K = 1 + \max [|q(x)| + 2|p(x)|]$, the maximum being taken over any finite closed interval $[a, b]$, we obtain

$$\sigma'(x) \leq K\sigma(x), \qquad K < +\infty$$

By Lemma 2 of Ch. 1, §10, it follows that $\sigma(x) = 0$ for all $x \in [a, b]$. Hence $u(x) \equiv 0$ and $v(x) \equiv w(x)$ on the interval, as claimed.

The Uniqueness Theorem just proved implies an important extension of the Superposition Principle stated in §1.

THEOREM 4.   *Let f and g be two solutions of the homogeneous second-order linear DE*

$$(8) \qquad\qquad u'' + p(x)u' + q(x)u = 0, \qquad p,q \in \mathcal{C}$$

For some $x = x_0$, let $(f(x_0), f'(x_0))$ and $(g(x_0), g'(x_0))$ be linearly independent vectors. Then every solution of this DE is equal to some linear combination $h(x) = cf(x) + dg(x)$ of f and g with constant coefficients c,d.

In other words, the *general solution* of the given homogeneous DE (8) is $cf(x)$ $+ dg(x)$, where c and d are arbitrary constants.

Proof.   By the Superposition Principle, any such $h(x)$ satisfies (8). Conversely, suppose the function $h(x)$ satisfies the given DE (8). Then, at the given point x_0, constants c and d can be found such that

$$cf(x_0) + dg(x_0) = h(x_0), \qquad cf'(x_0) + dg'(x_0) = h'(x_0)$$

In fact, the constants c and d are given by Cramer's Rule, as

$$c = (h_0 g_0' - g_0 h_0')/(f_0 g_0' - g_0 f_0')$$

$$d = (f_0 h' - h_0 f_0')/(f_0 g_0' - g_0 f_0')$$

where we have used the abbreviations $f_0 = f(x_0)$, $f_0' = f'(x_0)$, and so on. For this choice of c and d, the function

$$u(x) = h(x) - cf(x) - dg(x)$$

satisfies the given homogeneous DE by the Superposition Principle and the initial conditions $u(x_0) = u'(x_0) = 0$. Hence by the Uniqueness Theorem, $u(x)$ is the trivial solution, $u(x) \equiv 0$, of the given homogeneous DE; therefore $h = cf + dg$.

Two solutions, f and g, of a homogeneous linear second-order DE (8) with the property that every other solution can be expressed as a linear combination of them are said to be a *basis of solutions* of the DE.

5   THE WRONSKIAN

The question of whether two solutions of a homogeneous linear DE form a basis of solutions is easily settled by examining their Wronskian, a concept that we now define.

DEFINITION.   The *Wronskian* of any two differentiable functions $f(x)$ and $g(x)$ is

$$(9) \qquad W(f, g; x) = f(x)g'(x) - g(x)f'(x) = \begin{vmatrix} f(x) & f'(x) \\ g(x) & g'(x) \end{vmatrix}$$

THEOREM 5.   *The Wronskian* (9) *of any two solutions of* (8) *satisfies the identity*

$$(10) \qquad W(f, g; x) = W(f, g; a) \exp\left(-\int_a^x p(t)\, dt\right)$$

Proof.   If we differentiage (9) and write $W(f, g; x) = W(x)$ for short, a direct computation gives $W' = fg'' - gf''$. Substituting for g'' and f'' from (8) and cancelling, we have the linear homogeneous first-order DE

$$(11) \qquad W'(x) + p(x)W(x) = 0$$

Equation (10) follows from the first-order homogeneous linear DE (11) by Theorem 4 of Ch. 1, §6.

COROLLARY. *The Wronskian of any two solutions of the homogeneous linear* DE (8) *is identically positive, identically negative, or identically zero.*

We now relate the Wronskian of two functions to the concept of linear independence. In general, a collection of functions $f_1, f_2 \ldots, f_n$ is called *linearly independent* on the interval $a \leq x \leq b$ when no linear combination $c_1 f_1(x) + c_2 f_2(x) + \ldots + c_n f_n(x)$ of these functions gives the identically zero function for $a \leq x \leq b$, except the trivial linear combination where all coefficients vanish. Functions that are not linearly independent are called linearly dependent. If f and g are any two linearly dependent functions, then $cf + dg = 0$ for suitable constants c and d, not both zero. Hence $g = -(d/d)f$ or $f = -(c/c)g$; the functions f and g are proportional.

LEMMA. *If f and g are linearly dependent differentiable functions, then their Wronskian vanishes identically.*

Proof. Suppose that f and g are linearly dependent. Then there are two constants c and d, not both zero, which satisfy the *two* linear equations

$$cf(x) + dg(x) = 0, \qquad cf'(x) + dg'(x) = 0$$

identically on the interval of interest. Therefore, the determinant of the two equations, which is the Wronskian $W(f, g; x)$, vanishes identically.

The interesting fact is that when f and g are both solutions of a second-order linear DE, a strong *converse* of this lemma is also true.

THEOREM 6. *If f and g are two linearly independent solutions of the nonsingular second-order linear* DE (8), *then their Wronskian never vanishes.*

Proof. Suppose that the Wronskian $W(f, g; x)$ vanished at some point x_1. Then the vectors $[f(x_1), f'(x_1)]$ and $[g(x_1), g'(x_1)]$ would be linearly dependent and, therefore, proportional: $g(x_1) = kf(x_1)$ and $g'(x_1) = kf'(x_1)$ for some constant k. Consider now the function $h(x) = g(x) - kf(x)$. This function is a solution of the DE (8), since it is a linear combination of solutions. It also satisfies the initial conditions $h(x_1) = h'(x_1) = 0$. By the Uniqueness Theorem, this function must vanish identically. Therefore, $g(x) = kf(x)$ for all x, contradicting the hypothesis of linear independence of f and g.

Remark 1. The fact that the DE (8) is nonsingular is essential in Theorem 6. For example, the Wronskian x^4 of the two linearly independent solutions x^2 and x^3 of the DE $x^2 u'' - 4xu' + 6u = 0$ vanishes at $x = 0$. This is possible because the leading coefficient $p_0(x)$ of the DE vanishes there.

Remark 2. There is an obvious connection between the formula for the Wronskian of two functions and the formula for the derivative of their quotient:

$$\left(\frac{g}{f}\right)' = \frac{(fg' - gf')}{f^2} = \frac{W(f, g)}{f^2}$$

This suggests that the ratio of two functions is a constant if and only if their Wronskian vanishes identically. However, this need not be true if f vanishes: the ratio of the two functions x^3 and $|x|^3$ is *not* a constant, yet their Wronskian $W(x^3, |x|^3) \equiv 0$. (Note also that both functions satisfy the DEs $xu' = 3u$ and $3xu'' - 2u' = 0$.)

Nevertheless, the connection between $W(f,g)$ and g/f is a useful one. Thus, it allows one to construct a second solution $g(x)$ of (8) if one nontrivial solution is known. Namely, if $P(x) = \int p(x)dx$ is any indefinite integral of $p(x)$, then the function

(12)
$$g(x) = f(x) \int \left[\frac{e^{-P(x)}}{f^2(x)} \right] dx$$

is a second, linearly independent solution of (8) in any interval where $f(x)$ is nonvanishing. This is evident, since $(g/f)' = W(f,g)/f^2$, whence

$$\frac{g(x)}{f(x)} = \int \left[\frac{W(f,g)}{f^2} \right] dx = \int \left[\frac{e^{-P(x)}}{f^2(x)} \right] dx$$

For example, knowing that e^{3x} is one nontrivial solution of $u'' - 6u' + 9 = 0$, since $P(x) = -6x = \int p \, dx$, setting $e^{-P(x)} = e^{6x}$, we obtain the second solution

$$g(x) = e^{3x} \int \left[\frac{e^{6x}}{(e^{3x})^2} \right] dx = e^{3x} \int dx = xe^{3x}$$

Riccati Equation. Finally, consider the formula for the derivative of the ratio $v = u'/u$,† where u is any nontrivial solution of (8):

(13)
$$v' = \left(\frac{u'}{u} \right)' = \frac{u''}{u} - \frac{u'^2}{u^2} = -p(x)v - q(x) - v^2$$

The quadratic first-order DE (13) is called the Riccati equation associated with (8); its solutions form a one-parameter family. Conversely, if $v(x)$ is any solution of the Riccati equation (13) and if $u' = v(x)u$, then u satisfies (8). Hence, every solution $u(x)$ of (8) can be written in any interval where u does not vanish, in the form,

(14)
$$u(x) = C \exp \int v(x)dx$$

where $v(x)$ is some solution of the associated Riccati equation (13).

The Riccati substitution $v = u'/u$ thus reduces the problem of solving (8) to the integration of a first-order quadratic DE and a quadrature. For instance,

---

† Since $v = u'/u = d(\ln u)/dx$, this is called the *logarithmic* derivative of u.

the Riccati equation associated with the trigonometric equation $u'' + k^2 u = 0$ is $v' + v^2 + k^2 = 0$, whose general solution is $v = k \tan k(x_1 - x)$.

EXERCISES B

1. Show that all solutions of (8) have continuous second derivatives. Show also that this is not true for (1).

2. Find a formula expressing the fourth derivative u^{iv} of any solution u of (8) in terms of u, u', and the derivatives of p and q. What differentiability conditions must be assumed on the coefficients of (8) to justify this formula?

For the solution pairs of the DEs specified in Exs. 3–5 to follow, (a) calculate the Wronskian, and (b) solve the initial-value problem for the DE specified with each of the initial conditions $u(0) = 2$, $u'(0) = 1$, and $u(0) = 1$, $u'(0) = -1$ (or explain why there is no solution).

3. $f(x) = \cos x$, $g(x) = \sin x$ (solutions of $u'' + u = 0$).

4. $f(x) = e^{-x}$, $g(x) = e^{-3x}$ (solutions of $u'' + 4u' + 3u = 0$).

5. $f(x) = x + 1$, $g(x) = e^x$ (solutions of $xu'' - (1 + x)u' + u = 0$).

6. Let $f(x)$, $g(x)$, and $h(x)$ be any three solutions of (8). Show that

$$\begin{vmatrix} f & f' & f'' \\ g & g' & g'' \\ h & h' & h'' \end{vmatrix} \equiv 0$$

7. (a) Prove the Corollary of Theorem 5.
   (b) Prove that if $f(x)$ and $g(x)$ satisfy the hypotheses of Theorem 6, then
   $$p(x) = (gf'' - fg'')/W \quad \text{and} \quad q(x) = (f'g'' - g'f'')/W.$$

8. What is wrong with the following "proof" of Theorem 5: "Let $w(x) = \log W(x)$; then $w'(x) = -p(x)$. Hence, $w(x) = w(a) - \int_a^x p(x)\, dx$, from which (10) follows."

9. Construct second-order linear homogeneous DEs having the following bases of solutions; you may assume the result of Ex. 7:
   (a) x, $\sin x$,   (b) x^m, x^n,   (c) $\sinh x$, $\sin x$,   (d) $\tan x$, $\cot x$.
   For each of the examples of Ex. 9, determine the singular points of the resulting DE.

10. (a) Show that if $p,q \in \mathcal{C}^n$, then every solution of (8) is of class $\mathcal{C}^{n+2}$.
    (b) Show that if every solution of (8) is of class $\mathcal{C}^{n+2}$, then $p \in \mathcal{C}^n$ and $q \in \mathcal{C}^n$.

11. Let $f(x)$, $g(x)$, $h(x)$ be three solutions of the linear third-order DE

$$y''' + p_1(x)y'' + p_2(x)y' + p_3(x)y = 0$$

Derive a first-order DE satisfied by the determinant

$$w(x) = \begin{vmatrix} f & f' & f'' \\ g & g' & g'' \\ h & h' & h'' \end{vmatrix}$$

*12. Let $y'' + q(x)y = 0$, where $g(x)$ is "piecewise continuous" (i.e., continuous except for a finite number of finite jumps). Define a "solution" of such a DE as a function $y = f(x) \in \mathcal{C}^1$ that satisfies the DE except at these jumps.

(a) Show that any such solution has left- and right-derivatives at every point of discontinuity.

(b) Describe explicitly a basis of solutions for the DE $y'' + q(x)y = 0$, if

$$q(x) = \begin{cases} +1 & \text{when } x > 0 \\ -1 & \text{when } x < 0 \end{cases}$$

[N.B. The preceding function $q(x)$ is commonly denoted sgn x_i.]

6 SEPARATION AND COMPARISON THEOREMS

The Wronskian can also be used to derive properties of the graphs of solutions of the DE (8). The following result, the celebrated Sturm Separation Theorem, states that all nontrivial solutions of (8) have essentially the same number of oscillations, or *zeros*. (A "zero" of a function is a point where its value is zero; functions have two zeros in each complete oscillation.)

THEOREM 7. *If $f(x)$ and $g(x)$ are linearly independent solutions of the* DE (8), *then $f(x)$ must vanish at one point between any two successive zeros of $g(x)$. In other words, the zeros of $f(x)$ and $g(x)$ occur alternately.*

Proof. If $g(x)$ vanishes at $x = x_i$, then the Wronskian

$$W(f, g; x_i) = f(x_i)g'(x_i) \neq 0$$

since f and g are linearly independent; hence, $f(x_i) \neq 0$ and $g'(x_i) \neq 0$ if $g(x_i) = 0$. If x_1 and x_2 are two successive zeros of $g(x)$, then $g'(x_1)$, $g'(x_2)$, $f(x_1)$, and $f(x_2)$ are all nonzero. Moreover, the nonzero numbers $g'(x_1)$ and $g'(x_2)$ cannot have the same sign, because if the function is increasing at $x = x_1$, then it must be decreasing at $x = x_2$, and vice-versa. Since $W(f, g; x)$ has constant sign by the Corollary of Theorem 4, it follows that $f(x_1)$ and $f(x_2)$ must also have opposite signs. Therefore $f(x)$ must vanish somewhere between x_1 and x_2.

For instance, applied to the trigonometric DE $u'' + k^2 u = 0$, the Sturm Separation Theorem yields the well-known fact that the zeros of sin kx and cos kx must alternate, simply because these functions are two linearly independent solutions of the same linear homogeneous DE.

A slight refinement of the same reasoning can be used to prove an even more useful Comparison Theorem, also due to Sturm.

THEOREM 8. *Let $f(x)$ and $g(x)$ be nontrivial solutions of the* DEs $u'' + p(x)u = 0$ *and $v'' + q(x)v = 0$, respectively, where $p(x) \geq q(x)$. Then $f(x)$ vanishes at least once between any two zeros of $g(x)$, unless $p(x) \equiv q(x)$ and f is a constant multiple of g.*

Proof.   Let x_1 and x_2 be two successive zeros of $g(x)$, so that $g(x_1) = g(x_2) = 0$. Suppose that $f(x)$ failed to vanish in $x_1 < x < x_2$. Replacing f and/or g by their negative, if necessary, we could find solutions f and g positive on $x_1 < x < x_2$. This would make

$$W(f, g; x_1) = f(x_1)g'(x_1) \geq 0 \quad \text{and} \quad W(f, g; x_2) = f(x_2)g'(x_2) \leq 0$$

On the other hand, since $f > 0$, $g > 0$, and $p \geq q$ on $x_1 < x < x_2$, we have

$$\frac{d}{dx}[W(f, g; x)] = fg'' - gf'' = (p - q)fg \geq 0 \quad \text{on} \quad x_1 < x < x_2$$

Hence, W is nondecreasing, giving a contradiction unless

$$p - q \equiv W(f, g; x) \equiv 0$$

In this event, $f \equiv kg$ for some constant k by Theorem 4, completing the proof.

COROLLARY 1.   *If $q(x) \leq 0$, then no nontrivial solution of $u'' + q(x)\,u = 0$ can have more than one zero.*

The proof is by contradiction. By the Sturm Comparison Theorem, the solution $v \equiv 1$ of the DE $v'' = 0$ would have to vanish at least once between any two zeros of any nontrivial solution of the DE $u'' + q(x)u = 0$.

The preceding results show that the oscillations of the solutions of $u'' + q(x)u = 0$ are largely determined by the sign and magnitude of $q(x)$. When $q(x) \leq 0$, oscillations are impossible: no solution can change sign more than once. On the other hand, if $q(x) \geq k^2 > 0$, then any solution of $u'' + q(x)u = 0$ must vanish between any two successive zeros of any given solution $A \cos k(x - x_1)$ of the trigonometric DE $u'' + k^2u = 0$, hence in any interval of length π/k.

This result can be applied to solutions of the Bessel DE (**) of §1 (i.e., to the Bessel function of order n; see Ch. 4, §4). Substituting $u = v/\sqrt{x}$ into (**), we obtain the equivalent DE

$$(15) \qquad v'' + \left[1 - \frac{4n^2 - 1}{4x^2}\right]v = 0$$

whose solutions vanish when u does (for $x \neq 0$). Applying the Comparison Theorem to (15) and $u'' + u = 0$, we obtain the following.

COROLLARY 2.   *Each interval of length π of the positive x-axis contains at least one zero of any solution of the Bessel DE of order zero, and at most one zero of any nontrivial solution of the Bessel DE of order n if $n > \frac{1}{2}$.*

The fact that the oscillations of the solutions of $u'' + q(x)u = 0$ depend on the sign of $q(x)$ is illustrated by Figures 2.2 and 2.3, which depict sample solution curves for the cases $q(x) = 1$ and $q(x) = -1$, respectively.

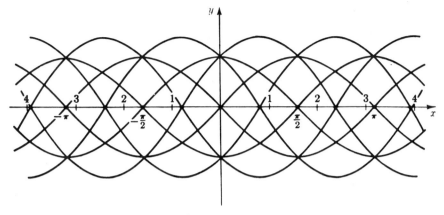

Figure 2.2   Solution curves of $u'' + u = 0$.

7   THE PHASE PLANE

In the theory of normal second-order DEs $u'' = F(x,u,u')$, linear or nonlinear, the two-dimensional space of all vectors (u,u') is called the *phase plane*. As was noted in §5, the points of this phase plane correspond to the *states* of any physical system whose behavior is modeled by such a DE.

Clearly, any solution $u(x)$ of the given DE determines a parametric curve or *trajectory* in this phase plane, which consists of all $[u(x),u'(x)]$ associated with this solution. [A trivial exception arises at *equilibrium states* at which $F(x,c,0) \equiv 0$, so that $u'(x) \equiv 0$ and $u(x) \equiv c$. Clearly, any such equilibrium point is necessarily on the *u*-axis, where $u' = 0$.]

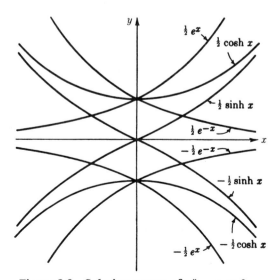

Figure 2.3   Solution curves of $u'' - u = 0$.

The trajectories just defined have some important general geometrical properties. For example, since u is increasing when $u' > 0$ and decreasing when $u' < 0$, the paths of solutions must go to the right in the upper half-plane and to the left in the lower half-plane. Furthermore, paths of solutions ("trajectories") must cut the u-axis $u' = 0$ orthogonally, except where $F = 0$.

We will treat in this chapter only homogeneous second-order *linear* DEs (8), deferring discussion of the nonlinear case to Chapter 5. Using the letter v to signifly u', this DE is obviously equivalent to the system

(16) $$\frac{du}{dx} = v, \qquad \frac{dv}{dx} = -p(x)v - q(x)u$$

which can also be written in vector form as

$$\frac{d}{dx}\begin{pmatrix} u \\ v \end{pmatrix} = \begin{bmatrix} 0 & 1 \\ -q(x) & -p(x) \end{bmatrix}\begin{pmatrix} u \\ v \end{pmatrix}$$

Note that if $q(x) < 0$, then $du'/dx = -q(x)u$ has the same sign as u on the u-axis. It follows that if $q(x)$ is negative, then any trajectory once trapped in the first quadrant can never leave it, because it can neither cross the u-axis into the fourth quadrant nor recross the u'-axis into the second quadrant. The same is true, for similar reasons, of trajectories trapped in the third quadrant.

Even more important, two nontrivial solutions of (8) are *linearly dependent* if and only if they are on the same straight line through the origin in the (u,v)-plane. It follows that each straight line through the origin moves as a unit. The preceding facts also become evident analytically, if we introduce *clockwise polar coordinates* in the phase plane, by the formulas

(17) $$u'(x) = r \cos \theta(x), \qquad u(x) = r \sin \theta(x)$$

(We adopt this clockwise orientation so that θ will be an *increasing* function on the u'-axis.) Differentiating the relation $\cot \theta = u'/u$, we then have the formulas

$$-(\csc^2 \theta)\theta' = (u''/u) - (u'/u)^2 = -p(u'/u) - q - (u'/u)^2$$
$$= -p \cot \theta - q - \cot^2 \theta$$

If we multiply through by $-\sin^2 \theta$, this equation gives

(18) $$d\theta/dx = \cos^2 \theta + p(x) \cos \theta \sin \theta + q(x) \sin^2 \theta$$

This *first*-order DE gives much information about the oscillations of u.

Differentiating $r^2(x) = u^2(x) + u'^2(x)$ as in the proof of Theorem 1, where $\sigma(x) = r^2(x)$, we get

$$rr' = uu' + u'u'' = u'(u - pu' - qu)$$
$$= r^2 \cos \theta[(1 - q(x)) \sin \theta - p(x) \cos \theta]$$

Dividing through by r^2 and simplifying, we obtain

(19) $$\frac{1}{r}\frac{dr}{dx} = -p(x)\cos^2\theta + (1 - q(x))\cos\theta\sin\theta$$

As in Theorem 1, it follows that the magnitude $|d(\ln r)/dx|$ of the logarithmic derivative of $r(x)$ is bounded by $|p|_{max} + (1 + |q|_{max})/2$.

Now, consider the graph of the multiple-valued function $\theta(x)$ in the (x, θ)-plane. Since $\cot\theta$ is periodic with period π, the graph at $\theta = $ arc $\cot(u'/u)$ for any solution of (8) consists of an infinite family of congruent curves, all obtained from any one by vertical translation through integral multiples of π. The curves that form the graphs of $\theta_1(x)$ and $\theta_2(x)$, for any two linearly independent solutions u_1, u_2 of (8), occur alternately. Moreover, by the uniqueness theorem of Ch. 1, they can never cross.

In (17), $u = 0$ precisely when $\sin\theta = 0$, that is, when $\theta \equiv 0 \pmod{\pi}$. Inspecting (18), we also see that

(20a)   When $\theta \equiv 0 \pmod{\pi}$, that is, $u = 0$, then $d\theta/dx > 0$

(20b)   When $\theta \equiv \pi/2 \pmod{\pi}$, $d\theta/dx$ has the sign of q

From (20a) it follows that, after the graph of any $\theta(x)$ has crossed the line $\theta = n\pi$, it can never recross it backwards. Where $u(x)$ next vanishes (if it does), we must have $\theta = (n + 1)\pi$; in other words, successive zeros of $u(x)$ occur precisely where θ increases from one integral multiple of π to the next!

After verifying that the right side of (18) satisfies a Lipschitz condition, we see that this inequality can never cease to hold; hence, in any interval where $\theta_1(x)$ increases from $n\pi$ to $n\pi + \pi$, $\theta_2(x)$ must cross the line $\theta = n\pi + \pi$ and so u_2 must vanish there. Sturm's Comparison Theorem follows similarly: if $q(x)$ is increased and $p(x)$ is left constant, the Comparison Theorem of Ch. 1, applied to (19), yields it as a corollary.

Oscillatory Solutions. The preceding considerations also enable one to extend some of the results stated in §3 for constant-coefficient DEs to second-order linear DEs with variable coefficients. When $q(x) > p^2(x)/4$, the quadratic form on the right side of (18) is *positive definite;* hence $d\theta/dx$ is identically positive. Unless $q(x)$ gets very near to $p^2(x)/4$, the zero-crossings of solutions occur with roughly uniform frequency, and so the DE (8) may be said to be of *oscillatory type.*

When $q(x) < 0$, the DE (8) is said to be of *positive* type. One can also say by (20a) and (20b), that once $\theta(x)$ has entered the first or third quadrant, it can never escape from this quadrant; it is trapped in it. Therefore, a given solution $u(x)$ of (8) can have at most one zero if $q(x) < 0$; solutions are *nonoscillatory.* Moreover, since $uu' > 0$ in the first and third quadrants, $u^2(x)$ and hence $|u(x)|$ are perpetually *increasing* after a solution has been trapped in one of these quadrants.

Using more care, one can show that when $q(x) < 0$ the limit as $\alpha\downarrow - \infty$ of the solutions $u_\alpha(x)$ satisfying $u_\alpha(0) = 1$ and $u_\alpha(\alpha) = 0$ is an everywhere *increasing* positive solution. Moreover, replacing x by $-x$, which reverses the sign of $u'(x)$, one can construct similarly an everywhere *decreasing* positive solution. These two monotonic solutions (e^x and e^{-x} for $u'' - u = 0$) are usually unique, up to constant positive factors, and provide a natural basis of solutions.

Focal, Nodal, and Saddle Points.   Even more interesting than Sturm's theorems are the qualitative differences between the behavior of solutions of different second-order DEs that become apparent when we look at the corresponding trajectories in the phase plane (their so-called phase portraits). We shall discuss these for *nonlinear* DEs in Chapter 5; here we shall discuss only the linear, constant-coefficient case. We have already discussed this case briefly in §§2–3, primarily from an algebraic standpoint.

In the linear constant-coefficient case, using the letter v to signify u', we obviously have

$$(21) \qquad \frac{du}{dx} = v, \frac{dv}{dx} = -pv - qu.$$

Deferring to Chapter 5, §5, the discussion of the possibilities $q = 0$ and $\Delta = p^2 - 4q = 0$, the original DE $u'' + pu' + qu = 0$ has a basis of solutions of one of the following three main kinds: A) if $p^2 < 4q$, $e^{\alpha x} \cos kx$ and $e^{\alpha x} \sin kx$, B) if $p^2 > 4q > 0$, functions $e^{\alpha x}$ and $e^{\beta x}$, where α and β have the same sign, C) if $p^2 > 0 > 4q$, functions $e^{\alpha x}$ and $e^{\beta x}$ where α and β have opposite signs. These three cases give very different-looking configurations of trajectories in the phase plane.

Note that Cases B and C are subcases of the "Case 1" discussed in §1, while Case A coincides with "Case 2" discussed there. As will be explained in Chapter 5, §5, most of the qualitative differences to be pointed out below have analogues for *nonlinear* DEs of the general form $\dfrac{dv}{du} = F(u,v)$, of which the form

$$(21') \qquad \frac{dv}{du} = \frac{-pv - qu}{v}$$

of (21) is a special case.

Case A.   By (18), writing $\gamma = \cot \theta$, we have

$$\theta' = d\theta/dx = (\sin^2\theta)(\gamma^2 + p\gamma + q) > 0 \text{ for all } \theta.$$

Hence, θ increases monotonically. In each half-turn around the origin, r is amplified or damped by a factor $e^{|\alpha|\pi}$, according as $\alpha > 0$ or $\alpha < 0$. In either case there are no invariant lines; the critical point at $(0,0)$ is said to be a *focal point*. Figure 2.4a shows the resulting phase portrait for $u'' + 0.2u' + 4.01u = 0$.

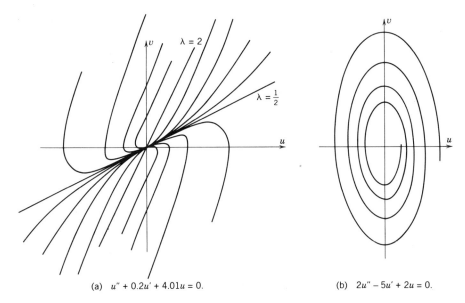

(a) $u'' + 0.2u' + 4.01u = 0.$ (b) $2u'' - 5u' + 2u = 0.$

Figure 2.4   Two phase portraits.

When $p^2 > 4q$ (i.e., in Cases B and C), the two lines $u' = \alpha u$ and $u' = \beta u$ in the phase plane are *invariant lines* (Ch. 1, §7). These lines, which correspond to the solutions $e^{\alpha x}$ and $e^{\beta x}$, divide the uu'-plane up into four sectors, in each of which θ' is of constant sign and so θ is monotonic. If $q = \alpha\beta > 0$, the two invariant lines lie in the same quadrant; if $q < 0$, they lie in adjacent quadrants.

Case B. In this case, the trajectories in each sector are all tangent at the origin to the same invariant line, and have an asymptotic direction parallel to the other invariant line at ∞. Fig. 2.4b depicts the phase portrait for $2u'' - 5u' + 2u = 0$. The lines $v = 2u$ and $u = 2v$ are the invariant lines of the corresponding linear fractional DE, $dv/du = (5v - u)/v$. In Case B, the origin is said to be a *nodal point*.

Case C. In the *saddle point* case that $p^2 > 0 > 4q$, the two invariant lines lie in different quadrants, and all trajectories are asymptotic to one of them as they come in from infinity, and to the other as they recede to it. Figure 1.5 depicts the phase portrait for the case $u'' = u$, with hyperbolic trajectories $u^2 - v^2 = 4AB$ in the phase plane given parametrically by $u = Ae^x + Be^{-x}$, $v = u' = Ae^x - Be^{-x}$. The invariant lines are the asymptotes $v = \pm u$.

EXERCISES C

1. (a) Show that if $g(x) = f'(x)$, then $g(x)$ vanishes at least once between any two zeros of $f(x)$.
   (b) Show how to construct, for any n, a function $f(x)$ satisfying $f(0) = f(1) = 0$, $f(x) \neq 0$ on $(0,1)$, yet for which $f'(x)$ vanishes n times on $(0,1)$.

2. Show that there is a zero of $J_1(x)$ between any two successive zeros of $J_0(x)$.

3. Show that every solution of $u'' + (1 + e^x)u = 0$ vanishes infinitely often on $(-\infty, 0)$, and also infinitely often on $(0;\infty)$.

4. Show that no nontrivial solution of $u'' + (1 - x^2)u = 0$ vanishes infinitely often.

*5. The Legendre polynomial $P_n(x)$ satisfies the DE $(1 - x^2)u'' - 2xu' + n(n + 1)u = 0$. Show that $P_n(x)$ must vanish $O(n)$ times on $[-1,1]$.

6. Apply numerical methods (Ch. 1, §8) to (18), to determine about how many times any solution of $u'' + xu = 0$ must vanish on $(0,100)$.

7. Same question for the Mathieu DE $u'' + [\pi^2 + 4.\cos 2x]u = 0$.

8. (a) Show that no normal second-order linear homogeneous DE can be satisfied by both some $\cos kx$ with $k \neq 0$, and some e^{ax}. [HINT: Consider the Wronskian.]
   (b) Find a normal third-order homogeneous linear DE that has as solutions both the oscillatory functions $\sin x$, $\cos x$, and the nonoscillatory function e^x.

8 ADJOINT OPERATORS; LAGRANGE IDENTITY

Early studies of differential equations concentrated on formal manipulations yielding solutions expressible in terms of familiar functions. Out of these studies emerged many useful concepts, including those of integrating factor and exact differential discussed in Ch. 1, §6. We will now extend these concepts to second-order *linear* DEs, and derive from them the extremely important notions of adjoint and self-adjoint equations.

DEFINITION. The second-order homogeneous linear DE

$$(22) \qquad L[u] = p_0(x)u''(x) + p_1(x)u'(x) + p_2(x)u(x) = 0$$

is said to be *exact* if and only if, for some $A(x), B(x) \in \mathcal{C}^1$,

$$(22') \qquad p_0(x)u'' + p_1(x)u' + p_2(x)u = \frac{d}{dx}[A(x)u' + B(x)u]$$

for all functions $u \in \mathcal{C}^2$. An *integrating factor* for the DE (22) is a function $v(x)$ such that $vL[u]$ is exact. [Here and later, it will be assumed that $p_0 \in \mathcal{C}^2$ and that $p_1, p_0 \in \mathcal{C}^1$ in discussing the DEs (22) and (22').]

If an integrating factor v for (22) can be found, then clearly

$$v(x)[p_0(x)u'' + p_1(x)u' + p_2(x)u] = \frac{d}{dx}[A(x)u' + B(x)u]$$

Hence, the solutions of the homogeneous DE (22) are those of the first-order inhomogeneous linear DE

$$(23) \qquad A(x)u' + B(x)u = C$$

where C is an arbitrary constant. Also, the solutions of the inhomogeneous DE $L[u] = r(x)$ are those of the first-order DE

$$(23') \qquad A(x)u' + B(x)u = \int v(x)r(x)\, dx + C$$

The DEs (23) and (23') can be solved by a quadrature (Ch. 1, §6). Hence, if an integrating factor of (22) can be found, we can reduce the solution $L[u] = r(x)$ to a sequence of quadratures.

Evidently, $L[u] = 0$ is exact in (22) if and only if $p_0 = A$, $p_1 = A' + B$, and $p_2 = B'$. Hence (22) is exact if and only if

$$p_2 = B' = (p_1 - A')' = p_1' - (p_0')'$$

This simple calculation proves the following important result.

LEMMA. *The DE (22) is exact if and only if its coefficient functions satisfy*

$$p_0'' - p_1' + p_2 = 0$$

COROLLARY. *A function $v \in \mathcal{C}^2$ is an integrating factor for the DE (22) if and only if it is a solution of the second-order homogeneous linear DE*

$$(24) \qquad M[v] = [p_0(x)v]'' - [p_1(x)v]' + p_2(x)v = 0$$

DEFINITION. The operator M in (24) is called the *adjoint* of the linear operator L. The DE (24), expanded to the DE

$$(24') \qquad p_0 v'' + (2p_0' - p_1)v' + (p_0'' - p_1' + p_2)v = 0$$

is called the *adjoint* of the DE (22).

Clearly, whenever a nontrivial solution of the adjoint DE (24) or (24') of a given second-order linear DE (22) can be found, every solution of any DE $L[u] = r(x)$ can be obtained by quadratures, using (23').

Lagrange Identity. The concept of the adjoint of a linear operator, which originated historically in the search for integrating factors, is of major importance because of the role which it plays in the theory of orthogonal and biorthogonal expansions. We now lay the foundations for this theory.

Substituting into (24), we find that the adjoint of the adjoint of a given second-order linear DE (20) is again the original DE (20). Another consequence of (24) is the identity, valid whenever $p_0 \in \mathcal{C}^2$, $p_i \in \mathcal{C}^1$,

$$vL[u] - uM[v] = (vp_0)u'' - u(p_0v)'' + (vp_1)u' + u(p_1v)'$$

Since $wu'' - uw' = (wu' - uw')'$ and $(uw)' = uw' + wu'$, this can be simplified to give the Lagrange identity

(25) $$vL[u] - uM[v] = \frac{d}{dx}[p_0(u'v - uv') - (p_0' - p_1)uv]$$

The left side of (25) is thus always an exact differential of a homogeneous bilinear expression in u,v, and their derivatives.

Self-Adjoint Equations.   Homogeneous linear DEs that coincide with their adjoint are of great importance; they are called *self-adjoint*. For instance, the Legendre DE of Example 2, §1, is self-adjoint. The condition for (22) to be self-adjoint is easily derived. It is necessary by (24') that $2p_0' - p_1 = p_1$, that is, $p_0' = p_1$. Since this relation implies $p_0'' - p_1' = 0$, it is also sufficient. Moreover, in this self-adjoint case, the last term in (25) vanishes. This proves the first statement of the following theorem.

THEOREM 9.   *The second-order linear* DE (22) *is self-adjoint if and only if it has the form*

(26) $$\frac{d}{dx}\left[p(x)\frac{du}{dx}\right] + q(x)u = 0$$

The DE (22) *can be made self-adjoint by multiplying through by*

(26') $$h(x) = \left[\exp\int(p_1/p_0)\,dx\right]/p_0.$$

To prove the second statement, first reduce (22) to normal form by dividing through by p_0, and then observe that the DE

$$hu'' + (ph)u' + (qh)u = 0$$

is self-adjoint if and only if $h' = ph$, or $h = \exp(\int p\,dx)$.

For example, the self-adjoint form of the Bessel DE of Example 1 is

$$(xu')' + [x - (n^2/x)]u = 0$$

For self-adjoint DEs (26), the Lagrange identity simplifies to

(26'') $$vL[u] - uL[v] = \frac{d}{dx}[p(x)(u'v - uv')]$$

EXERCISES D

1. Show that if $u(x)$ and $v(x)$ are solutions of the self-adjoint DE

$$(pu')' + q(x)u = 0$$

then $p(x)[uv' - vu']$ is a constant (Abel's identity).

2. Reduce the following DEs to self-adjoint form:
   (a) $(1 - x^2)u'' - xu' + \lambda u = 0$ (Chebyshev DE)
   (b) $x^2u'' + xu' + u = 0$ (c) $u'' + u' \tan x = 0$

3. For each of the following DEs, $y = x^3$ is one solution; use (12) to find a second, linearly independent solution by quadratures.
   (a) $x^2y'' - 4xy' + 6y = 0$ (b) $xy'' + (x - 2)y' - 3y = 0$

4. Show that the substitution $y = e^{\int p\,dx/2}u$ replaces (8) by

$$y'' + I(x)y = 0, \qquad I(x) = q - \frac{p^2}{4 - p'/2}$$

*5. Show that two DEs of the form (8) can be transformed into each other by a change of dependent variable of the form $y = v(x)u, v \neq 0$, if and only if the function $I(x) = q(x) - p^2(x)/4 - p'(x)/2$ is the same for both DEs [$I(x)$ is called the *invariant* of the DE].

6. Reduce the self-adjoint DE $(pu')' + qu = 0$ to normal form, and show that, in the notation of Ex. 5, $I(x) = (p'^2 - 2pp'' + 4pq)/4p^2$.

7. (a) Show that, for the normal form of the Legendre DE $[(1 - x^2)u']' + \lambda u = 0$

$$I(x) = \frac{(1 + \lambda - \lambda x^2)}{(1 - x^2)^2} \qquad \text{(Use Ex. 6.)}$$

   (b) Show that, if $\lambda = n(n + 1)$, then every solution of the Legendre equation has at least $(2n + 1)/\pi$ zeros on $(-1, 1)$.

8. Let $u(x)$ be a solution of $u'' = q(x)u$, $q(x) > 0$ such that $u(0)$ and $u'(0)$ are positive. Show that uu' and $u(x)$ are increasing for $x > 0$.

9. Let $h(x)$ be a nonnegative function of class $\mathcal{C}^1$. Show that the change of independent variable $t = \int_a^x h(s)\,ds$, $u(x) = v(t)$, changes (8) into $v'' + p_1(t)v' + q_1(t)v = 0$, where $p_1(t) = [p(x)h(x) + h'(x)]/h(x)^2$ and $q_1(t) = q(x)/h(x)^2$.

10. (a) Show that a change of independent variable $t = \pm \int |q(x)|^{1/2}\,dx$, $q \in \mathcal{C}^1$ changes the DE (8) into one whose normal form is

(*) $$\frac{d^2u}{dt^2} + \left(\frac{q' + 2pq}{2|q|^{3/2}}\right)\frac{du}{dt} \pm u = 0$$

   (b) Show that no other change of independent variable makes $|q| = 1$

*11. Using Ex. 10, show that Eq. (8) is equivalent to a DE with constant coefficients under a change of independent variable if and only if $(q' + 2pq)/q^{3/2}$ is a constant.

12. Making appropriate definitions, show that $p_0u''' + p_1u'' + p_2u' + p_3u = 0$ is an exact DE if and only if $p_0''' - p_1'' + p_2' - p_3 = 0$.

9   GREEN'S FUNCTIONS

The *inhomogeneous* linear second-order DE in normal form,

$$(27) \qquad L[u] = \frac{d^2u}{dx^2} + p(x)\frac{du}{dx} + q(x)u = r(x)$$

differs from the homogeneous linear DE

$$(27') \qquad L[u] = \frac{d^2u}{dx^2} + p(x)\frac{du}{dx} + q(x)u = 0$$

by the nonzero function $r(x)$ on the right side. In applications to electrical and dynamical systems, the function $r(x)$ is called the *forcing term* or *input function*. By the Uniqueness Theorem of §4 and Lemma 2 of §4, it is clear that the solution $u(x)$ of $L[u] = r(x)$ for given homogeneous initial conditions such as $u(0) = u'(0) = 0$ depends linearly on the forcing term. We will now determine the exact nature of this linear dependence.

Given the inhomogeneous linear DE (27), we will show that there exists an integral operator G,

$$(28) \qquad G[r] = \int_a^x G(x, \xi)r(\xi)\, d\xi$$

such that $G[r] = u$. In fact, one can always find a function G that makes $G[r]$ satisfy given homogeneous boundary conditions, provided that the latter define a well-set problem.

The kernel $G(x, \xi)$ of Eq. (28) is then called the *Green's function*† associated with the given boundary value problem. In operator notation, it is defined by the identity $L[G[r]] = r$ (G is a "right-inverse" of the linear operator L) and the given boundary conditions.

Green's functions can be defined for linear differential operators of any order, as we will show in Ch. 3, §9. To provide an intuitive basis for this very general concept, we begin with the simplest, first-order case. In this example, the independent variable will be denoted by t and should be thought of as representing time.

Example 3.   Suppose that money is deposited continuously in a bank account, at a continuously varying rate $r(t)$, and that interest is compounded continuously at a constant rate p ($=100p\%$ per annum). As a function of time, the amount

---

† To honor the British mathematician George Green (1793–1841), who was the first to use formulas like (28) to solve boundary value problems. Cauchy and Fourier used similar formulas earlier to solve DEs in infinite domains.

$u(t)$ in the account satisfies the DE

$$\frac{du}{dt} = pu + r(t)$$

If the account is opened when $t = 0$ and initially has no money: $u(0) = 0$, then one can calculate $u(T)$ at any later time $T > 0$ as follows. Each infinitesimal deposit $r(t)\, dt$, made in the time interval $(t, t + dt)$, increases through compound interest accrued during the time interval from t to T by a factor $e^{p(T-t)}$ to the amount $e^{p(T-t)}r(t)\, dt$. Hence the account should amount, at time T, to the integral (limit of sums)

$$(29) \qquad u(T) = \int_0^T e^{p(T-t)}r(t)\, dt = e^{pT} \int_0^T e^{-pt}r(t)\, dt$$

This plausible argument is easily made rigorous. It is obvious that $u(0) = 0$ in (29). Differentiating the product in the final expression of (29), we obtain

$$u'(T) = pe^{pT} \int_0^T e^{-pt}r(t)\, dt + e^{pT}e^{-pT}r(T) = pu(T) + r(T)$$

where the derivative of the integral is evaluated by the Fundamental Theorem of the Calculus.

Example 4.   Consider next the motion of a mass m on an elastic spring, which we model by the DE $u'' + pu' + qu = r(t)$. Here p signifies the damping coefficient and q the restoring force; $r(t)$ is the forcing function; we will assume that $q > p^2/4$. Finally, suppose that the mass is at rest up to time t_0, and is then given an impulsive (that is, instantaneous) velocity v_0 at time t_0.

The function f describing the position of the mass m as a function of time under such conditions is continuous, but its derivative f' is not defined at t_0, because of the sudden jump in the velocity. However, the left-hand derivative of f at the point t_0 exists and is equal to zero, and the right-hand derivative also exists and is equal to v_0, the impulsive velocity. For $t > t_0$, the function f is obtained by solving the constant-coefficient DE $u'' + pu' + qu = 0$. Since $q > p^2/4$, the roots of the characteristic equation are complex conjugate, and we obtain an oscillatory solution

$$u(t) = \begin{cases} 0 & t < t_0 \\ (v_0/\nu)e^{-\mu(t-t_0)} \sin \nu(t - t_0) & t \geq t_0 \end{cases}$$

where $\mu = p/2$ and $\nu = \sqrt{q - p^2/4}$.

Now suppose the mass is given a sequence of small velocity impulses $\Delta v_k = r(t_k)\,\Delta t$, at successive instants t_0, $t_1 = t_0 + \Delta t, \ldots, t_k = t_0 + k\,\Delta t, \ldots$. Summing the effects of these over the time interval $t_0 \leq t \leq T$, and passing to the limit as $\Delta t \to 0$, we are led to conjecture the formula

$$(30) \qquad u(T) = \int_{t_0}^{T} \frac{1}{\nu} e^{-\mu(T-t)} \sin \nu(T - t) r(t)\, dt$$

This represents the *forced* oscillation associated with the DE

$$(31) \qquad u'' + pu' + qu = r(t), \qquad q > p^2/4$$

having the forcing term $r(t)$.

Variation of Parameters. The conjecture just stated can be verified as a special case of the following general result, valid for all linear second-order DEs with continuous coefficients.

THEOREM 10. *Let the function $G(t, \tau)$ be defined as follows:*
(i) $G(t, \tau) = 0$, *for* $a \leq t \leq \tau$,
(ii) *for each fixed* $\tau \geq a$ *and all* $t > \tau$, $G(t, \tau)$ *is that particular solution of the* DE
  $L[G] = G_{tt} + p(t)G_t + q(t)G = 0$ *which satisfies the initial conditions* $G = 0$
  and $G_t = 1$ *at* $t = \tau$.
Then G is the Green's function of the operator L for the initial value problem on $t \leq a$.

Proof. We must prove that, for any continuous function r, the definite integral

$$(32) \qquad u(t) = \int_{a}^{t} G(t, \tau) r(\tau)\, d\tau = \int_{a}^{\infty} G(t, \tau) r(\tau)\, d\tau \qquad \text{[by (i)]}$$

is a solution of the second-order inhomogeneous linear DE (27), which satisfies the initial conditions $u(a) = u'(a) = 0$.

The proof is based on Leibniz' Rule for differentiating definite integrals.†
This rule is: For any continuous function $g(t, \tau)$ whose derivative $\partial g/\partial t$ is piecewise continuous, we have

$$\frac{d}{dt} \int_{a}^{t} g(t, \tau)\, d\tau = g(t, t) + \int_{a}^{t} \frac{\partial g}{\partial t} (t, \tau)\, d\tau$$

---

† Kaplan, *Advanced Calculus*, p. 219. In our applications, $\partial g/\partial t$ has, at worst, a simple jump across $t = \tau$.

Applying this rule twice to the right side of formula (32), we obtain successively, since $G(t, t) = 0$,

$$u'(t) = G(t, t)r(t) + \int_a^t G_t(t, \tau)r(\tau) \, d\tau = \int_a^t G_t(t, \tau)r(\tau) \, d\tau$$

$$u''(t) = G_t(t, t)r(t) + \int_a^t G_{tt}(t, \tau)r(\tau) \, d\tau$$

By assumption (ii), the last equation simplifies to

$$u''(t) = r(t) + \int_a^t G_{tt}(t, \tau)r(\tau) \, d\tau$$

Here the subscripts indicate partial differentiation with respect to t. Thus

$$L[u] = u''(t) + p(t)u'(t) + q(t)u(t)$$

$$= r(t) + \int_a^t [G_{tt}(t, \tau) + p(t)G_t(t, \tau) + q(t)G(t, \tau)]r(\tau) \, d\tau = r(t)$$

completing the proof.

The reader can easily verify that the function $\nu^{-1}e^{-\mu(t-\tau)} \sin \nu(t - \tau)$ in (30) satisfies the conditions of Theorem 10, in the special case of Example 4.

To construct the Green's function $G(t, \tau)$ of Theorem 9 explicitly, it suffices to know two linearly independent solutions $f(t)$ and $g(t)$ of the reduced equation $L[u] = 0$. Namely, to compute $G(t, \tau)$ for $t > \tau$ write $G(t, \tau) = c(\tau)f(t) + d(\tau)g(t)$, by Theorem 3. Solving the simultaneous linear equations $G = 0$, $G_t = 1$, at $t = \tau$ specified in condition (ii) of Theorem 10

$$c(\tau)f(\tau) + d(\tau)g(\tau) = 0, \qquad c(\tau)f'(\tau) + d(\tau)g'(\tau) = 1$$

we get the formulas

$$c(\tau) = -g(\tau)/W(f, g; \tau), \qquad d(\tau) = f(\tau)/W(f, g; \tau)$$

where $W(f, g; \tau) = f(\tau)g'(\tau) - g(\tau)f'(\tau)$ is the Wronskian. This gives for the Green's function the formula

$$G(t, \tau) = [f(\tau)g(t) - g(\tau)f(t)]/[f(\tau)g'(\tau) - g(\tau)f'(\tau)]$$

Substituting into (32), we obtain our final result.

COROLLARY.   *Let $f(t)$ and $g(t)$ be any two linearly independent solutions of the linear homogeneous DE (27′). Then the solution of $L[u] = r(t)$ for the initial conditions*

$u(a) = u'(a) = 0$ *is the function*

(33) $$u(t) = \int_a^t \frac{f(\tau)g(t) - g(\tau)f(t)}{W[f(\tau),g(\tau)]} r(\tau)\, d\tau$$

Consequently, if we define the functions $\phi(t)$ and $\psi(t)$ as the following definite integrals:

$$\phi(t) = \int_a^t \frac{f(\tau)}{W(f,g)} r(\tau)\, d\tau, \qquad \psi(t) = -\int_a^t \frac{g(\tau)}{W(f,g)} r(\tau)\, d\tau$$

we can write the solution of $L[u] = r(t)$, in the form

(33') $$u(t) = \phi(t)g(t) + \psi(t)f(t)$$

In textbooks on the elementary theory of DEs, formula (33) is often derived formally by posing the question: What must $c(\tau)$ and $d(\tau)$ be in order that the function

$$G(t, \tau) = c(\tau)f(t) + d(\tau)g(t)$$

when substituted into (28), will give a solution of the inhomogeneous DE $L[u] = r(t)$? Since $c(\tau)$ and $d(\tau)$ may be regarded as "variable parameters," which vary with τ, formula (33) is said to be obtained by the method of *variation of parameters*.

EXERCISES E

1. Integrate the following DEs by using formula (33):
   (a) $y'' - y = x^n$ (b) $y'' + y = e^t$
   (c) $y'' - qy' + y = 2xe^x$ (d) $y'' + 10y' + 25y = \sin x$

2. Show that the general solution of the inhomogeneous DE $y'' + k^2y = R(x)$ is given by $y = (1/k)[\int_a^t \sin k(x - t)R(t)\, dt] + c_1 \sin kx + c_2 \cos kx$.

3. Solve $y'' + 3y' + 2y = x^3$ for the initial conditions $y(0) = y'(0) = (0)$.

4. Show that any second-order inhomogeneous linear DE which is satisifed by both x^2 and $\sin^2 x$ must have a singular point at the origin.

5. Construct Green's functions for the initial-value problem, for the following DEs:
   (a) $u'' = 0$ (b) $u'' = u$ (c) $u'' + u = 0$
   (d) $x^2u'' + (x^2 + 2x)u' + (x + 2)u = 0$ [HINT: x is a solution.]

6. Find the general solutions of the following inhomogeneous Euler DEs:
   (a) $x^2y'' - 2xy' + 2y = x^2 + px + q$ (b) $x^2y'' + 3xy' + y = R(x)$
   [HINT: Any homogeneous Euler DE has a solution of the form x^r.]

7. (a) Construct a Green's function for the inhomogeneous first-order DE

$$du/dt = p(t)u + r(t)$$

(b) Interpret in terms of compound interest (cf. Example 3).

(c) Relate to formula (8′) of Ch. 1.

8. Show that, if $q(t) < 0$, then the Green's function $G(t, \tau)$ of $u_{tt} + q(t)u = 0$ for the initial-value problem is positive and convex upward for $t > \tau$.

*10 TWO-ENDPOINT PROBLEMS

So far, we have considered only "initial conditions." That is, in considering solutions of second-order DEs such as $y'' = -p(x)y' - q(x)y$, we have supposed y and y' both given at the same point a. That is natural in many dynamical problems. One is given the initial position and velocity, and a general law relating the acceleration to the instantaneous position and velocity, and then one wishes to determine the subsequent motion from this data, as in Example 4.

In other problems, *two-endpoint* conditions, at points $x = a$ and $x = b$, are more natural. For instance, the DE $y'' = 0$ characterizes straight lines in the plane, and one may be interested in determining the straight line joining two given points (a, c) and (b, d). That is, the problem is to find the solution $y = f(x)$ of the DE $y'' = 0$ which satisfies the *two* endpoint conditions $f(a) = c$ and $f(b) = d$.

Many two-endpoint problems for second-order DEs arise in the calculus of variations. Here a standard problem is to find, for a given function $F(x,y,y')$, the curve $y = f(x)$ which minimizes the integral

$$(34) \qquad I(f) = \int_a^b F(x,y,y')\, dx$$

By a classical result of Euler,† the line integral (34) is an extremum (maximum, minimum, or minimax), relative to all curves $y = f(x)$ of class $\mathcal{C}^2$ satisfying $f(a) = c$ and $f(b) = d$, if and only if $f(x)$ satisfies the Euler-Lagrange variational equation

$$(34') \qquad \frac{d}{dx}\left(\frac{\partial F}{\partial y'}\right) = \left(\frac{\partial F}{\partial y}\right)$$

For example, if $F(x,y,y') = \sqrt{1 + y'^2}$ so that $I(f)$ is the length of the curve, Eq. (34′) gives zero curvature:

$$\frac{d}{dx}\left(\frac{y'}{\sqrt{1 + y'^2}}\right) = y''\left[\frac{1}{(1 + y'^2)^{1/2}} - \frac{y'^2}{(1 + y'^2)^{3/2}}\right] = \frac{y''}{(1 + y'^2)^{3/2}} = 0$$

---

† See for example Courant and John, Vol. 2, p. 743.

as the condition for the length to be an extremum. This is equivalent to $y'' = 0$, whose solutions are the straight lines $y = cx + d$.

It is natural to ask: Under what circumstances does a second-order DE have a unique solution, assuming given values $f(a) = c$ and $f(b) = d$ at two given endpoints a and $b > a$? When this is so, the resulting two-endpoint problem is called *well-set*. Clearly, the two-endpoint problem is always well-set for $y'' = 0$.

Example 5.   Now consider, for given $p, q, r \in C^1$, the curves that minimize the integral (34) for $F = \frac{1}{2}[p(x)y'^2 + 2q(x)yy' + r(x)y^2]$. For this F, the Euler–Lagrange DE is the second-order linear self-adjoint DE $(py')' + (q' - r)y = 0$. The question of when the two-endpoint problem is well-set in this example is partially answered by the following result.

THEOREM 10.   *Let the second-order linear homogeneous* DE

$$(35) \qquad p_0(x)u'' + p_1(x)u' + p_2(x)u = 0 \qquad p_0(x) \neq 0$$

with continuous coefficient-functions have two linearly independent solutions.† Then the two-endpoint problem defined by (35) *and the endpoint conditions* $u(a) = c$, $u(b) = d$ *is well-set if and only if no nontrivial solution satisfies the endpoint conditions*

$$(36) \qquad u(a) = u(b) = 0$$

Proof.   By Theorem 2, the general solution of the DE (35) is the function $u = \alpha f(x) + \beta g(x)$, where f and g are a basis of solutions of the DE (35), and α, β are arbitrary constants. By the elementary theory of linear equations, the equations

$$\alpha f(a) + \beta g(a) = c \qquad \alpha f(b) + \beta g(b) = d$$

have one and only one solution vector (α, β) if and only if the determinant $f(a)g(b) - g(a)f(b) \neq 0$. The alternative $f(a)g(b) = f(b)g(a)$ is, however, the condition that the *homogeneous* simultaneous linear equations

$$(37) \qquad \alpha f(a) + \beta g(a) = \alpha f(b) + \beta g(b) = 0$$

have a nontrivial solution $(\alpha, \beta) \neq (0, 0)$. This proves Theorem 11.

When the DE (35) has a nontrivial solution satisfying the homogeneous endpoint conditions $u(a) = u(b) = 0$, the point $(b, 0)$ on the x-axis is called a *conjugate point* of the point $(a, 0)$ for the given homogeneous linear DE (35) or for a variational problem leading to this DE. In general, such conjugate points exist

---

† In Ch. 6, it will be shown that this hypothesis is unnecessary; a basis of solutions always exists for continuous $p_i(x)$.

for DEs whose solutions *oscillate* but not for those of nonoscillatory type, such as $u'' = q(x)u$, $q(x) > 0$.

Thus, in Example 5, let $p = 1$, $q = 0$, and $r = -k^2 < 0$. Then the general solution of (35) for the initial condition $u(a) = 0$ is $u = A \sin [k(x - a)]$. For $u(b) = 0$ to be compatible with $A \neq 0$, it is necessary and sufficient that $b = a + (n\pi/k)$. The conjugate points of a are spaced periodically. On the other hand, the DE $y'' - \lambda y = 0$, corresponding to the choice $p = 1$, $q = 0$, $r = \lambda$ in Example 5, admits no conjugate points if $\lambda = r$ is positive.

*11   GREEN'S FUNCTIONS, II

We now show that, except in the case that a and b are conjugate points for the reduced equation $L[u] = 0$, the inhomogeneous linear DE (27) can be solved for the boundary conditions $u(a) = u(b) = 0$ by constructing an appropriate Green's function $G(x, \xi)$ on the square $a \leq x, \xi \leq b$ and setting

$$(38) \qquad u(x) = \int_a^b G(x, \xi)r(\xi) \, d\xi = G[r]$$

Note that G is an *integral operator* whose kernel is the Green's function $G(x, \xi)$.

The existence of a Green's function for a typical two-endpoint problem is suggested by simple physical considerations, as follows.

Example 6.   Consider a nearly horizontal taut string under constant tension T, supporting a continuously distributed load $w(x)$ per unit length. If $y(x)$ denotes the vertical displacement of the string, then the load $w(x) \, \Delta x$ supported by the string in the interval $(x_0, x_0 + \Delta x)$ is in equilibrium with the net vertical component of tension forces, which is

$$T\{y'(x_0 + \Delta x) - y'(x_0)\}$$

in the nearly horizontal ("small amplitude") approximation.† Dividing through by Δx and letting $\Delta x \downarrow 0$, we get $Ty''(x) = w(x)$.

The displacement $y(x)$ depends linearly on the load, by Lemma 2 of §4. This suggests that we consider the load as the sum of a large number of point-concentrated loads $w_i = w(\xi_i) \, \Delta \xi_i$ at isolated points ξ_i. For each individual such load, the taut string consists of two straight segments, the slope jumping by w_i/T at ξ_i. Thus, if the string extends from 0 to 1, the vertical displacement is

$$w_i G(x, \xi_i) = \begin{cases} \epsilon_i(\xi_i - 1)x & 0 < x \leq \xi_i \\ \epsilon_i \xi_i (x - 1) & \xi_i \leq x \leq 1 \end{cases}$$

---

† For a more thorough discussion, see J. L. Synge and B. A. Griffith, *Principles of Mechanics*, McGraw-Hill, 1949, p. 99.

where ϵ_i is set equal to w_i/T in order to give a jump in slope of w_i/T at the point $x = \xi_i$. Passing to the limit as the $\Delta\xi_i \downarrow 0$, we are led to guess that

$$y(x) = \int_0^1 G(x, \xi)w(\xi)\,d\xi$$

where

$$G(x, \xi) = \begin{cases} (\xi - 1)x/T & 0 \le x \le \xi \\ \xi(x - 1)/T & \xi \le x \le 1 \end{cases}$$

These heuristic considerations suggest that, in general, the Green's function $G(x, \xi)$ for the two-endpoint problem is determined for each fixed ξ by the following four conditions:

(i) $L[G] = 0$ in each of the intervals $a \le x \le \xi$ and $\xi \le x \le b$.

(ii) $G(a, \xi) = G(b, \xi) = 0$.

(iii) $G(x, \xi)$ is continuous across the diagonal $x = \xi$ of the square $a \le x, \xi \le b$ over which $G(x, \xi)$ is defined.

(iv) The derivative $\partial G/\partial x$ jumps by $1/p_0(x)$ across this diagonal. To fulfill these conditions for any given ξ, let $f(x)$ and $g(x)$ be any nontrivial solutions of $L[u] = 0$ that satisfy $f(a) = 0$ and $g(b) = 0$, respectively. Then for any factor $\epsilon(\xi)$, the function

$$G(x, \xi) = \begin{cases} \epsilon(\xi)f(x)g(\xi) & a \le x \le \xi \\ \epsilon(\xi)f(\xi)g(x) & \xi \le x \le b \end{cases}$$

will satisfy $L[G] = 0$ in the required intervals because $L[f] = L[g] = 0$; it will satisfy (ii) because $f(a) = g(b) = 0$; and it approaches the same limit $\epsilon(\xi)f(\xi)g(\xi)$ from both sides of the diagonal $x = \xi$; hence it is continuous there. For the factor $\epsilon(\xi)$ to give $\partial G/\partial x$ a jump of $1/p_0(x)$ across $x = \xi$, a direct computation gives the condition

$$(39) \qquad \frac{\partial G}{\partial x}(\xi^+, \xi) - \frac{\partial G}{\partial x}(\xi^-, \xi) = \epsilon(\xi)\{f(\xi)g'(\xi) - g(\xi)f'(\xi)\} = 1/p_0(\xi)$$

We are therefore led to try the kernel

$$(39') \qquad G(x, \xi) = \begin{cases} f(x)g(\xi)/p_0(\xi)W(\xi) & a \le x \le \xi \\ f(\xi)g(x)/p_0(\xi)W(\xi) & \xi \le x \le b \end{cases}$$

where $W = fg' - gf'$ is the Wronskian of f and g. Observe again that since $f(a) = g(b) = 0$, $G(a, \xi) = G(b, \xi) = 0$ *for all* $\xi \in [a, b]$.

THEOREM 11′. *For any continuous function r(x) on [a, b], the function u(x) ∈ $\mathcal{C}^2$ defined by (39) and (39′) is the solution of $p_0 u'' + p_1 u' + p_2 u = r$ that satisfies the boundary conditions of u(a) = u(b) = 0, provided that W(f, g) ≠ 0, i.e., that there is no nontrivial solution of (35) satisfying the same boundary conditions.*

The proof is similar to that of Theorem 10; the existence of two linearly independent solutions of (35) is again assumed. Rewriting (38) in the form

$$u(x) = \int_a^x G(x, \xi) r(\xi)\, d\xi + \int_x^b G(x, \xi) r(\xi)\, d\xi$$

and differentiating by Leibniz' Rule, we have

$$u'(x) = \int_a^x G_x(x, \xi) r(\xi)\, d\xi + \int_x^b G_x(x, \xi) r(\xi)\, d\xi$$

The endpoint contributions cancel since $G(x, \xi)$ is continuous for $x = \xi$. Differentiating again, we obtain

$$u''(x) = \int_a^x G_{xx}(x, \xi) r(\xi)\, d\xi + G_x(x, x^-) r(x^-)$$
$$+ \int_x^b G_{xx}(x, \xi) r(\xi)\, d\xi - G_x(x, x^+) r(x^+)$$

where $f(x^+)$ signifies the limit of $f(\xi)$ as ξ approaches x from above, and $f(x^-)$ the limit as ξ approaches x from below. The two terms corresponding to the contributions from the endpoints come from the sides $\xi < x$ and $\xi > x$ of the diagonal; since r is continuous, $r(x^-) = r(x^+)$. Hence their difference is $[G_x(x^+, x) - G_x(x^-, x)] r(x)$, which equals $r(x/p_0(x))$ by (39). Simplifying, we obtain

$$u''(x) = \int_a^b G_{xx}(x, \xi) r(\xi)\, d\xi + \frac{r(x)}{p_0(x)}$$

From this identity, we can calculate $L[u]$. It is

$$L[u] = \int_a^b L[G(x, \xi)] r(\xi)\, d\xi + r(x) = r(x)$$

since $L[G(x, \xi)] = 0$ except at $x \equiv \xi$. Here the operator L acts on the variable x in $G(x, \xi)$; though G is not in $\mathcal{C}^2$, the expression $L[G]$ is meaningful for one-sided derivatives and the above can be justified. This gives the identity (38).

Since $G(x, \xi)$, considered as a function of x for fixed ξ, satisfies the boundary conditions $G(a, \xi) = G(b, \xi) = 0$, it follows from (38) that $u(a) = u(b) = 0$, completing the proof of the theorem.

Delta-Function Interpretation.   The ideas underlying the intuitive discussion for Examples 4 and 5 can be given the following heuristic interpretation. Let the symbolic function $\delta(x)$ stand for the limit of nonnegative "density" functions $p(x)$ concentrated in a narrow interval $(-\epsilon, \epsilon)$ near $x = 0$, with total mass $\int_{-\epsilon}^{\epsilon} p(x)\, dx = 1$, as $\epsilon \downarrow 0$. Likewise, $\delta(x - \xi)$ stands for the density of a unit mass (or charge) concentrated at $x = \xi$.

For any $f \in C[a, b]$ $a < 0 < b$ and any such p with support $(-\epsilon, \epsilon) \subset [a, b]$, we will have by the Second Law of the Mean for integrals

$$\int_a^b f(x)p(x)\, dx = \int_{-\epsilon}^{\epsilon} f(x)p(x)\, dx = f(x_1) \int_{-\epsilon}^{\epsilon} p(x)\, dx = f(x_1)$$

where $-\epsilon < x_1 < \epsilon$. Letting ϵ approach zero, we get in the limit

$$(40) \qquad \int_a^b f(x)\delta(x)\, dx = f(0)$$

Translating through ξ, we have similarly

$$(40') \qquad \int_a^b \delta(x - \xi)f(x)\, dx = f(\xi), \qquad \xi \in (a, b)$$

In particular, setting $f(x)$ equal to one, we get

$$(41) \qquad \int_a^b \delta(x - \xi)\, dx = \begin{cases} 1 & \text{if } \xi \in (a, b) \\ 0 & \text{if } \xi \notin [a, b] \end{cases}$$

Finally, the Green's function of a differential operator L and given *homogeneous* linear initial or boundary conditions satisfies the symbolic equation

$$(42) \qquad L_x G(x, \xi) = \delta(x - \xi)$$

and the same initial or boundary conditions (in x). Now consider the function

$$(43) \qquad u(x) = \int G(x, \xi)r(\xi)\, d\xi$$

Extending heuristically the Superposition Principle to integrals (considered as limits of sums), we are led to the good guess that $u(x)$ satisfies the same initial (resp. boundary) conditions and also

$$(44) \qquad L[u] = L_x\left[\int G(x, \xi)r(\xi)\, d\xi \right] = \int \delta(x - \xi)r(\xi)\, d\xi = r(x)$$

EXERCISES F

In Exs. 1–3, (a) construct Green's functions for the two-endpoint problem defined by the DE specified and the boundary conditions $u(0) = u(1) = 0$, and (b) solve for $r(x) = x^2$:

1. $u'' - u = r(x)$　　　　　　　　　　2. $u'' + 4u = r(x)$

3. $u'' - \dfrac{4x}{2x - 1} u' + \dfrac{4}{2x - 1} u = r(x)$

In Exs. 4–5, find the conjugate points nearest to $x = 0$ for the DE specified.

4. $u'' + 2u' + 10u = 0$

5. $(x^2 - x + 1)u'' + (1 - 2x)u' + 2u = 0$ [HINT: Look for polynomial solutions.]

*6. $u_{tt} - u_t + e^{2t}u = 0$

7. (a) Show that, for two-endpoint problems containing no pairs of conjugate points, Green's function is always negative.
   (b) Show that, if $q(x) < 0$, then the Green's function for $u'' + q(x)u = 0$ in the two-endpoint problem is always negative and convex (concave downward), with negative slope where $x < \xi$ and positive slope where $x > \xi$.

*8. Set $F(x, y, y') = y'^2(1 - y')^2$ in (34), and find the curves joining $(0, 0)$ and $(1, \tfrac{1}{2})$ which minimize $I(f)$.

9. Show that the Euler-Lagrange DE for $F(x, y, y') = g\rho(x)y + \tfrac{1}{2} Ty'^2$ (g, T constants) is $Ty'' = g\rho(x)$. Relate to the sag of a loaded string under tension T.

ADDITIONAL EXERCISES

1. Show that the ratio $v = f/g$ of any two linearly independent solutions of the DE $u'' + q(x)u = 0$ is a solution of the third-order nonlinear DE

(*)　　　　　　　$$S[v] = \frac{v''}{v'} - \frac{3}{2}\left(\frac{v''}{v'}\right)^2 = 2q(x)$$

2. The *Schwarzian* $S[v]$ of a function $v(x)$ being defined by the middle term of (*), show that $S[(av + b)/(cv + d)] = S[\infty]$ for any four constants, a, b, c, d with $ad \neq be$.

*3. Prove that, if v_0, v_1, v_2, v_3 are any four distinct solutions of the Riccati DE, their cross ratio is constant: $(v_0 - v_1)(v_3 - v_2)/(v_0 - v_2)(v_3 - v_1) = c$.

4. Find the general solutions of the following inhomogeneous Euler DEs:
   (a) $x^2y'' - 2xy' + 2y = x^2 + px + q$　　(b) $x^2y'' + 3xy' + y = R(x)$

5. (a) Show that, if f and g satisfy $u'' + q(x)u = 0$, the product $fg = y$ satisfies the DE $2yy'' = (y')^2 - 4y^2g(x) + c$ for some constant c.
   (b) As an application, solve $2yy'' = (y')^2 - (x + 1)^{-2}y^2$.

6. Show that, if u is the general solution of the DE (1) of the text, and $W_{ij} = p_i p_j' - p_i' p_j$, then $v = u'$ is the general solution of

$$p_0 p_2 v'' + (p_1 p_2 - W_{02})v' + (p_2^2 - W_{12})v = W_{23}.$$

7. (a) Show that the Riccati equation $y' = 1 + x^2 + y^2$ has no solution on the interval $(0, \pi)$.
   (b) Show that the Riccati equation $y' = 1 + y^2 - x^2$ has a solution on the interval $(-\infty, +\infty)$.

8. Let $\phi_k(x) = \dfrac{1}{\pi}\dfrac{d}{dx}$ (arctan $kx = k/\pi(1 + k^2x^2)$. Show that, if $f(x)$ is any continuous function bounded on $(-\infty, \infty)$, then $\lim_{k\uparrow\infty}\int_{-\infty}^{\infty}\phi_k(x - c)f(x)\,dx = f(c)$.

9. For the DE $u'' + (B/x^2)u = 0$, show that every solution has infinitely many zeros on $(1, +\infty)$ if $B > \frac{1}{4}$ and a finite number if $B < \frac{1}{2}$. [HINT: The DE is an Euler DE.]

10. For the DE $u'' + q(x)u = 0$, show that every solution has a finite number of zeros on $(1, +\infty)$ if $q(x) < \frac{1}{4}x^2$, and infinitely many if $q(x) > B/x^2$, $B > \frac{1}{4}$.

*11. For the DE $u'' + q(x)u = 0$, show that every solution has infinitely many zeros on $(1, +\infty)$ if

$$\int_1^\infty \left[xq(x) - \frac{1}{4x} \right] dx = +\infty.$$

*12. Show that, if $p, q \in \mathcal{C}^2$, we can transform the DE (8) to the form $d^2z/d\xi^2 = 0$ in some neighborhood of the y-axis by transformations of the form $z = f(x)y$ and $d\xi = h(x)\,dx$. [HINT: Transform a basis of solutions to $y_1 = 1, y_2 = \xi$.]

CHAPTER 3

LINEAR EQUATIONS WITH CONSTANT COEFFICIENTS

1 THE CHARACTERISTIC POLYNOMIAL

So far, we have discussed only first- and second-order DEs, primarily because so few DEs of higher order can be solved explicitly in terms of familiar functions. However, general algebraic techniques make it possible to solve *constant-coefficient* linear DEs of arbitrary order and to predict many properties of their solutions, including especially their *stability* or instability.

This chapter will be devoted to explaining and exploiting these techniques. In particular, it will exploit complex algebra and the properties of the complex exponential function, which will be reviewed in this section. It will also apply polynomial algebra to linear differential operators with constant coefficients, using principles to be explained in §2.

The nth order linear DE with constant coefficients is

$$(1) \qquad L[u] = u^{(n)} + a_1 u^{(n-1)} + a_2 u^{(n-2)} + \cdots + a_n u = r(x)$$

Here $u^{(k)}$ stands for the kth derivative $d^k u/dx^k$ of the unknown function $u(x)$; $a_1, \ldots, a_n$ are arbitrary constants; and $r(x)$ can be any continuous function. As in Ch. 2, §1, the letter L in (1) stands for a (homogeneous) *linear operator*. That is, $L[\alpha u + \beta v] = \alpha L[u] + \beta L[v]$ for any functions u and v of class $\mathcal{C}^n$ and any constant α and β.

As in the second-order case treated in Chapter 2, the solution of linear DEs of the form (1) is best achieved by expressing its general solution as the sum $u = u_p + u_h$ of some *particular* solution $u_p(x)$ of (2), and the *general solution* $u_h(x)$ of the "reduced" (homogeneous) equation

$$(2) \qquad L[u] = u^{(n)} + a_1 u^{(n-1)} + a_2 u^{(n-2)} + \cdots + a_n u = 0$$

obtained by setting the right-hand side of (1) equal to 0.

Solutions of (2) can be found by trying the *exponential substitution* $u = e^{\lambda x}$, where λ is a real or complex number to be determined. Since $d^n(e^{\lambda x})/dx^n = \lambda^n e^{\lambda x}$, this substitution reduces (2) to the identity

$$L[e^{\lambda x}] = (\lambda^n + a_1 \lambda^{n-1} + a_2 \lambda^{n-2} + \cdots + a_n)e^{\lambda x} = 0$$

This is satisfied if and only if λ is a (real or complex) root of the *characteristic polynomial* of the DE (1), defined as

$$(3) \qquad p(\lambda) = p_L(\lambda) = \lambda^n + a_1\lambda^{n-1} + \cdots + a_{n-1}\lambda + a_n$$

For the second-order DE $u'' + pu' + qu = 0$, the roots of the characteristic polynomial are $\lambda = \frac{1}{2}(-p \pm \sqrt{p^2 - 4q})$. In Ch. 2, §2, it was shown by a special method that, when $p^2 < 4q$ so that the characteristic polynomial has *complex* roots $\lambda = -p/2 \pm iv$ $(v = \sqrt{4q - p^2})$, the real functions $e^{-px/2}\begin{Bmatrix}\cos \\ \sin\end{Bmatrix} vx$ form a basis of solutions. We will now show how to apply the exponential substitution $u = e^{\lambda x}$ to solve the general DE (2), beginning with the second-order case.

Loosely speaking, when the characteristic polynomial $p(\lambda)$ has n distinct roots $\lambda_1, \ldots, \lambda_n$, the functions $\phi_j(x) = e^{\lambda_j x}$ form a basis of *complex* solutions of the DE (2). By this we mean that for any "initial" $x = x_0$ and specified (complex) numbers $u_0, u_0', \ldots, u_0^{(n-1)}$, there exist unique numbers $c_1, \ldots, c_n$ such that the solution $f(x) = u_h(x) = \sum_{j=1}^n c_j\phi_j(x)$ satisfies $f(x_0) = u_0, f'(x_0) = u_0', \ldots,$ $f^{(n-1)}(x) = u_0^{(n-1)}$.

Moreover, the complex roots λ_j of $p(\lambda)$ occur in pairs $\mu_j \pm iv_j$, just as in the second-order case treated in Ch. 2. Therefore, the real functions $e^{\mu_j x}\cos v_j x$, $e^{\mu_j x}\sin v_j x$ together with the $e^{\lambda_j x}$ corresponding to *real* roots of $p(\lambda) = 0$, form a basis of *real* solutions of (2).

Initial Value Problem.   By the "initial value problem" for the nth order DE (1) is meant finding, for specified x_0 and numbers $u_0, u_0', \ldots, u_0^{(n-1)}$, a solution $u(x)$ of (1) that satisfies $u(x_0) = u_0$, and $u^{(j)}(x_0) = u_0^{(j)}$ for $j = 1, \ldots, n - 1$. If a basis of solutions $\phi_j(x)$ of the "reduced" DE (2) is known, together with one "particular" solution $u_p(x)$ of the *inhomogeneous* DE (1), then the sum $u(x) = u_p(x) + \sum c_j\phi_j(x)$, with the c_j chosen to make $u_h(x) = \sum c_j\phi_j(x)$ satisfy $u_h(x_0) = u_0 - u_p(x_0), u_h'(x_0) = u_0' - u_p'(x_0), \ldots, u_h^{(n-1)}(x_0) = u_0^{(n-1)} - u_p^{(n-1)}(x_0)$, constitutes one solution of the stated initial value problem. In §4, we will prove that this is the *only* solution (a uniqueness theorem), so that the stated initial value problem is "well-posed."

2   COMPLEX EXPONENTIAL FUNCTIONS

When the characteristic polynomial of $u'' + pu' + qu = 0$ has complex roots $\lambda = -p/2 \pm iv$, as before, the exponential substitution gives two *complex exponential functions* as formal solutions, namely

$$e^{-px/2 \pm ivx} = e^{-px/2}\{\cos vx \pm i \sin vx\}$$

From these complex solutions, the *real* solutions $e^{-px/2}\begin{Bmatrix}\cos \\ \sin\end{Bmatrix} vx$ obtained by a special method in Ch. 2 can easily be constructed as linear combinations. The present section will be devoted to explaining how similar constructions can be applied to arbitrary homogeneous linear constant-coefficient DEs (2).

The first consideration that must be invoked is the so-called Fundamental Theorem of Algebra.† To apply this theorem effectively, one must also be familiar with the basic properties of the complex exponential function. We shall now take these up in turn.

The Fundamental Theorem of Algebra states that any real or complex polynomial $p(\lambda)$ can be uniquely factored into a product of powers of distinct linear factors $(\lambda - \lambda_j)$:

$$(4) \qquad p(\lambda) = (\lambda - \lambda_1)^{k_1}(\lambda - \lambda_2)^{k_2} \cdots (\lambda - \lambda_m)^{k_m}$$

Clearly, the roots of the equation $p(\lambda) = 0$ are the λ_j. The exponent k_j in (4) is called the *multiplicity* of the root λ_j; evidently the sum of the k_j is the degree of p. When all λ_j are distinct (i.e., all $k_j = 1$ so that $m = n$), the DE has a *basis* of complex exponential solutions $\phi_j(x) = e^{\lambda_j x}$, $j = 1, 2, \ldots, n$; see §4 for details.

Example 1. For the fourth-order DE $u^{iv} = u$, the characteristic polynomial is $\lambda^4 = 1$, with roots ± 1, $\pm i$. Therefore, a basis of *complex* solutions is provided by e^x, e^{-x}, e^{ix}, and e^{-ix}. From these we can construct a basis of four *real* solutions

$$e^x, \quad e^{-x}, \quad \cos x = (e^{ix} + e^{-ix})/2i, \quad \sin x = (e^{ix} - e^{-ix})/2i$$

Complex Exponentials. In this chapter and in Ch. 9, properties of the complex exponential function e^z will be used freely, and so we recall some of them. The exponent $z = x + iy$ is to be thought of as a point in the (x,y)-plane, which is also referred to as the complex z-plane. The complex "value" $w = e^z$ of the exponential function is evidently a vector in the complex w-plane with magnitude $|e^z| = e^x$, which makes an angle y with the u-axis. (Here $w = u + iv$, so that $u = e^x \cos y$ and $v = e^x \sin y$ if $w = e^z$.)

Because $e^{i\theta} = \cos \theta + i \sin \theta$, one also often writes $z = x + iy$ as $z = re^{i\theta}$, where $r = \sqrt{x^2 + y^2}$ and $\theta = \arctan(y/x)$ are *polar coordinates* in the (x,y)-plane. In this notation, the inverse of the complex exponential function e^z is the complex "natural logarithm" function

$$\ln z = \ln(x + iy) = \ln r + i\theta$$

Since θ is defined only modulo 2π, $\ln z$ is evidently a *multiple-valued* function.

In the problems treated in this chapter, the coefficients a_j of the polynomial (2) will usually all be *real*. Its roots λ_j will then all be either real or complex conjugate in pairs, $\lambda = \mu \pm iv$. Thus, for the second-order DE $u'' + pu' + qu = 0$ discussed in Chapter 2, the roots are $\lambda = \frac{1}{2}(-p \pm \sqrt{p^2 - 4q})$. They are real when $p^2 > 4q$, and complex conjugate when $p^2 < 4q$.

---

† Birkhoff and MacLane, p. 113.

In this chapter, the independent variable x will also be considered as real. Now recall that if $\lambda = \mu + i\nu$, where μ, ν are real, then we have for real x

(5) $$e^{\lambda x} = e^{\mu x + i \nu x} = e^{\mu x}(\cos \nu x + i \sin \nu x)$$

Hence, if $\lambda = \mu + i\nu$ and $\lambda^* = \mu - i\nu$ are both roots of $p_L(\lambda) = 0$ in (3), the functions $e^{\mu x}(\cos \nu x \pm i \sin \nu x)$ are both solutions of (2). Since $|e^{i\nu x}| = 1$ for all real ν, x, it also follows that, where $e^{\lambda x}$ is considered as a *complex*-valued function of the real independent variable x

(5′) $$|e^{\lambda x}| = e^{\mu x}, \qquad \arg\{e^{\lambda x}\} = \nu x, \qquad \lambda = \mu + i\nu$$

Example 1′. For the DE $u^{iv} + 4u = 0$, the characteristic polynomial $\lambda^4 + 4$ has the roots $\pm 1 \pm i$. Hence it has a basis of real solutions $e^x \cos x$, $e^x \sin x$, $e^{-x}\cos x$, $e^{-x}\sin x$. (An equivalent basis is provided by the functions $\cosh x \cos x$, $\cosh x \sin x$, $\sinh x \cos x$, $\sinh x \sin x$.)

Euler's Homogeneous DE. The homogeneous linear DE

(6) $$x^n u^{(n)} + b_1 x^{n-1} u^{(n-1)} + b_2 x^{n-2} u^{(n-2)} + \cdots + b_n u = 0$$

is called Euler's homogeneous differential equation. It can be reduced to the form (2) on the positive semi-axis $x > 0$, by making the substitutions

$$t = \ln x, \qquad \frac{d}{dt} = x \frac{d}{dx}, \qquad e^{\lambda t} = x^\lambda$$

Corresponding to the real solutions $e^{\lambda_j t}$, $t e^{\lambda_j t}$, ... of (2), we have real solutions x^{λ_j}, $x^{\lambda_j} \ln x$, of (6).

Moreover, these can easily be found by substituting x^λ for u in (6). This substitution yields an equation of the form $I(\lambda)x^\lambda = 0$, where $I(\lambda)$ is a polynomial of degree n, called the indicial equation. Any λ for which $I(\lambda) = 0$ gives a solution x^λ of (6); if λ is a double root, then x^λ and $x^\lambda \ln x$ are both solutions, and so on.

For example, when $n = 2$, Euler's homogeneous DE is

(7) $$x^2 u'' + pxu' + qu = 0, \qquad p, q \text{ real constants}$$

Trying $u = x^\lambda$, we get the *indicial equation* of (8):

(7′) $$\lambda(\lambda - 1) + p\lambda + q = 0$$

Alternatively, making the change of variable $x = e^t$, we get

(8) $$\frac{d^2 u}{dt^2} + (p - 1)\frac{du}{dt} + qu = 0, \qquad t = \ln x$$

since

$$\frac{d^2}{dt^2} = x \frac{d}{dx}\left(x \frac{d}{dx}\right) = x^2 \frac{d^2}{dx^2} + x \frac{d}{dx}$$

If $(p-1)^2 > 4q$, the indicial equation has two distinct real roots $\lambda = \alpha$ and $\lambda = \beta$, and so the DE (7) has the two linearly independent *real* solutions x^α and x^β, defined for *positive* x. For positive *or* negative x, we have the solutions $|x|^\alpha$ and $|x|^\beta$ since the substitution of $-x$ for x does not affect (7). Note that the DE (7) has a singular point at $x = 0$ and that $|x|$ has discontinuous slope there if $\alpha \leq 1$.

When the discriminant $(p-1)^2 - 4q$ is negative, the indicial equation has two conjugate complex roots $\lambda = \mu \pm i\nu$, where $\mu = (1-p)/2$ and $\nu = [4q - (p-1)^2]^{1/2}/2$. A basis of real solutions of (8) is then $e^{\mu t} \cos \nu t$ and $e^{\mu t} \sin \nu t$; the corresponding solutions of the second order Euler homogeneous DE (7) are $x^\mu \cos(\nu \ln x)$ and $x^\mu \sin(\nu \ln x)$. These are, for $x > 0$, the real and imaginary parts of the *complex power* function

$$x^\lambda = x^{\mu \pm i\nu} = e^{(\mu \pm i\nu)\ln x} = x^\mu[\cos(\nu \ln x) \pm i \sin(\nu \ln x)]$$

as in (5). For $x < 0$, we can get real solutions by using $|x|$ in place of x. But for $x < 0$, the resulting real solutions of (7) are no longer the real and imaginary parts of x^λ, because $\ln(-x) = \ln x \pm i\pi$; cf. Ch. 9, §1.

General Case.   The general nth-order case can be treated in the same way. We can again make the change of independent variable

$$x = e^t, \qquad t = \ln x, \qquad x \frac{d}{dx} = \frac{d}{dt}$$

This reduces (6) to a DE of the form (2), whose solutions $t^r e^{\lambda t}$ give a basis of solutions for (6) of the form $(\ln x)^r x^\lambda$.

EXERCISES A

In Exs. 1–4, find a basis of real solutions of the DE specified.

1. $u'' + 5u' + 4u = 0$ 2. $u''' = u$

3. $u^{iv} = u$ *4. $u^{iv} + u = 0$

In Exs. 5–6, find a basis of complex exponential solutions of the DE specified.

5. $u'' + 2iu' + 3u = 0$ 6. $u'' - 2u' + 2u = 0$

In Exs. 7–10, find the solution of the initial value problem specified.

7. $u'' + 5u' + 4u = 0$, $u(0) = 1$, $u'(0) = 0$

8. $u''' = u$, $u(0) = u''(0) = 0$, $u'(0) = 1$

*9. $u^{iv} = u$, $u(0) = u''(0) = 0$, $u'(0) = u'''(0) = 1$

*10. $u'' - 2u' + 2u = 0$, $u(0) = 1$, $u'(0) = 0$

In Exs. 11 and 12, find a basis of solutions of the Euler DE.

11. $x^2u'' + 5xu' + 3u = 0$ 12. $x^2u'' + 2ixu - 3u = 0$

13. Describe the behavior of the function z^i of the complex variable $z = x + iy$ as z traces the unit circle $z = e^{i\theta}$ around the origin.

14. Do the same as in Ex. 13 for the function $z^i e^{iz}$.

3 THE OPERATIONAL CALCULUS

We have already explained the general notion of a linear operator in Ch. 2, §2. Obviously, any *linear combination* $M = c_1L_1 + c_2L_2$ of linear operators L_1 and L_2, defined by the formula $M[u] = c_1L_1[u] + c_2L_2[u]$, is itself a linear operator, in the sense that $M[au + bv] = aM[u] + bM[v]$ for all u,v to which L_1 and L_2 are applicable. Moreover the same is true of the (left-) *composite* L_2L_1 of L_1 and L_2, defined by the formula $L_2[L_1[u]]$.

For linear operators with constant coefficients, one can say much more. In the first place, they are *permutable,* in the sense of the following lemma.

LEMMA. *Linear operators with constant coefficients are permutable: for any constants a_j,b_k, if $p(D) = \Sigma a_jD^j$ and $q(D) = \Sigma b_kD^k$, then $p(D)q(D) = q(D)p(D) = \Sigma a_jb_kD^{j+k}$.*

Proof. Iterate the formula $D[b_kD[u]] = b_kD^2[u]$. It follows that, for any two constants a_j and b_k and any two positive integers j and k, we have $a_jD^jb_kD^k = a_jb_kD^{j+k} = b_kD^ka_jD^j$.

This is not true of linear differential operators with variable coefficients. Thus, since

$$Dxf = (xf)' + xf' + f = (xD + I)f, \quad \text{for} \quad f \in \mathcal{C}^1[a,b]$$

we have $Dx = xD + I$. This shows that the differentiation operator D is not permutable with the operator "multiply by x." Likewise $(x^2D)(xD) = x^3D^2 + x^2D$, whereas $(xD)(x^2D) = x^3D^2 + 2x^2D$.

Because *constant-coefficient* linear differential operators are permutable, we can fruitfully apply *polynomial algebra* to them. As an immediate application, we have

THEOREM 1. *If λ is a root of multiplicity k of the characteristic polynomial (3), then the functions $x^re^{\lambda x}$ $(r = 0, \ldots, k-1)$ are solutions of the linear DE (2).*

Proof. An elementary calculation gives, after cancelation, $(D - \lambda)[e^{\lambda x}f(x)] = e^{\lambda x}f'(x)$ for any differentiable function $f(x)$. By induction, this implies $(D - \lambda)^k[f(x)e^{\lambda x}] = e^{\lambda x}f^{(k)}(x)$ for any $f \in \mathcal{C}^k$. Since the kth derivative of x^r is zero when $k > r$, it follows that

$$(D - \lambda)^k[x^re^{\lambda x}] = 0, \quad \text{if } k > r$$

Moreover, the operators $(D - \lambda_i)^{k_i}$ being permutable, we can write, for any i

$$L[u] = q_i(D)(D - \lambda_i)^{k_i}, \qquad \text{where} \qquad q_i(D) = \prod_{j \neq i} (D - \lambda_j)^{k_j}$$

Hence $L[x^r e^{\lambda_i x}] = 0$ for each λ_i and $r = 0, 1, \ldots, k_i - 1$, as stated.

Real and Complex Solutions. Theorem 1 holds whether the coefficients a_k of the DE (2) are real or complex. Indeed, although the *independent* variable x will be interpreted as *real* in this chapter (especially in discussing stability), the operational calculus just discussed above, and the solutions constructed with it, are equally applicable to functions of the complex variable $z = x + iy$.

However, when all the coefficients a_k are real numbers, more detailed information can be obtained about the solutions, as follows.

LEMMA. *Let the complex-valued function* $w(x) = u(x) + iv(x)$ *satisfy a homogeneous linear* DE (1) *with real coefficients. Then the functions* $u(x)$ *and* $v(x)$ *[the real and imaginary parts of* $w(x)$*] both satisfy the* DE.

Proof. The complex conjugate† $w^*(x) = u(x) - iv(x)$ of $w(x)$ satisfies the complex conjugate of the given DE (2), obtained by replacing every coefficient a_k by its complex conjugate a_k^*, because $L^*[w^*] = \{L[w]\}^* = 0$. If the a_k are real, then $a_k^* = a_k$, and so $w^*(x)$ also satisfies (2). Hence, the linear combinations

$$u(x) = \frac{[w(x) + w^*(x)]}{2} \qquad \text{and} \qquad v(x) = \frac{[w(x) - w^*(x)]}{2i}$$

also satisfy (2), as stated.

This result is also valid for DEs with variable coefficients $a_k(x)$.

COROLLARY 1. *If the* DE (2) *has real coefficients and* $e^{\lambda x}$ *satisfies* (2), *then so does* $e^{\lambda^* x}$. *The nonreal roots of the characteristic polynomial* (3) *thus occur in conjugate pairs* $\lambda_j = \mu_j \pm iv_j$, *having the same multiplicity* k_j.

Now, using formula (5), we obtain the following.

COROLLARY 2. *Each pair of complex conjugate roots* λ_j, λ_j^* *of* (3) *of multiplicity* k_j *gives real solutions of* (2) *of the form*

(9) $$x^r e^{\mu_j x} \cos v_j x, \qquad x^r e^{\mu_j x} \sin v_j x, \qquad r = 0, \ldots, k_j - 1$$

These solutions differ from the solutions $e^{\lambda x}$ with real λ in that they have infinitely many zeros in any infinite interval of the real axis: that is, they are oscillatory. This proves the following result.

---

† The complex conjugate w^* of a complex number $w = u + iv$ is $u - iv$. Some authors use $\overline{w}$ instead of w^* to denote the complex conjugate of w.

THEOREM 2.   *If the characteristic polynomial* (3) *with real coefficients has* $2r$ *non-real roots, then the* DE (2) *has* $2r$ *distinct oscillatory real solutions of the form* (9).

4   SOLUTION BASES

We now show that *all* solutions of the real homogeneous linear DE (2) are linear combinations of the special solutions described in Corollary 2 above. The proof will appeal to the concept of a *basis* of solutions of a general nth order linear homogeneous DE

$$(10) \qquad L[u] = u^{(n)} + p_1(x)u^{(n-1)} + \cdots + p_n(x)u = 0$$

The coefficient-functions $p_k(x)$ in (10) may be variable, but they must be real and continuous.

DEFINITION.   A *basis* of solutions of the DE (10) is a set of solutions $u_k(x)$ of (10) such that every solution of (10) can be uniquely expressed as a linear combination $c_1u_1(x) + \cdots + c_nu_n(x)$.

The aim of this section is to prove that the special solutions described in Corollary 2 form a basis of real solutions of the DE (2). The fact that every nth order homogeneous linear DE *has* a basis of n solutions is, of course, a theorem to be proved.

First, as in Ch. 2, §2, we define a set of n real or complex functions f_1, $f_2, \ldots, f_n$ defined on an interval (a,b) to be *linearly independent* when no linear combination of the functions with constant coefficients not all zero can vanish identically: that is, when $\Sigma_{k=1}^n c_k f_k(x) \equiv 0$ implies $c_1 = c_2 = \cdots = c_n = 0$. A set of functions that is not linearly independent is said to be *linearly dependent*.

There are two notions of linear independence, according as we allow the coefficients c_k to assume only *real values,* or also *complex* values. In the first case, one says that the functions are linearly independent over the *real field;* in the second case, that they are linearly independent over the *complex field.*

LEMMA 1.   *A set of real-valued functions on an interval* (a,b) *is linearly independent over the complex field if and only if it is linearly independent over the real field.*

Proof.   Linear dependence over the real field implies linear dependence over the complex field, *a fortiori.* Conversely, the $f_j(x)$ being real, suppose that $\Sigma c_j f_j(x) \equiv 0$ for $a < x < b$. Then $[\Sigma c_j f_j(x)]^* \equiv 0$, and hence $\Sigma c_j^* f_j(x) \equiv 0$. Subtracting, we obtain $\Sigma [(c_j - c_j^*)/i]f_j(x) \equiv 0$. If all c_j are real, there is nothing to prove. If some c_j is not real, some real number $(c_j - c_j^*)/i$ will not vanish, and we still have a vanishing linear combination with real coefficients.

A set of functions that is linearly dependent on a given domain may become linearly independent when the functions are extended to a larger domain. However, a linearly independent set of functions clearly remains linearly independent when the functions are so extended.

LEMMA 2. *Any set of functions of the form*

(11) $$f_{rj}(x) = x^r e^{\lambda_j x}, \qquad j = 1, \ldots, n$$

where the r are nonnegative integers and the λ_j complex numbers, is linearly indepen-dent on any nonvoid open interval, unless two or more of the functions are identical.

Proof. Suppose that $\Sigma c_{rj} f_{rj}(x) \equiv 0$. For any given λ_j, choose R to be the larg-est r such that $c_{rj} \neq 0$. Form the operator

$$q(D) = (D - \lambda_j)^R \prod_{i \neq j} (D - \lambda_i)^{k_i+1}$$

where for each $i \neq j$, k_i is the largest r such that $x^r e^{\lambda_i x}$ is a member of the set of functions in (11). It follows that $q(D)[f_{ri}] = 0$ unless $i = j$, and that $q(D)[f_{rj}] = 0$ for $r < R$. Hence, we have

$$q(D)[\Sigma c_{ri} f_{ri}(x)] = c_{Rj} q(D)[x^R e^{\lambda_j x}]$$

On the other hand, as in the proof of Theorem 1, we see that

$$q(D)[x^R e^{\lambda_i x}] = (R!) \prod_{i \neq j} (\lambda_j - \lambda_i)^{k_i+1} e^{\lambda_j x} \neq 0$$

Therefore, substituting back, we find that $c_{Rj} = 0$. Since we assumed that $c_{Rj} \neq 0$, this gives a contradiction unless all $c_{rj} = 0$, proving linear independence.

From Theorem 1 we obtain the following corollary.

COROLLARY 1. *The DE (2) has at least n linearly independent, real or complex solutions of the form $x^r e^{\lambda x}$.*

The analogous result for *real* solutions of DE of the form (2) with *real* coef-ficients can be proved as follows. For any two conjugate complex roots $\lambda = \mu + iv$ and $\lambda^* = \mu - iv$ of the characteristic equation of (2), the real solutions $x^r e^{\mu x} \cos vx$ and $x^r e^{\mu x} \sin vx$ are complex linear combinations of $x^r e^{\lambda x}$ and $x^r e^{\lambda^* x}$, and conversely. Hence, they can be substituted for $x^r e^{\lambda x}$ and $x^r e^{\lambda^* x}$ in any set of solutions without affecting their linear independence. Since linear indepen-dence over the complex field implies linear independence over the real field, this proves the following.

COROLLARY 2. *A linear DE (2) with constant real coefficients a_k has a set of n solutions of the form $x^r e^{\mu x}$ or (9), which is linearly independent over the real field in any nonvoid interval.*

We now show that *all* solutions of the real homogeneous linear DE (2) are linear combinations of the special solutions described in Corollary 2. (The proof will be extended to the case of *complex* coefficient-functions in Ch. 6, §11.) To this end, we first prove a special uniqueness lemma for the more general homogeneous linear DE (10),

$$L[u] = u^{(n)} + p_1(x)u^{(n-1)} + \cdots + p_n(x)u = 0$$

with real and continuous coefficient-functions $p_k(x)$.

LEMMA 3. *Let $f(x)$ be any real or complex solution of the nth order homogeneous linear DE* (10) *with continuous real coefficient-functions in the closed interval $[a,b]$. If $f(a) = f'(a) = \cdots = f^{(n-1)}(a) = 0$, then $f(x) \equiv 0$ on $[a,b]$.*

Proof. We first suppose $f(x)$ *real*. The function

$$\sigma(x) = f(x)^2 + f'(x)^2 + \cdots + f^{(n-1)}(x)^2 \geq 0$$

satisfies the initial condition $\sigma(a) = 0$. Differentiating $\sigma(x)$, we find, since $\sigma(x)$ is real, that

$$\sigma'(x) = 2[f(x)f'(x) + f'(x)f''(x) + \cdots + f^{(n-1)}(x)f^{(n)}(x)]$$

Using the inequality $|2\alpha\beta| \leq \alpha^2 + \beta^2$ repeatedly $n - 1$ times, we have

$$\sigma'(x) \leq (f^2 + f'^2) + (f'^2 + f''^2) + \cdots + ([f^{(n-2)}]^2 + [f^{(n-1)}]^2) + 2f^{(n-1)}f^{(n)}$$

Since $L[f] = 0$, it follows that $f^{(n)} = -\sum_{k=1}^{n} p_k f^{(n-k)}$. Hence, the last term can be rewritten in the form

$$f^{(n-1)}f^{(n)} = -\sum_{k=1}^{n} p_k f^{(n-1)} f^{(n-k)}$$

Applying the inequality $|2\alpha\beta| \leq \alpha^2 + \beta^2$ again, we obtain

$$2|f^{(n-1)}f^{(n)}| \leq \sum_{k=1}^{n} |p_k|([f^{(n-k)}]^2 + [f^{(n-1)}]^2)$$

Substituting and rearranging terms, we obtain

$$\sigma'(x) \leq (1 + |p_n|)f^2 + (2 + |p_{n-1}|)f'^2 + (2 + |p_{n-2}|)f''^2$$
$$+ \cdots + (2 + |p_2|)[f^{(n-2)}]^2 + \left(1 + |p_1| + \sum_{k=1}^{n} |p_k|\right)[f^{(n-1)}]^2$$

Now let $K = 2 + \max|p_1(x)| + \max_{a \leq x \leq b} \sum_{k=1}^{n} |p_k(x)|$. Then it follows from the last inequality that $\sigma'(x) \leq K\sigma(x)$. From this inequality and the initial condition

$\sigma(a) = 0$, the identity $\sigma(x) \equiv 0$ follows by Lemma 2 of Ch. 1, §11. Hence, we have $f(x) \equiv 0$.

If $h(x) = f(x) + ig(x)$ (f,g real) is a *complex* solution of (10), then $f(x)$ and $g(x)$ satisfy (10) by Lemma 2 of §3. Moreover, $h(a) = h'(a) = \cdots = h^{(n-1)}(a) = 0$ implies the corresponding equalities on f and g. Hence, by the preceding paragraph, we have $h = f + ig \equiv 0 + 0 = 0$, completing the proof.

We now show that any n linearly independent solutions of (10) form a basis of solutions.

THEOREM 3. *Let $u_1, \ldots, u_n$ be n linearly independent real solutions of the nth order linear homogeneous DE* (10) *with real coefficient-functions. Then, given arbitrary real numbers a, u_0, u_0', $\ldots$, $u_0^{(n-1)}$, there exist unique constants $c_1, \ldots, c_n$ such that $u(x) = \Sigma c_k u_k(x)$ is a solution of* (10) *satisfying*

$$(12') \qquad u(a) = u_0, \; u'(a) = u_0', \ldots, u^{(n-1)}(a) = u_0^{(n-1)}$$

The functions $u_k(x)$ are a basis of solutions of (10).

Theorem 3 follows readily from the lemma. Suppose that, for some a, u_0, u_0', $\ldots, u_0^{(n-1)}$, there were no linear combination $\Sigma c_k u_k(x)$ satisfying the given initial conditions (12'). That is, suppose the n vectors

$$[u_k(a), u_k'(a), \ldots, u_k^{(n-1)}(a)], \qquad k = 1, \ldots, n$$

were *linearly dependent*. Then there would exist constants $y_1, \ldots, y_n$, not all zero, such that

$$\sum_{k=1}^{n} \gamma_k u_k(a) = 0, \qquad \sum_{k=1}^{n} \gamma_k u_k'(a) = 0, \ldots, \qquad \sum_{k=1}^{n} \gamma_k u^{(n-1)}(a) = 0$$

That is, the function $\phi(x) = \gamma_1 u_1(x) + \cdots + \gamma_n u_n(x)$ would satisfy

$$\phi(a) = \phi'(a) = \cdots = \phi^{(n-1)}(a) = 0$$

From this it would follow, from the lemma, that $\phi(x) \equiv 0$.

Recapitulating, we can find either $c_1, \ldots, c_n$ not all zero such that

$$u(x) = c_1 u_1(x) + \cdots + c_n u_n(x)$$

satisfies (12'), or $\gamma_1, \ldots, \gamma_n$ not all zero such that

$$\phi(x) = \gamma_1 u_1(x) + \cdots + \gamma_n u_n(x) \equiv 0$$

The second alternative contradicts the hypothesis of linear independence in Theorem 3, which proves the first conclusion there.

To prove the second conclusion, let $v(x)$ be *any* solution of (10). By the first conclusion, constants $c_1, \ldots, c_n$ can be found such that

$$u(x) = c_1 u_1(x) + \cdots + c_n u_n(x)$$

satisfies $u(a) = v(a)$, $u'(a) = v'(a), \ldots, u^{(n-1)}(a) = v^{(n-1)}(a)$. Hence the difference $f(x) = u(x) - v(x)$ satisfies the hypotheses of Lemma 3. Using the lemma, we obtain $u(x) \equiv v(x)$ and $v(x) = \Sigma c_k u_k(x)$, proving the second conclusion of Theorem 3.

COROLLARY 1.   *Let* $\lambda_1, \ldots, \lambda_m$ *be the roots of the characteristic polynomial of the real*† *DE* (2) *with multiplicities* $k_1, \ldots, k_m$. *Then the functions* $x^r e^{\lambda_j x}$, $r = 0, \ldots, k_j - 1$, *are a basis of complex solutions of* (2).

Referring back to Theorem 2, we have also the following.

COROLLARY 2.   *If the coefficients of the DE* (2) *are real, then it has a basis of real solutions of the form* $x^r e^{\lambda x}$, $x^r e^{\mu x} \cos \nu x$, *and* $x^r e^{\mu x} \sin \nu x$, *where* λ, μ, *and* ν *are real constants.*

EXERCISES B

1. Solve the following initial-value problems:
    (a) $u^{iv} - u = 0$, $u(0) = u'(0) = u'''(0) = 0$, $u''(0) = 1$
    (b) $u^{iv} = 0$, $u'(0) = u'''(0) = 0$, $u(0) = 1$, $u''(0) = -2$
    (c) $u^{iv} + u'' = 0$, $u''(0) = u'''(0) = 0$, $u(0) = u'(0) = 1$

2. (a) Find a DE $L[u] = 0$ of the form (2) having e^{-t}, te^{-t}, and e^t as a basis of solutions.
    (b) For this linear operator L, find a basis of solutions of the sixth-order DE $L^2[u] = 0$ and the ninth-order DE $L^3[u] = 0$.

3. Find bases of solutions for the following DEs:
    (a) $u^{vi} = u$ (b) $u^{iv} - 3u'' + 2u = 0$
    (c) $u''' + 6u'' + 12u' + 8u = 0$
    (d) $u''' + 6u'' + 12u' + (8 + i)u = 0$

4. Knowing bases of solutions $L_1[u] = 0$ and $L_2[u] = 0$ of the form given by Theorem 1, find a basis of solutions of $L_1[L_2[u]] = 0$.

5. Show that in every *real* DE of the form (2), L can be factored as $L = L_1 L_2 \ldots L_r$, where $L_j = D_j + b_j$ or $L_j = D^2 + p_j D + q_j$, with all b_j, p_j, q_j real.

6. Extend Lemma 2 of §4 to the case where the r are arbitrary complex numbers.

*7. State an analog of Corollary 2 of §4 for Euler's homogeneous DE, and prove your statement without assuming Corollary 2.

8. Prove that the DE of Ex. A5 has no nontrivial real solution.

---

† The preceding result can be proved more generally for linear DEs with constant complex coefficients, by similar methods; see Ch. 6, §11.

5 INHOMOGENEOUS EQUATIONS

We now return to the nth order *inhomogeneous* linear DE with constant coefficients,

$$(13) \qquad L[u] = \frac{d^n u}{dt^n} + a_1 \frac{d^{n-1}u}{dt^{n-1}} + \cdots + a_{n-1}\frac{du}{dt} + a_n u = r(t)$$

already introduced in §1. As in the second-order case of Ch. 2, §8, the function $r(t)$ in (13) may be thought of as an "input" or "source" term, and $u(t)$ as the "output" due to $r(t)$. We first describe a simple method for finding a particular solution of the DE (13) in closed form, in the special case that $r(t) = \Sigma p_k(t)e^{\lambda k^t}$ is a linear combination of products of polynomials and exponentials.

We recall that, by Lemma 2 of §3,

$$(D - \lambda)[e^{\lambda t}f(t)] = e^{\lambda t}f'(t)$$

As a corollary, since every polynomial of degree s is the derivative $r(t) = q'(t)$ of a suitable polynomial $q(t)$ of degree $s + 1$, we obtain the following result.

LEMMA 1. *If $r(t)$ is a polynomial of degree s, then $(D - \lambda)[u] = e^{\lambda t}r(t)$ has a solution of the form $u = e^{\lambda t}q(t)$, where $q(t)$ is a polynomial of degree $s + 1$.*

More generally, one easily verifies the identity

$$(D - \lambda_1)[e^{\lambda t}f(t)] = e^{\lambda t}[(\lambda - \lambda_1)f(t) + f'(t)]$$

If $\lambda \neq \lambda_1$, and $f(t)$ is a polynomial of degree s, then the right side of the preceding identity is a polynomial of degree s times $e^{\lambda t}$. This proves another useful algebraic fact:

LEMMA 2. *If $r(t)$ is a polynomial of degree s and $\lambda \neq \lambda_1$, then*

$$(D - \lambda_1)[u] = e^{\lambda t}r(t)$$

has a solution of the form $u = e^{\lambda t}q(t)$, where $q(t)$ is a polynomial of degree s.

Applying the two preceding lemmas repeatedly to the factors of the operator

$$L = p_L(D) = (D - \lambda_1)^{k_1}(D - \lambda_2)^{k_2} \cdots (D - \lambda_m)^{k_m}$$

we get the following result.

THEOREM 4. *The DE $L[u] = e^{\lambda t}r(t)$, here $r(t)$ is a polynomial, has a particular solution of the form $e^{\lambda t}q(t)$,where $q(t)$ is also a polynomial. The degree of $q(t)$ equals that* of $r(t)$ unless $\lambda = \lambda_j$ is a root of the characteristic polynomial $p_L(\lambda) = \Sigma(\lambda - \lambda_j)^{k_j}$ of L. If $\lambda = \lambda_j$ is a k-fold root of $p_L(\lambda)$, then the degree of $q(t)$ exceeds that of $r(t)$ by k.

Knowing the form of the answer, we can solve for the coefficients b_k of the unknown polynomial $q(t) = \Sigma b_k t^k$ by the method of undetermined coefficients. Namely, applying $p(D)$ to $u(t) = e^{\lambda t}(\Sigma b_k t^k)$, one can compute the numbers P_{kl} in the formula

$$p(D)[u] = e^{\lambda t}\Sigma(P_{kl}b_l)t^k$$

using formulas for differentiating elementary functions. One does not need to factor p_L. The simultaneous linear equations $\Sigma P_{kl}b_l = c_k$ can then be solved for the b_k, given $r(t) = \Sigma c_k t^k$, by elementary algebra. Theorem 4 simply states how many unknowns b_k must be used to get a compatible system of linear equations.

Example 2. Find the solution of the DE

$$L[u] = u''' + 3u'' + 2u' = 12te^t$$

that satisfies the initial conditions $u(0) = -17/3$, $u'(0) = u''(0) = 1/3$. First, since the two-dimensional subspace of functions of the form $(\alpha + \beta t)e^t$ is mapped into itself by differentiation, the constant-coefficient DE (*) may be expected to have a particular solution of this form. And indeed, substituting $u = (\alpha + \beta t)e^t$ into (*) and evaluating, we get

(*) $$L[u] = [(6\alpha + 11\beta) + 6\beta t]e^t = 12te^t$$

Comparing coefficients, we find a *particular* solution $u = (-11/3 + 2t)e^{-t}$ of (*).

Second, the *reduced* DE $u''' + 3u'' + 2u' = 0$ of (*) has 1, e^{-t}, and e^{-2t} as a basis of solutions. The *general* solution of (*) is therefore

$$u = a + be^{-t} + ce^{-2t} + (-11/3 + 2t)e^t$$

The initial conditions yield three simultaneous linear equations in a, b, c whose solution is $a = 1$, $b = -4$, $c = 1$. Hence the solution of the specified initial value problem is

(**) $$u = 1 - 4e^{-t} + e^{-2t} + (-11/3 + 2t)e^t$$

EXERCISES C

In Exs. 1–4, find a particular solution of the DE specified. In Exs. 1–3, find the solutions satisfying (a) $u(0) = 0$, $u'(0) = 1$ and (b) $u(0) = 0$, $u'(0) = 1$.

1. $u'' = te^t$ 2. $u'' + u = te^t$

3. $u'' - u = te^t$ 4. $u^{iv} = t^5e^t$

In each of Exs. 5–8, find a particular solution of the DE specified.

5. $u^{vi} + 4u = \sin t$ 6. $u''' + 2u'' + 3u' + 6u = \cos t$

7. $u^{iv} + 5u'' + 4u = e^t$ 8. $u'' + iu = \sin 2t, \quad i = \sqrt{-1}$

In each of Exs. 9–12, find (a) the *general* solution of the DE specified four exercises earlier, and (b) the particular solution satisfying the initial condition specified.

9. $u^{(v)}(0) = 0$ for $v = 0, 1, 2, 3, 4$ $u^v(0) = 1$

10. $u(0) = u'(0), \quad u''(0) = 1$

11. $u(0) = 10, \quad u'(0) = u''(0) = u'''(0) = 0$

12. $u(0) = 0, \quad u'(0) = i$

6 STABILITY

An important physical concept is that of the stability of equilibrium. An equilibrium state of a physical system is said to be *stable* when small departures from equilibrium remain small with the lapse of time, and *unstable* when arbitrarily small initial deviations from equilibrium can ultimately become quite large.

In considering the stability of equilibrium, it is suggestive to think of the independent variable as standing for the time t. Accordingly, one rewrites the DE (2) as

$$(14) \qquad \frac{d^n u}{dt^n} + a_1 \frac{d^{n-1}u}{dt^{n-1}} + a_2 \frac{d^{n-2}u}{dt^{n-2}} + \cdots + a_{n-1}\frac{du}{dt} + a_n u = 0$$

For such constant-coefficient *homogeneous linear* DEs, the trivial solution $u \equiv 0$ represents an equilibrium state, and the possibilities for stable and unstable behavior are relatively few. They are adequately described by the following definition (cf. Ch. 5, §7 for the nonlinear case).

DEFINITION. The *homogeneous linear* DE (14) is *strictly stable* when every solution tends to zero as $t \to \infty$; it is *stable* when every solution remains bounded as $t \to \infty$; when not stable, it is called *unstable*.

Evidently, a homogeneous linear DE is strictly stable if and only if it has a finite *basis* of solutions tending to zero, and stable if and only if it has a basis of bounded solutions. The reason for this is that every finite linear combination of bounded functions is bounded, as is easily shown. Hence Theorem 3, Corollary 6 gives algebraic tests for stability and strict stability of the DE (14). Take a basis of solutions of the form $t^r e^{\mu t}$, $t^r e^{\mu t} \sin vt$, $t^r e^{\mu t} \cos vt$. Such a solution tends to zero if and only if $\mu < 0$ and remains bounded as $t \to \infty$ if and only if $\mu < 0$ or $\mu = r = 0$. This gives the following result.

THEOREM 5. *A given* DE (14) *is strictly stable if and only if every root of its characteristic polynomial has a negative real part. It is stable if and only if every mul-*

tiple root λ_i [*with* $k_i > 1$ *in* (4)] *has a negative real part, and no simple root* (*with* $k_i = 1$) *has a positive real part.*

Polynomials all of whose roots have negative real parts are said to be of *stable type*.† There are algebraic inequalities, called the *Routh–Hurwitz conditions,* on the coefficients of a real polynomial, which are necessary and sufficient for it to be of stable type. Thus, consider the quadratic characteristic polynomial of the DE of the second-order DE (5) of Ch. 2, §1. An examination of the three cases discussed in §2 above shows that the real DE

$$\ddot{u} + a_1\dot{u} + a_2 u = 0, \qquad \ddot{u} = \frac{d^2u}{dt^2}, \qquad \dot{u} = \frac{du}{dt}$$

is strictly stable if and only if a_1 and a_2 are both positive (positive damping and positive restoring force). That is, when $n = 2$, the Routh-Hurwitz conditions are $a_1 > 0$ and $a_2 > 0$.

To make it easier to correlate the preceding results with the more informal discussion of stability and oscillation found in Ch. 2, §2, we can rewrite the DE discussed there as $\ddot{u} + p\dot{u} + qu = 0$. We have just recalled that this DE is strictly stable if and only if $p > 0$ and $q > 0$. It is oscillatory if and only if $q > p^2/4$, so that its characteristic polynomial $\lambda^2 + p\lambda + q$ has complex roots $-(p \pm \sqrt{p^2 - 4q})/2$.

In the case of a third-order DE ($n = 3$), the test for strict stability is provided by the inequalities $a_j > 0$ ($j = 1, 2, 3$) and $a_1a_2 > a_3$. When $n = 4$, the conditions for strict stability are $a_j > 0$ ($j = 1, 2, 3, 4$), $a_1a_2 > a_3$, and $a_1a_2a_3 > a_1^2a_4 + a_3^2$.

When $n > 2$, there are no equally simple conditions for solutions to be oscillatory or nonoscillatory. Thus, the characteristic polynomial of the DE $\dddot{u} + \ddot{u} + \dot{u} + u = 0$ is $(\lambda + 1)(\lambda^2 + 1)$; hence its general solution is

$$a \cos t + b \sin t + ce^{-t}$$

Unless $a = b = 0$, this solution will become oscillatory for large positive t, but will be nonoscillatory for large negative t. Other illustrative examples are given in Exercises C.

7   THE TRANSFER FUNCTION

Inhomogeneous linear DEs (13) are widely used to represent electric alternating current networks or *filters*. Such a filter may be thought of as a "black

---

† See Birkhoff and MacLane, p. 122. For polynomials of stable type of higher degree, see F. R. Gantmacher, *Applications of Matrices*, Wiley–Interscience, New York, 1959.

box" into which an electric current or a voltage is fed as an *input* $r(t)$ and out of which comes a resulting *output* $u(t)$.

Mathematically, this amounts to considering an *operator* transforming the function r into a function u, which is the solution of the inhomogeneous linear DE (13). Writing this operator as $u = F[r]$, we easily see that $L[F[r]] = r$. Thus, such an input–output operator is a right-*inverse* of the operator L.

Since there are many solutions of the inhomogeneous DE (13) for a given input $r(t)$, the preceding definition of F is incomplete: the preceding equations do not define $F = L^{-1}$ unambiguously.

For input–output problems that are unbounded in time, this difficulty can often be resolved by insisting that $F[r]$ be in the class $B(-\infty, +\infty)$ of *bounded* functions; in §§7–8, we will make this restriction. For, in this case, for any two solutions u_1 and u_2 of the inhomogeneous DE $L[u] = r$, the difference $v = u_1 - u_2$ would have to satisfy $L[v] = 0$. Unless the characteristic polynomial $p_L(\lambda) = 0$ has pure imaginary roots, this implies $v = 0$. Hence, in particular, the DE $L[u] = r$ has at most one bounded solution if the DE $L[u] = 0$ is *strictly stable*— an assumption which corresponds in electrical engineering to a passive electrical network with dissipation. Moreover, the effect of initial conditions is "transient": it dies out exponentially.

For initial value problems and their Green's functions, it is more appropriate to define F by restricting its values to functions that satisfy $u(0) = u'(0) = \cdots = u^{(n-1)}(0) = 0$; this also defines F unambiguously, by Theorem 3.

We now consider bounded solutions of (13) for various input functions, without necessarily assuming that the homogeneous DE is strictly stable.

Sinusoidal input functions are of the greatest importance; they represent alternating currents and simple musical notes of constant pitch. These are functions of the form

$$A \cos (kt + \alpha) = \text{Re} \{ce^{ikt}\}, \qquad A = |c|, \qquad x = \arg c$$

A is called the *amplitude*, $k/2\pi$ the *frequency*, and α the *phase constant*. The frequency $k/2\pi$ is the reciprocal of the *period* $2\pi/k$.

Except in the case $p_L(ik) = 0$ of perfect resonance, there always exists a unique *periodic solution* of the DE (13), having the same period as the given input function $r(t) = ce^{ikt}$. This output function $u(t)$ can be found by making the substitution $u = C(k)ce^{ikt}$, where $C(k)$ is to be determined. Substituting into the inhomogeneous DE (14), we see that $L[C(k)ce^{ikt}] = ce^{ikt}$ if and only if

$$(15) \qquad\qquad C(k) = 1/p_L(ik), \qquad p_L(ik) \neq 0$$

where $p_L(\lambda)$ is the characteristic polynomial defined by (3).

DEFINITION. The complex-valued function $C(k)$ of the real variable k defined by (15) is called the *transfer function* associated with the linear, time-independent operator L. If $C(k) = \rho(k)e^{-ij(k)}$, then $\rho = |C(k)|$ is the *gain function*, and $\gamma(k) = -\arg C(k) = \arg p_L(ik)$ is the *phase lag* associated with k.

The reason for this terminology lies in the relationship between the real part of $u(t)$ and that of the input $r(t)$. Clearly,

$$\text{Re } \{u(t)\} = \text{Re } \{C(k)ce^{ikt}\} = |C(k)| \cdot |c| \cos (kt + \alpha - \gamma)$$

This shows that the amplitude of the output is $\rho(k)$ times the amplitude of the input, and the phase of the output lags $\gamma = -\arg C$ behind that of the input at all times.

In the *strictly stable* case, the particular solution of the inhomogeneous linear DE $L[u] = ce^{ikt}$ found by the preceding method is the only bounded solution; hence $F[ce^{ikt}] = C(k)ce^{ikt}$ describes the effect of the input–output operator F on sinusoidal inputs. Furthermore, since every solution of the homogeneous DE (14) tends to zero as $t \to +\infty$, every solution of $L[u] = ce^{ikt}$ approaches $C(k)ce^{ikt}$ exponentially.

Example 3. Consider the forced vibrations of a lightly damped harmonic oscillator:

(*) $$[L[u] = u'' + \epsilon u' + p^2 u = \sin kt, \qquad \epsilon \ll 1]$$

The transfer function of (*) is easily found using the *complex* exponential trial function e^{ikt}. Since

$$L[ce^{ikt}] = (-k^2 + \epsilon ik + p^2)ce^{ikt}$$

we have $C(k) = 1/[(p^2 - k^2) + \epsilon ik]$. De Moivre's formulas give from this the gain function $\rho = 1/[(p^2 - k^2)^2 + \epsilon^2 k^2]^{1/2}$ and the phase lag

$$\gamma = \arctan \left[\frac{\epsilon k}{(p^2 - k^2)} \right]$$

The solution of (*) is therefore $\rho \sin (kt - \gamma)$, where ρ and γ are as stated.

One can also solve (*) in real form. Since differentiation carries functions of the form $u = a \cos kt + b \sin kt$ into functions of the same form, we look for a periodic solution of (*) of this form. An elementary computation gives for u as before:

$$L[u] = [(p^2 - k^2)a + \epsilon kb] \cos kt + [(p^2 - k^2)b - \epsilon ka] \sin kt$$

To make the coefficient of $\cos kt$ in (*) vanish, it is necessary and sufficient that $a/b = \epsilon k/(k^2 - p^2)$, the tangent of the phase *advance* (negative phase lag). The gain can be computed similarly; we omit the tedious details.

Finally, note that the characteristic polynomial of any real DE (2) can be factored into real linear and quadratic factors

$$p_L(\lambda) = \prod_{j=1}^{r} (\lambda + b_j) \prod_{l=1}^{s} (\lambda^2 + p_l\lambda + q_l), \qquad r + 2s = n$$

and since all b_j, p_l, and q_l are positive in the strictly stable case that all roots of $p_L(\lambda) = 0$ have negative parts. Therefore

$$j(k) = \sum_{j=1}^{r} \arg (b_j + ik) + \sum_{l=1}^{g} \arg (q_l + ikp_l - k^2)$$

increases *monotonically* from 0 to $n\pi/2$ as k increases from 0 to ∞. This is evident since each $\arg (b_j + ik)$ increases from 0 to $\pi/2$, while $\arg (q_l + ikp_l - k^2)$ increases from 0 to π, as one easily sees by visualizing the relevant parametric curves (straight line or parabola). Theorem 6 below will prove a corresponding result for complex constant-coefficient DEs.

Resonance. The preceding method fails when the characteristic polynomial $p_L(\lambda)$ has one or more purely imaginary roots $\lambda = ik_j$ (in electrical engineering, this occurs in a "lossless passive network").

Thus, suppose that ik is a root of the equation $p_L(\lambda) = 0$ and that we wish to solve the inhomogeneous DE $L[u] = e^{ikt}$. From the identity (cf. §6)

$$L[te^{\lambda t}] = L\left[\frac{\partial}{\partial\lambda} e^{\lambda t}\right] = \frac{\partial}{\partial\lambda} L[e^{\lambda t}] = \frac{\partial}{\partial\lambda} [p_L(\lambda)e^{\lambda t}] = p_L'(\lambda)e^{\lambda t} + p_L(\lambda)te^{\lambda t}$$

we obtain, setting $\lambda = ik$,

$$L[ie^{ikt}] = p_L'(ik)e^{ikt}$$

If ik is a simple root of the characteristic equation, then $p_L'(ik) \neq 0$. Hence a solution of $L[u] = e^{ikt}$ is $u(t) = [1/p_L'(ik)]te^{ikt}$. The amplitude of this solution is $[1/|p_L'(ik)|]t$, and it increases to infinity as $t \to \infty$. This is the phenomenon of *resonance*, which arises when a nondissipative physical system Σ is excited by a force whose period equals one of the periods of free vibration of Σ.

A similar computation can be made when ik is a root of multiplicity n of the characteristic polynomial, using the identity $L[t^n e^{ikt}] = p_L^{(n)}(ik)e^{ikt}$, which is proved in much the same way. In this case the amplitude of the solution again increases to infinity.

Periodic Inputs. The transfer function gives a simple way for determining periodic outputs from any periodic input function $r(t)$ in (13). Changing the time unit, we can write $r(t + 2\pi) = r(t)$. We can then expand $r(t)$ in a Fourier series, getting

(16)
$$L[u] = \frac{a_0}{2} + \sum_{k=1}^{\infty} (a_k \cos kt + b_k \sin kt)$$

or, in complex form,

(16′)
$$2L[u] = \sum_{k=-\infty}^{\infty} c_k e^{ikt}; \qquad c_0 = a_0, \qquad c_k = c^*_{-k} = a_k - ib_k$$

summed over all integers k.

Applying the superposition principle to the Fourier components of $c_k e^{ikt}$ of $r(t)$ in (16′), we obtain, as at least a formal solution,

(17)
$$u(t) = \sum_{k=-\infty}^{\infty} \left[\frac{c_k}{p_L(ik)} \right] e^{ikt}$$

provided that no $p_L(ik)$ vanishes. The series (17) is absolutely and uniformly convergent, since $p_L(ik) = 0(k^{-n})$ for an nth-order DE. We leave to the reader the proof and the determination of sufficient conditions for term-by-term differentiability.

EXERCISES D

In Exs. 1–4, test the DE specified for stability and strict stability.

1. $u'' + 5u' + 4u = 0$ 2. $u''' + 6u'' + 12u' + 8t = 0$

3. $u''' + 6u'' + 11u' + 6u = 0$ 4. $u^{iv} + 4u''' + 4u'' = 0$

5. For which n is the DE $u^{(n)} + u = 0$ stable?

In Exs. 6–9, plot the gain and transfer functions of the operator specified (I denotes the identity operator):

6. $D^2 + 4D + 4I$ 7. $D^3 + 6D^2 + 12D + 8I$

8. $D^2 + 2D + 101I$ 9. $D^4 - I$

10. For a strictly stable $L[u] = u'' + au' + bu = r(t)$, calculate the outputs (the *responses*) to the inputs $r(t) = 1$ and $r(t) = t$ for $a^2 > 4b$ and $a^2 < 4b$.

*8   THE NYQUIST DIAGRAM

The transfer function $C(k) = 1/p_L(ik)$ of a linear differential equation with constant coefficients $L[u] = 0$ is of great help in the study of the inhomogeneous DE (13). To visualize the transfer function, one graphs the logarithmic gain $\ln \rho(k)$ and phase lag $\gamma(k)$ as functions of the frequency $k/2\pi$. If $\lambda_1, \ldots,$

λ_n are the roots of the characteristic polynomial, we have

(18a) $\ln \rho(k) = - \sum\limits_{j=1}^{n} \ln |ik - \lambda_j| = -\frac{1}{2} \Sigma \ln [(k - \nu_j)^2 + \mu_j^2]$

(18b) $\gamma(k) = \sum\limits_{j=1}^{n} \arg (ik - \lambda_j) = \Sigma \arctan \left[\dfrac{(k - \nu_j)}{\mu_j} \right]$

from which these graphs are easily plotted. Figure 3.1*a* depicts the gain function and phase lag of the DE

$$u^{iv} + 0.8u''' + 5.22u'' + 1.424u' + 4.1309u = 0$$

whose characteristic polynomial has the roots $\lambda_j = -0.1 \pm i, -0.3 \pm 2i$.

We now compute how the phase lag $\gamma(k)$ changes as the frequency $k/2\pi$ increases from $-\infty$ to $+\infty$. By (18b), it suffices to add the changes in the functions $\arg (ik - \lambda_j)$ for each λ_j. If Re $\{\lambda_j\}$ is negative, then the vertical straight line $ik - \lambda_j (-\infty < k < \infty)$ lies in the right half of the complex plane; hence, $\arg (ik - \lambda_j)$ increases by π as k increases from $-\infty$ to $+\infty$. If Re $\{\lambda_j\}$ is positive, then $\arg (ik - \lambda_j)$ decreases by π for a similar reason. Hence, if there are no roots with zero real part, the change in $\gamma(k)$ is $(m - p)\pi$, where m is the number of λ_j with negative real part, and p is the number with positive real part.† If there are no purely imaginary roots, then $m + p = n$, and we obtain a useful test for strict stability.

THEOREM 6.   *The DEL $[u] = r(t)$ of order n is strictly stable if and only if the phase lag $\gamma = -\arg C(k)$ increases by $n\pi$ as k increases from $-\infty$ to $+\infty$. In this case, the phase lag increases monotonically with k.*

If the differential operator L has *real* coefficients, then $\gamma(-k) = -\gamma(k)$ and $\rho(-k) = \rho(k)$, since the complex roots λ_j occur in conjugate pairs $\mu_j \pm i\nu_j$. In particular, we have $\gamma(0) = 0$, and the change in $\gamma(k)$ as k increases from 0 to ∞ is $(m - p)\pi/2$. This proves the following specialization of Theorem 6.

COROLLARY 1.   *A linear DE of order n with constant real coefficients is strictly stable if and only if the phase lag increases from 0 to $n\pi/2$ as k increases from 0 to ∞.*

If all roots λ_j of the characteristic polynomial are real, then one easily verifies that all $\ln |ik - \lambda_j|$ increase monotonically as k increases from 0 to ∞. Hence,

---

† For purely imaginary roots, the change of argument of λ is undefined (it could be π or $-\pi$). In this case, we make the convention that the change in the argument is zero. The following theorem is true with the proviso that, whenever the argument is undefined, the change is taken to be zero.

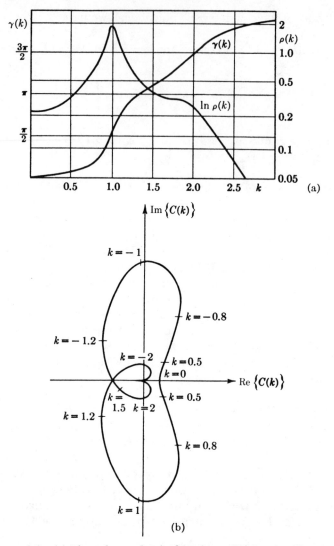

Figure 3.1   (a) Phase-lag and gain functions, (b) Nyquist diagram.

in this case, the gain $\rho(k)$ is a monotonically decreasing function of the frequency. In the case of complex roots $\lambda_j = \mu_j \pm i\nu_j$ with μ_j very small, the gain function $\rho(k)$ is very large near $k = \nu_j$; this is due to near *resonance*, as in Example 3 above.

Another useful way to visualize the transfer function is to plot the curve $z = C(k)$ in the *complex plane* as k ranges through all real values. The curve thus obtained is called the *Nyquist diagram* of DE (14). Figure 3.1b depicts the Nyquist diagram of the DE below (18b).

Since $C(k)$ is the inverse of a polynomial, it tends to the origin as $k \to \pm\infty$, that is, it "starts" and "ends" at the origin. It is a continuous curve except when the characteristic equation has one or more imaginary roots $\lambda_j = ik_j$. From Theorem 6 we obtain the following *Nyquist Stability Criterion*.

COROLLARY 2.   *The equation $L[u] = 0$ is strictly stable if and only if the Nyquist diagram for $C(k)$ turns through $-n\pi$ radians as k increases from $-\infty$ to $+\infty$.*

If L is real, then $C(-k) = C^*(k)$, and it suffices to plot half of the Nyquist diagram. The operator L is strictly stable if and only if the Nyquist diagram turns through $n\pi/2$ radians as k increases from 0 to ∞.

*9   THE GREEN'S FUNCTION

The concept of the Green's function for initial value problems was introduced in Ch. 2, §9. For any inhomogeneous linear DE $L[u] = r(t)$, it is a function $G(t, \tau)$ such that

$$(19) \qquad u = f(t) = \int_a^t G(t, \tau)r(\tau)d\tau$$

satisfies $L[u] = r(t)$ if $t \geq a$, for any continuous function r. We now state a generalization of Theorem 10 of Ch. 2, §9, which describes the Green's function of a linear operator of arbitrary order.

THEOREM 7.   *The Green's function for the initial value problem of the nth order real linear differential operator with continuous coefficients*

$$(20)\ \ L = \frac{d^n}{dt^n} + p_1(t)\frac{d^{n-1}}{dt^{n-1}} + \cdots + p_{n-1}(t)\frac{d}{dt} + p_n(t), \qquad a \leq t \leq b$$

is zero if $t < \tau$. For $t \geq \tau$, it is that solution of the DE $L[G] = 0$ (for fixed τ and variable t) which satisfies the initial conditions

$$G = \partial G/\partial t = \cdots = \partial^{n-2}G/\partial t^{n-2} = 0, \qquad \partial^{n-1}G/\partial t^{n-1} = 1 \qquad for \qquad t = \tau$$

In the case $p_i(t) = a_i$ of linear DEs with constant real coefficients, the existence of such a solution follows from the results of §3. Green's function is easily computed as a sum of polynomials times exponentials. Thus, if (20) is $u''' + 3u'' + 3u' + u$, the Green's function is

$$G(t, \tau) = \begin{cases} 0 & \text{if } t < \tau \\ (t - \tau)^2 e^{\tau - t}/2 & \text{if } t \geq \tau \end{cases}$$

For variable coefficient-functions, existence will follow from the results of the next chapter.

We omit the proof of Theorem 7. It follows exactly the proof for second-order differential operators, given in Ch. 2, §9. This can be easily extended to the present case: one simply applies Leibniz's rule for differentiation under the integral sign n times instead of twice.

The computation of Green's functions for linear DEs with constant coefficients is most easily performed and its significance best understood by using the following result.

THEOREM 8. *The Green's function for the initial value problem of any linear differential operator with constant coefficients is a function $G(t, \tau) = \Gamma(t - \tau)$ depending only on the difference $t - \tau$.*

Proof. Let $\Gamma(t) = G(t, 0)$; then $\Gamma(t) = 0$ if $t < 0$. If $t \geq 0$, the function $\Gamma(t)$ is the solution of the DE $L[\Gamma] = 0$, which satisfies the initial conditions

$$\Gamma(0) = \Gamma'(0) = \cdots = \Gamma^{(n-2)}(0) = 0, \qquad \Gamma^{(n-1)}(0) = 1$$

We now remark that, if $u(t)$ is a solution of the DE $L[u] = 0$, for each fixed τ the function $u(t + \tau)$ of the variable t is also a solution of the DE. It follows that the function $F(t) = G(t + \tau, \tau)$ (for fixed τ) is a solution of the DE. This function satisfies the same initial conditions as the function Γ, because of the way Green's functions are defined. By the uniqueness theorem (Theorem 3), it follows that $\Gamma(t) = G(t + \tau, \tau)$. Hence, setting $s = t + \tau$, we obtain $\Gamma(s - \tau) = G(s, \tau)$, q.e.d.

Referring to Theorems 1 and 3, we obtain the next corollary.

COROLLARY 1. *In Theorem 8, the function $\Gamma(t - \tau)$ is of class $\mathscr{C}^{n-2}$. It satisfies $\Gamma(s) = 0$ for $s < 0$, while $\Gamma(s)$ is a linear combination of functions $s^r e^{\lambda_j s}$ for $s > 0$.†*

Changing variables in (19), we have also the following corollary.

COROLLARY 2. *If $r(t)$ vanishes for $t < a$, and is bounded and continuous for $a \leq t$, then the function*

$$(21a) \qquad f(t) = \int_{-\infty}^{\infty} \Gamma(t - \tau) r(\tau) \, d\tau = \int_{-\infty}^{\infty} \Gamma(s) r(t - s) \, ds$$

is a solution of the inhomogeneous linear DE with constant coefficients (13) on $[a, \infty]$, which satisfies $f(a) = f'(a) = \cdots = f^{(n-1)}(a) = 0$. (Note that, unless $r(a) = 0$, $f^{(n)}(a)$ does not exist.)

---

† The function $\Gamma(s)$ is, of course, *also* expressible for $s > 0$ as a *real* linear combination of functions of the form (9).

Indeed, since $r(\tau)$ vanishes for $\tau < a$ and $\Gamma(t - \tau)$ vanishes for $\tau > t$, the integral (21a) can be written as

(21b) $$f(t) = \int_a^t \Gamma(t - \tau) r(\tau) \, d\tau$$

and is equal to (19), by Theorem 8.

THEOREM 9.   *If L is strictly stable, formulas (21a) remain valid for any bounded continuous function r(t) defined for* $-\infty < t < \infty$; *for such a function, f (t) is the only solution of the DE (13) which is bounded for* $-\infty < t < \infty$.

Proof.   By Corollary 1 above, $\Gamma(s)$ is equal (for $s \geq 0$) to a linear combination of the form

(22) $$\Gamma(s) = \sum_{j=1}^n c_j s^{r_j} e^{\lambda_j x}, \qquad s \geq 0$$

If the equation $L[u] = 0$ is strictly stable, then the real parts of all λ_j are negative. Let $-m$ be the largest of these real parts. Then $-m < 0$, and

$$|e^{ms/2} \Gamma(s)| \leq \sum_{j=1}^n |c_j s^{r_j} e^{(\lambda_j + (m/2))s}|$$

Since Re $\{\lambda_j\} + (m/2) < 0$ for $1 \leq j \leq n$, the right side remains bounded for $0 \leq s \leq \infty$. Let M be an upper bound for the right side. Then we obtain

$$|\Gamma(s)| \leq M e^{-ms/2}, \qquad 0 \leq s < \infty$$

We next show that the integrals in (21a) are well-defined for any bounded continuous function r. The first integral can be rewritten in the form

$$\int_{-\infty}^t \Gamma(t - \tau) r(\tau) \, d\tau$$

since $\Gamma(t - \tau) = 0$ for $t < \tau$. Using the foregoing bound for $\Gamma(s)$, and letting R be an upper bound for $|r(\tau)|$ on $-\infty < \tau < \infty$, we obtain

$$\left| \int_{-\infty}^t \Gamma(t - \tau) r(\tau) \, d\tau \right| \leq \int_{-\infty}^t |\Gamma(t - \tau)| \, |r(\tau)| \, d\tau$$

$$\leq RM \int_{-\infty}^t e^{-m(t-\tau)/2} \, d\tau = \frac{2RM}{m} < +\infty$$

for all t. Hence the integral is well-defined and defines a bounded function $f(t)$. To show that f is a solution of the DE, we can argue as in Theorem 10 of Ch. 2, provided we can carry out the differentiation under the integral sign. This can indeed be justified‡; instead, however, we shall give a direct argument.

Consider the sequence of functions $r_k(t)$ defined by the formulas

(23) $$r_k(t) = \begin{cases} r(t) & \text{if } t \geq -k \\ 0 & \text{if } t < -k \end{cases} \qquad k = 1, 2, \ldots$$

Then the functions

(23′) $$f_k(t) = \int_{-k}^{t} \Gamma(t - \tau) r_k(\tau) \, d\tau = \int_{-\infty}^{\infty} \Gamma(t - \tau) r_k(\tau) \, d\tau$$

are solutions of the DE (14). We shall show that, for t ranging over any interval $a \leq t \leq b$, the functions $f_k(t)$, as well as their derivatives of orders up to n, converge uniformly to the derivatives of the function $f(t)$. This will also prove that $f(t)$ is a solution of the DE (14).

From the expression (22) for $\Gamma(s)$ as a linear combination of functions of the form $s^{r_j} e^{\lambda_j t}$, we see that all derivatives of $\Gamma(s)$ are also linear combinations of functions of the same form, for different r_j, but with the same sequence of exponents λ_j. That is, for the derivative of order ℓ, we have

$$\Gamma^{(\ell)}(s) = \sum_{j=1}^{n} p_j(s) e^{\lambda_j s},$$

where the p_j are polynomials in the variable s, depending on the order of differentiation ℓ.† It follows, as before, that

$$|\Gamma^{(\ell)}(s)| \leq M_\ell e^{-ms/2}, \qquad \ell = 1, 2, \ldots$$

Now, from the expression

$$f_k^{(\ell)}(t) = \int_{-k}^{t} \Gamma^{(\ell)}(t - \tau) r_k(\tau) \, d\tau$$

we find, for sufficiently large k and j, where $k \geq j$

$$|f_k^{(\ell)}(t) - f_j^{(\ell)}(t)| \leq \int_{-k}^{-j} |\Gamma^{(\ell)}(t - \tau)| \, |r_k(\tau) - r_j(\tau)| \, d\tau$$

$$\leq \int_{-k}^{-j} |\Gamma^{(\ell)}(t - \tau)| \, |r_k(\tau)| \, d\tau$$

$$\leq RM_\ell \int_{-k}^{-j} e^{-m(t-\tau)/2} \, d\tau$$

---

‡ Courant, Vol. 2, p. 312.
† This can be easily seen by applying Leibniz's rule; cf. Courant and John, p. 203.

and the last integral clearly tends to zero as j, $k \to \infty$, uniformly for $a \le t \le b$. Therefore, $|f_k^{(\ell)}(t) - f_j^{(\ell)}(t)| < \epsilon$ for sufficiently large k, j, uniformly for $a \le t \le b$. This completes the proof of the fact that f is a solution of the DE.

Lastly, one can easily see, by the following argument, that f thus defined is the only bounded solution. If f_1 were another bounded solution, then $f - f_1$ would be a bounded solution of the homogeneous DE. But, since the DE is *strictly* stable as $t \to \infty$, no nontrivial solution of the homogeneous DE can remain bounded as $t \to -\infty$ (cf. Theorem 4). Hence, $f - f_1 \equiv 0$, and the proof is complete.

Convolution. The preceding results have very simple interpretations in terms of the important notion of convolution. The *convolution* of two functions $f(t)$ and $g(t)$ defined for all real t is defined by the formula

$$(24) \qquad h(x) = \int_{-\infty}^{\infty} f(x - t)g(t)\, dt$$

whenever the integral is finite. If the functions f and g are identically zero for $t < 0$, this formula simplifies to

$$(24') \qquad h(x) = f*g(x) = \int_0^x f(t)g(x - t)\, dx$$

In many ways this operation is analogous to the multiplication of two infinite series. It is commutative and associative, as can easily be seen.

Corollary 2 of Theorem 8 states that the solution of the DE $L[u] = r(t)$ for the initial conditions $u(0) = u'(0) = \cdots = u^{(n-1)}(0) = 0$ is the convolution $r * \Gamma$ of r and the Green's function Γ for the same initial conditions, provided that $r \equiv 0$ for $t < 0$.

Theorem 9 can also be interpreted in terms of the convolution operation. It asserts that, for strictly stable L, the solution of the "input–output" problem $L[u] = r(t)$ is $r * \Gamma$ for any uniformly bounded "input" $r(t)$.

EXERCISES E

In Exs. 1–6, construct the Green's function for the initial value problem of the DE indicated.

1. $d^3u/dt^3 = r(t)$ 2. $d^nu/dt^n = r(t)$

3. $u^{iv} - u = r(t)$ 4. $u''' + u = r(t)$

5. $u^{iv} + u = r(t)$ *6. $u^{iv} - u = r(t)$

7. Find Green's function of the DE $d^nu/dt^n = r(t)$.

8. Carry out in detail the proof of Theorem 7 for $n = 3$, performing all differentiations under the integral sign explicitly.

*9. Show that, if (19) is defined for all t and the Green's function $G(t, \tau) = \Gamma(t - \tau)$, then all coefficients $p_k(t)$ are constant.

*10. Show that, if $u(t + \tau)$ is a solution of (19) with $r(t) = 0$ whenever $u(t)$ is, then the coefficients $p_k(t)$ are all constants.

11. (a) Show that in Ex. D1, the values of $p_L(ik)$ traverse the parabola $x = 4 - (y/5)^2$.
    (b) Verify Theorem 6 in this case.

12. For $u'' + au' + bu = ce^{ikt}$, make graphs of the gain function versus the dimensionless frequency $k/\sqrt{b}$ for the values $\eta = 0, \eta = 1/2, \eta = 1/\sqrt{2}, \eta = 2$ of the parameter $\eta = a/2\sqrt{b}$ $(b > 0)$.

13. Show that, if $p_L(ik) = p_L'(ik) = \cdots = p_L^{(n-1)}(ik) = 0$ but $p_L^{(n)}(ik) \neq 0$, then a solution of $L[u] = e^{ikt}$ is $u(t) = (1/p_L^{(n)}(ik)t^n e^{ikt}$.

14. A DE (13) is stable at $t \to -\infty$ when all solutions of the DE remain bounded as $t \to -\infty$. Find necessary and sufficient conditions for stability at $-\infty$.

15. In Ex. 14, find necessary and sufficient conditions for strict stability at $-\infty$.

16. Show that no DE (13) can be strictly stable at both ∞ and $-\infty$.

17. Show that a DE (13) is stable at both ∞ and $-\infty$ if and only if every root of the characteristic equation is simple and a purely imaginary number.

CHAPTER 4

POWER SERIES SOLUTIONS

1 INTRODUCTION

In the preceding chapters, we have constructed many explicit solutions of DEs and initial value problems in terms of the so-called elementary functions. These are functions that can be built up using the four rational operations (addition, subtraction, multiplication, and division) from the exponential and trigonometric functions and their inverses (e.g., the inverse $\ln x$ of the exponential function x). Since $x^{\nu} = e^{\nu \ln x}$, fractional powers and nth roots $\sqrt[n]{x} = x^{1/n}$ of *positive* (real) elementary functions can also be considered as "elementary."

In these earlier chapters, functions like $P(x) = \int p(t)\, dt$ defined symbolically as *indefinite integrals* were also used freely. Indeed, accurate tables of such functions are easily computed using Simpson's rule (Ch. 1, §8). However, one should also realize that the indefinite integrals of most "elementary" functions are *not* themselves "elementary" in the sense defined above. This basic fact explains why formulas for such expressions as

$$(1) \qquad \int \frac{\sin x}{x}\, dx \qquad \int \frac{e^{-x}}{x}\, dx \qquad \text{and} \qquad \int \frac{d\theta}{\sqrt{1 - k^2 \sin^2 \theta}}$$

are conspicuously missing from tables of integrals.†

This chapter will introduce a much more powerful method for constructing solutions of DEs. This method consists in admitting sums of infinite *power series* as defining *functions* in any domain where they *converge*. Such functions are called "analytic." All sums, differences, products, and (except where the denominator vanishes) quotients of analytic functions are analytic. Moreover, as is almost evident, since

$$\int (a_0 + a_1 t + a_2 t^2 + \cdots)\, dt = C + a_0 x + \frac{a_1 x^2}{2} + \frac{a_3 x^3}{3} + \cdots$$

any indefinite integral of any analytic function is also expressible as an analytic function. For example, since $(\sin x)/x = 1 - (x^2/3!) + (x^4/5!) + \dots$, termwise

---

† Cf. Dwight. A valuable collection of analogous explicit solutions of ordinary DEs may be found in Kamke, pp. 293–660.

integration gives the sine integral function

$$(2) \qquad \text{Si}(x) = \int_0^x \frac{\sin \xi}{\xi} d\xi = x - \frac{x^3}{18} + \frac{x^5}{600} - \cdots$$

(Since $E_1(x) = \int_x^\infty \frac{e^{-\xi}}{\xi} d\xi$ is logarithmically infinite near $x = 0$, to express it as an analytic function is, however, more involved; cf. §3.)

Although most of the techniques introduced in this chapter are applicable with little change to *complex* analytic functions, we shall defer the discussion of these to Chapter 9. Instead, this chapter will introduce a second and even more fundamental idea than that of considering functions as defined by power series. This is the idea of considering *functions as defined by differential equations* and appropriate initial or boundary conditions, by using the defining DE itself to determine the properties of the function.

This approach is especially easy to carry out for first-order *linear* DEs. For example, we can use it to derive the key properties of the exponential function $E(x)$ (or e^x) as the (unique) solution of the DE $E'(x) = E(x)$ that satisfies the initial condition $E(0) = 1$. For any a, $f(x) = E(a + x)$ must satisfy $f(0) = E(a)$ and $f'(x) \equiv E'(a + x) = E(a + x) = f(x)$. Since $f(x)$ and $E(x)$ are solutions of the same first-order linear homogeneous DE, it follows that $f(x) = f(0)E(x) = E(a)E(x)$, giving the formula $e^{a+x} = e^a e^x$. In particular, $E(a)$ can never vanish and is always positive, together with $E'(a) = E(a)$, $E''(a) = E(a)$, $\ldots$, which shows that $E(x)$ is increasing and convex. Finally, its Maclaurin series is

$$(3) \qquad E(0) + E'(0)x + (E''(0)x^2/2!) + (E'''(0)x^3/3!) + \cdots$$

giving the exponential series $e^x = \sum_{k=0}^\infty x^k/(k!)$.

In Ch. 2, §§6–7, we have seen how one can derive many oscillation and non-oscillation properties of solutions of linear second-order DEs

$$(4) \qquad u'' + p(x)u' + q(x)u = 0$$

When the coefficient-functions

$$p(x) = p_0 + p_1 x + p_2 x^2 + \ldots, \qquad q(x) = q_0 + q_1 x + q_2 x^2 + \ldots,$$

can be expressed as sums of convergent power series, we will show in this chapter how to find a basis of solutions of the DE (4) having the same form. The functions so constructed include many of the special functions most commonly used in applied mathematics.

Namely, by substituting a formal power series

$$(4') \qquad u = a_0 + a_1 x + a_2 x^2 + a_3 x^3 + \cdots = \sum_{k=0}^\infty a_k x^k$$

into the DE (4), and using Cauchy's product formula (18'),

$$(4'') \qquad p(x)u(x) = \sum_{k=0}^{\infty} c_k x^k, \qquad c_k = \sum_{j=0}^{\infty} p_j a_{k-j}$$

we shall first show how to compute the unknown coefficients a_k in (4'). We shall then show that the radius of convergence of the resulting formal power series (4') is at least as great as the lesser of the radii of convergence of the series for $p(x)$ and $q(x)$.

We will then give a similar construction for the solutions of normal first-order DEs of the form $y' = F(x,y)$, where

$$(5) \qquad F(x,y) = b_{00} + b_{10}x + b_{01}y + b_{20}x^2 + b_{11}xy + b_{02}y^2 + \cdots$$

is also assumed to be analytic. Finally, we will estimate the radius of convergence of this series.

2 METHOD OF UNDETERMINED COEFFICIENTS

The class $\mathcal{A}(D)$ of functions *analytic* in a domain D is defined as the class of those functions that can be expanded in a power series around any point of D, which is *convergent* in some neighborhood of that point. By a translation of coordinates, one can reduce the consideration of such power series to power series having the origin as the center of expansion.

Many of the special functions commonly used in applied mathematics have simple power series expansions. This is especially true of functions defined as definite integrals of functions with known power series expansions: one simply integrates termwise! Thus, from (2), we obtain

$$\mathrm{Si}(x) = \int_0^x \sum_{k=0}^{\infty} (-1)^k \frac{\xi^{2k} \, d\xi}{(2k+1)!} = \sum_{k=0}^{\infty} (-1)^k \frac{x^{2k+1}}{(2k+1)^2(2k)!}$$

Expanding the integrand in power series, we can obtain similarly the first few terms of the series expansion for the elliptic integral of the first kind,

$$F(k,\sin^{-1}x) = \int_0^x (1 - \xi^2)^{1/2}(1 - k^2\xi^2)^{-1/2} \, d\xi$$

$$= x + \frac{1 + k^2}{6} x^3 + \frac{3 + 2k^2 + 3k^3}{40} x^5 - \cdots$$

with a little more effort.

Likewise, consider the exponential function e^x, defined as the solution of $y' = y$ and the initial condition $y(0) = 1$. Differentiating this DE n times, we get $y^{(n+1)} = y^{(n)}$. Substitution into Taylor's formula then gives the familiar exponential series $e^x = \sum_{k=0}^{\infty} x^k/(k!)$.

We now give some other applications of the same principle to some special functions familiar in applied mathematics, defined as solutions of second-order linear homogeneous DEs (4) with analytic coefficient-functions.

Example 1. The Legendre DE (Ch. 2, §1) is usually written

(6)
$$\frac{d}{dx}\left[(1-x^2)\frac{du}{dx}\right] + \lambda u = 0$$

where λ is a parameter.

Substituting the series (4′) into the linear DE (6), and equating to zero the coefficients of $1 = x^0, x, x^2, \ldots$, we get an infinite system of linear equations

$$0 = 2a_2 + \lambda a_0 = 6a_3 + (\lambda - 2)a_1 = 12a_4 + (\lambda - 6)a_2 = \cdots$$

The kth equation of this system is the *recurrence relation*

(6′)
$$a_{k+2} = \frac{k(k+1) - \lambda}{(k+1)(k+2)} a_k$$

This relation defines for each λ two linearly independent solutions, one consisting of *even* powers of x, and the other of *odd* powers of x. These solutions are power series whose *radius of convergence* is unity by the Ratio Test,† unless $\lambda = n(n+1)$ for some nonnegative integer n. When $\lambda = n(n+1)$, the Legendre DE has a polynomial solution which is an even function if n is even and an odd function if n is odd. These polynomial solutions are the *Legendre polynomials* $P_n(x)$.

Graphs of the Legendre polynomials $P_0(x), \ldots, P_4(x)$ are shown in Figure 4.1. Note how the number of their oscillations increases with $\lambda = n(n+1)$, as predicted by the Sturm Comparison Theorem.

In general, to construct a solution of an ordinary DE in the form of a power series, one first writes down a *symbolic* power series with letters as coefficients, $u = a_0 + a_1x + a_2x^2 + \ldots$. To determine the a_k numerically, one substitutes this series for u in the DE, differentiating it term by term. One then collects the coefficients of x^k that results, for each k, and equates their sum to zero.

This is called the Method of Undetermined Coefficients. For it to be applicable to the second-order linear DE $u'' + p(x)u' + q(x)u = 0$, p and q must be analytic near the origin. That is (cf. §1), they can be expanded into power series

(7)
$$p(x) = p_0 + p_1x + p_2x^2 + \cdots$$
$$q(x) = q_0 + q_1x + q_2x^2 + \cdots$$

convergent for sufficiently small x.

---
† Courant and John, p. 520; Widder, p. 288.

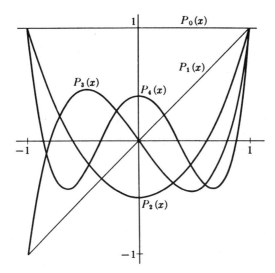

Figure 4.1   Graphs of $P_n(x)$; Legendre polynomials.

To compute the solution, one assumes that

(7') $$u = a_0 + a_1 x + a_2 x^2 + a_3 x^3 + \cdots = \sum_{k=0}^{\infty} a_k x^k$$

Term-by-term differentiation gives

$$u'' = 2a_2 + 6a_3 x + 12a_4 x^2 + \cdots + (n + 1)(n + 2)a_{n+2} x^n + \cdots$$

$$pu' = p_0 a_1 + (2p_0 a_2 + p_1 a_1)x + \cdots + \left[\sum_{k=1}^{n} (n + 1 - k)p_k a_{n+1-k} \right] x^n + \cdots$$

$$qu = q_0 a_0 + (q_0 a_1 + q_1 a_0)x + \cdots + \left[\sum_{k-1}^{n} q_k a_{n-k} \right] x^n + \cdots$$

Substituting into (4) and equating the coefficients of $1, x, \ldots, x^{n-1}, \ldots$ to zero, we get successively

$$a_2 = -\tfrac{1}{2}(p_0 a_1 + q_0 a_0), \qquad a_3 = -\tfrac{1}{6}(2p_0 a_2 + p_1 a_1 + q_0 a_1 + q_1 a_0)$$

and so on. The general equation is

(8) $$a_{n+1} = - \frac{\left[\displaystyle\sum_{k=0}^{n} (n - k)p_k a_{n-k} + \sum_{k=0}^{n} q_k a_{n-1-k} \right]}{n(n + 1)}, \qquad n \geq 1$$

Given $a_0 = f(0)$ and $a_1 = f'(0)$, a unique power series is determined by (8), which formally satisfies the DE (4). We have proved the following theorem.

THEOREM 1.   *Given a linear homogeneous second-order* DE (4) *with analytic coefficient functions* (5), *there exists a unique power series* (6) *that formally satisfies the* DE, *for each choice of a_0 and a_1.*

Example 2.   The Hermite DE is

$$(9) \qquad\qquad u'' - 2xu' + \lambda u = 0$$

Applying the Method of Undetermined Coefficients to (9), we obtain the recursion formula

$$(9') \qquad\qquad a_{k+2} = \frac{2k - \lambda}{(k + 1)(k + 2)} a_k$$

which again gives, for each λ, one power series in the even powers x^{2k} of x and another in the odd powers x^{2k+1}. These power series are convergent for all x; if $\lambda = 2n$ is a nonnegative even integer, one series is a polynomial of degree n, the *Hermite polynomial $H_n(x)$.*

Caution.   We have not stated or proved that the formal power series (7') converges or that it represents an analytic function. To see the need for proving this, consider the DE $x^2 u' = u - x$, which has the everywhere *divergent* formal power series solution

$$x + x^2 + (2!)x^3 + (3!)x^4 + \cdots + (n - 1)!x^n + \cdots.$$

For *normal* second-order linear DEs (4), the *convergence* of the power series defined by (7') to an *analytic solution* will be proved in §6. First we treat some more special cases, in which convergence is easily verified.

EXERCISES A

1. (a) Prove that the Legendre DE has a polynomial solution if and only if $\lambda = n(n + 1)$.
   (b) Prove that the radius of convergence of every nonpolynomial solution of the Legendre DE is one.

2. Find a recurrence relation like (8') for the DE $(1 + x^2)y'' = y$, and compute expansions through terms in x^{10} for a basis of solutions.

3. (a) Find power series expansions for a basis of solutions of Airy's DE $u'' + xu = 0$.
   (b) Prove that the radius of convergence of both solutions is infinite.
   (c) Show that any solution of Airy's DE vanishes infinitely often on $(0,\infty)$, but at most once on $(-\infty,0)$.
   (d) Reduce $u'' + (ax + b)u = 0$ to $d^2 u/dt^2 + tu = 0$ by a suitable change of independent variable.

4. Show that $u(x)$ satisfies the Hermite DE (9) if and only if $v = e^{-x^2/2}u$ satisfies $v'' + (\lambda + 1 - x^2)v = 0$.

5. (a) Find a basis of power series solutions for the DE $u'' + x^2 u = 0$.
   (b) Do the same for $u''' + xu = 0$.

6. In spherical coordinates, the Laplacian of a function $F(r,\theta)$ is

$$\nabla^2 F = F_{rr} + \frac{2}{r} F_r + \frac{1}{r^2} [F_{\theta\theta} + (\cos \theta) F_0].$$

Show that $F(r,\theta) = r^n P(\cos \theta)$ satisfies $\nabla^2 F = 0$ if and only if $P(x)$ satisfies the Legendre DE with $\lambda = n(n + 1)$.

7. (a) Show that in spherical coordinates, $U(r,\theta) = (1 - 2r \cos \theta + r^2)^{-1/2}$ satisfies $\nabla^2 U = 0$. [HINT: Consider the potential of a charge at $(1,0,0)$.]

   (b) Infer that $(1 - 2r \cos \theta + r^2)^{-1/2} = \sum_0^\infty r^n P_n(\cos \theta)$, where $P_n(x)$ is a solution of the Legendre equation (8) with $\lambda = n(n + 1)$.

*8. Show that, for any positive integer n, the polynomial $d^n[(x^2 - 1)^n]/dx^n$ is a solution of the Legendre DE with $\lambda = n(n + 1)$. (This is Rodrigues' formula.)

3   MORE EXAMPLES

Many other famous "higher transcendental functions" of classical analysis are best defined as solutions of linear homogeneous second-order DEs. Among these the Bessel functions have probably been most exhaustively studied.†

Example 3.   The Bessel function $J_n(x)$ of order n can be defined, for $n = 0$, 1, 2, . . . , as an (analytic) solution of the Bessel DE of order n,‡

(10)
$$u'' + \frac{1}{x} u' + \left(1 - \frac{n^2}{x^2}\right) u = 0$$

The coefficients of this DE have a singular point $x = 0$, but are analytic at all other points (see §5).

Though the Bessel DE (10) of integral order n has a singular point at the origin, it has a nontrivial analytic solution there. The power series expression for this solution can be computed by the Method of Undetermined Coefficients. For example, when $n = 0$, the DE (10) reduces to

(10′)
$$(xu')' + xu = 0$$

Substituting $u = \sum a_k x^k$ into (10′), we get the recursion relation $k^2 a_k = -a_{k-2}$. For $a_0 = 1$, $a_1 = 0$, (10′) has the analytic solution

(11)
$$J_0(x) = 1 - \left(\frac{x}{2}\right)^2 + \left(\frac{x^2}{2 \cdot 4}\right)^2 - \left(\frac{x^3}{2 \cdot 4 \cdot 6}\right)^2 + \cdots$$

$$= 1 - \frac{x^2}{4} + \frac{x^4}{64} - \frac{x^6}{2304} + \cdots + (-1)^r \frac{x^{2r}}{[2^r(r!)]^2} + \cdots$$

---

† See G. N. Watson, *A Treatise on Bessel Functions*, Cambridge University Press, 1926.

‡ Note that the Bessel DE of "order" n is still a "second-*order*" DE.

This series defines the Bessel function of order zero. It is convergent for all x (by the Ratio Test), and converges rapidly if $|x| < 2$.

Similar calculations give, for general n, the solution

$$(12) \qquad J_n(x) = \left(\frac{x}{2}\right)^n \left\{ \frac{1}{n!} - \frac{x^2}{4(n+1)!} + \frac{x^4}{32(n+2)!} - \cdots \right.$$

$$\left. + \frac{(-1)^r \left(\dfrac{x}{2}\right)^{2r}}{(r!)(n+r)!} + \cdots \right\}$$

whose coefficients $b_k = a_{k+n}$ satisfy the relation $(n^2 - k^2)b_k = b_{k-2}$. The series (12) is also everywhere convergent. Hence, the Bessel function $J_n(x)$ of any integral order n is an *entire* function, analytic for all finite x. Graphs of $J_0(x)$ and $J_1(x)$ are shown as Figure 4.2.

By comparing coefficients, one can easily verify the relations

$$J_1 = -J_0', \qquad xJ_2 = J_1 - xJ_1' = 2J_1 - xJ_0, \qquad xJ_3 = 2J_2 - xJ_2'$$

For general n, one can verify, in the same way, the *recursion formula*

$$(13) \qquad\qquad xJ_{n+1} = nJ_n - xJ_n' = 2nJ_n - xJ_{n-1}$$

Clearly, formula (13) defines $J_1, J_2, J_3, \ldots$ recursively from J_0.

Example 4.   A special case of the *Jacobi DE* is the following

$$(14) \qquad\qquad (1 - x^2)u'' - xu' + \lambda u = 0$$

where λ is a parameter.

If we set $u = \sum_{k=0}^{\infty} a_k x^k$, substitute into Eq. (14), and collect the terms in x^n, we get

$$(14') \qquad\qquad (n+1)(n+2)a_{n+2} - (n^2 - \lambda)a_n = 0$$

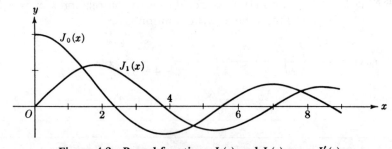

Figure 4.2   Bessel functions $J_0(x)$ and $J_1(x) = -J_0'(x)$.

For most values of λ, this leads to two solutions of (10), one an even function and the other odd. However, when $\lambda = n^2$ is a square, we have instead a *polynomial* solution.

Even simpler to solve by power series is the *Airy DE*

$$(15) \qquad\qquad u'' + xu = 0$$

One easily derives from (15) the recursion relation $n(n - 1)a_n + a_{n-3}$ for the coefficients. A basis of power series solutions therefore consists of the functions

$$Ai(x) = 1 - \frac{x^3}{6} + \frac{x^6}{180} - \frac{x^9}{12960} + \cdots$$

and

$$Bi(x) = x - \frac{x^4}{12} + \frac{x^7}{504} + \frac{x^{10}}{45360} + \cdots$$

4   THREE FIRST-ORDER DEs

The three homogeneous, linear, *second*-order DEs just treated were quite similar to each other. In this section, we will derive power series expansions for three *first*-order DEs having much less in common.

Example 5.   To obtain a useful power series expansion for the "exponential integral" function $E_1(x)$ defined by (2), one must supplement simple substitution into the exponential series, which gives only

$$\frac{e^{-x}}{x} = x^{-1} - 1 + \frac{x}{2!} - \frac{x^2}{3!} + \frac{x^3}{4!} - \cdots$$

from which termwise integration yields

$$(16) \qquad\qquad E_1(x) = -\gamma - \ln x - \sum_{k=1}^{\infty} \frac{(-x)^k}{[k \cdot (k!)]}$$

where γ is an unknown constant of integration. The fact that $\gamma = .5772156649$, discovered by Euler, requires additional analysis.

Moreover, to evaluate $E_1(x)$ when $x \geqq 10$, one should replace (16) by the asymptotic formula

$$(16') \quad E_1(x) = \frac{e^{-x}}{x} \int_x^{\infty} \frac{e^{-t}\, dt}{1 + (t/x)} \sim \left[\frac{e^{-x}}{x}\right] \left(1 - \frac{1}{x} + \frac{2}{x^2} - \frac{6}{x^3} + \cdots\right)$$

The final series of negative powers is *divergent* for all x: it is a so-called *asymptotic series* (cf. Ch. 7, §7). However, the partial sum of the first n terms has a relative error of less than 1%, and an absolute error less than $10^{-6.5}$ for $x \geq 10$.

Example 6. Pearson's DE is

(*)
$$(A + Bx + Cx^2)y' = (D + Ex)y, \qquad A \neq 0$$

Its solution by power series is straightforward, after dividing the equation through by A.

Setting $A = 1$, clearly $y = \sum_{k=0}^{\infty} a_k x^k$ implies

$$(1 + Bx + Cx^2)y' = a_1 + (2a_2 + Ba_1)x +$$
$$\sum_{k=3}^{\infty} [(k - 2)a_{k-2}C + (k - 1)a_{k-1}B + ka_k]x^{k-1}$$

and

$$(D + Ex)y = Da_0 + \sum_{k=1}^{\infty} (Da_k + Ea_{k-1})x^k$$

The two sides of the preceding equations are equal if and only if

$$a_1 = Da_0, \qquad a_2 = \frac{[Ea_0 + (D - Ba_1)]}{2}$$

$$ka_k = [D - (k - 1)B]a_{k-1} + [E - (k - 2)C]a_{k-2}$$

For given numerical values of A, B, C, D, E, one thus obtains a basic solution of (*) in the form of the power series

$$y = 1 + Dx + \frac{[E + D(1 - B)]x^2}{2} + \sum_{k=3}^{\infty} a_k x^k$$

where the a_k ($k \geq 3$) are computed recursively from the last previously displayed formula.

Example 7. We will next consider the function $y = \tan x$ as the solution of the *nonlinear* DE $y' = 1 + y^2$ for the initial value $y(0) = 0$.

By successive differentiation of this DE, we easily obtain formulas for the second and third derivatives:

$$y'' = (1 + y^2)' = 2yy' = 2y(1 + y^2)$$
$$y''' = 2y' + 6y^2y' = 2(1 + 3y^2)(1 + y^2)$$

Since $(1 + 3y^2)(1 + y^2) = 1 + 4y^2 + 3y^4$, we have further

$$y^{iv} = 2(8y + 12y^3)y' = 8y(2 + 3y^2)(1 + y^2)$$

In particular, setting $x = y = 0$, we get $y'(0) = 1$, $y''(0) = 0$, $y'''(0) = 2$, and $y^{iv}(0) = 0$, whence $\tan x = x + x^3/3 + O(x^5)$.

Note that since $1 + y^2$ is positive, the function $\tan x$ is always increasing. Also, since y'' has the same sign as y, the graph of $y = \tan x$ is concave upward in the upper half-plane and concave downward in the lower half-plane. Likewise, y''' is always positive, and so y'' is increasing.

Again, setting $t = -x$ and $z = -y$, we obtain the formulas

$$\frac{dz}{dt} = \frac{dy}{dx} = 1 + y^2 = 1 + z^2 \qquad \text{and} \qquad z(0) = 0$$

Hence $z(t)$ is also a solution of the initial-value problem of Example 7. Therefore, by the uniqueness theorem (Ch. 1, Theorem 6′) for first-order DEs, $z = \tan t$. This proves that $-\tan(-x) = \tan x$: $\tan x$ is an *odd function*.

Consequently, the Taylor series expansion of $\tan x$ contains only terms of odd order:

$$y = x + a_3 x^3 + a_5 x^5 + a_7 x^7 + a_9 x^9 + \cdots$$

Substituting into the defining DE, we obtain

$$1 + 3a_3 x^2 + 5a_5 x^4 + 7a_7 x^6 + 9a_9 x^8 + \cdots$$
$$= 1 + x^2 + 2a_3 x^4 + (2a_5 + a_3^2)x^6 + 2(a_7 + a_3 a_5)x^8 + \cdots$$

Equating coefficients of like powers of x, we get

$$3a_3 = 1, \qquad 5a_5 = 2a_3, \qquad 7a_7 = 2a_5 + a_3^2, \qquad 9a_9 = 2a_7 + 2a_3 a_5$$

and so on. Solving recursively, we get the first few terms of the power series expansion for $\tan x$,

$$(**) \qquad \tan x = x + \frac{x^3}{3} + \frac{2x^5}{15} + \frac{17}{315}x^7 + \frac{62}{2835}x^9 + \cdots$$

The radius of convergence of this power series is $\pi/2$; this follows from the formula $\tan x = \sin x/\cos x$ and the results of §3.

By differentiating the DE $y' = 1 + y^2$ repeatedly, one obtains similarly $y'' = 2yy'$, $y''' = (2 + 3y^2)(1 + y^2)$, $y^{iv} = 2(5y + 6y^2)(1 + y^2)$, and so on. The Taylor series expansion for $y_{n+1} = \tan(x_n + h)$ through terms in h^4 for given $y_n = \tan x_n$ is therefore

$$(*) \qquad y_{n+1} = y_n + h(1 + y_h^2)Q_n + O(h^5)$$

where

$$Q_n = 1 + 2y_n h + (2 + 3y_n^2)h^2 + (5 + 6y_n^2)y_n h^3$$

By neglecting the $O(h^5)$ term in (*), one obtains a formula

(**) $$Y_{n+1} = Y_n + h(1 + Y_n^2)Q_n(Y_n, h)$$

for computing a table of *approximate* values Y_n of $y_n = \tan (nh)$ from the *initial value* $Y_0 = y_0 = 0$, which is much more accurate than that given by the "Taylor series method" of Ch. 1, §8.

Order of Accuracy. We have used in (*) the convenient notation $O(h^5)$ to signify the fact that the remainder is bounded by Mh^5, where M is independent of h. As a result, one expects that the error per step h will be roughly proportional to h^5. Since the number of steps is proportional to h^{-1}, one therefore expects the cumulative error to be proportional to h^4. This contrasts with the simpler Taylor series method of Ch. 1, §8, which has only $O(h^2)$ cumulative accuracy.

EXERCISES B

1. Derive formula (13) by comparing the coefficients of the appropriate power series.
2. (a) Show that the function $(\sin r)/r$ satisfies the DE $u_{rr} + (2/r)u_r + u = 0$ and the initial conditions $u(0) = 1$, $u'(0) = 0$.
   (b) Find another, linearly independent solution of this DE.
3. Show that the DE $(Ax^2 + B)u'' + Cxu' + Du = 0$ has a solution that is a polynomial of degree n if and only if $An^2 + (C - A)n + D = 0$.
4. Show that the change of independent variable $x = \cos \theta$ transforms the Legendre DE (8) of the text into $u_{\theta\theta} + (\cos \theta)u_\theta + \lambda u = 0$. What is the self-adjoint form of this equation?
*5. Find conditions on the constants $A, \ldots, F$ necessary and sufficient for the DE $(Ax^2 + Bx + C)u'' + (Dx + E)u' + Fu = 0$ to have a polynomial solution of degree n.
6. Show that if $y' = 1 + y^2$, then $y'' = 2y(1 + y^2)$, $y''' = 2(1 + y^2)(1 + 3y^2)$, and $y^{iv} = 8y(1 + y^2)(2 + 3y^2)$.
7. Show that any function that satisfies $y' = 1 + y^2$ is an *increasing* function, and that its graph is convex upward in the upper half-plane. [HINT: Use Ex. 6.]
8. Derive the coefficients $1/3$, $2/15$, $17/315$, and $62/2835$ of the series (**) of the text.
9. Show that, if $y' = 1 + y^2$, $y^v = 8y(2 + 15y^2 + 15y^4)(1 + y^2)$.

5 ANALYTIC FUNCTIONS

A function is called *analytic* in a domain D when, near any point of D, it can be expanded in a power series that is *convergent* in some neighborhood of that point. For instance, a real function $p(x)$ of one real variable is analytic in the open interval (x_1, x_2) when, given any x_0 in this interval (i.e., satisfying $x_1 < x_0 < x_2$), there exist coefficients $p_0, p_1, p_2, \ldots$ and a positive number δ such that

$$(17) \qquad p(x) = \sum_{k=0}^{\infty} p_k(x - x_0)^k \qquad \text{if} \qquad |x - x_0| < \delta, \qquad \delta > 0$$

The numerical values of δ and the coefficients p_k will, of course, depend on x_0.

Likewise, a real function $F(x,y)$ is analytic in a domain of D of the real xy-plane when, given $(x_0, y_0) \in D$, there exist constants b_{jk} $(j,k = 0, 1, 2, \ldots)$ and $\delta > 0$ such that

$$F(x,y) = \sum_{j=0}^{\infty} \sum_{k=0}^{\infty} b_{jk}(x - x_0)^j(y - y_0)^k \qquad \text{if} \qquad |x - x_0| + |y - y_0| < \delta$$

An example of such a series expansion is the double geometric series

$$(18) \qquad G(x,y) = \frac{M}{\left(1 - \dfrac{x}{H}\right)\left(1 - \dfrac{y}{k}\right)} = \sum_{j=0}^{\infty} \sum_{k=0}^{\infty} \frac{M}{H^j K^k} x^j y^k$$

This series converges in the rectangle $|x| < H$, $|y| < K$ and defines an analytic function in this rectangle.

Analytic functions of three and more variables are defined similarly.

Domain of Convergence.   Crucial for work with any power series is an understanding of its domain of convergence. Within this domain, any power series is absolutely convergent, and it can be differentiated or integrated termwise any number of times. It follows that each coefficient of any power series is uniquely determined by the analytic function which the power series defines. This is because, if $u = F(x,y) = \Sigma_{j,k} \, b_{jk}(x - x_0)^j(y - y_0)^k$ is convergent in some neighborhood of (x_0, y_0), then

$$b_{jk} = \frac{(j!)(k!)\partial^{j+k}F}{\partial x^j \partial y^k (x_0, y_0)}$$

If F and $G(x,y) = \Sigma_{jk} \, c_{jk}(x - x_0)^j(y - y_0)^k$ are any two power series expansions about the same "center" (x_0, y_0), moreover, then their power series can be added, subtracted, and multiplied termwise within the intersection of their domains of convergence. Worth noting is Cauchy's formula for the product $h(x) = f(x)g(x)$ of two analytic functions $f(x) = \Sigma_{k=0}^{\infty} a_k x^k$ and $g(x) = \Sigma_{k=0}^{\infty} b_k x^k$. This formula is

$$(18') \qquad h(x) = \sum_{k=0}^{\infty} c_k x^k, \qquad \text{where} \qquad c_k = \sum_{j=0}^{k} a_j b_{k-j}$$

By a well-known theorem of the calculus, this will be absolutely convergent to $f(x)g(x)$ whenever the series for $f(x)$ and $g(x)$ are both absolutely convergent.†

In dealing with *linear* differential equations such as (4), it is sufficient to consider analytic functions of *one* variable. For these, the key concept is that of the radius of convergence. The *radius of convergence* R of the power series (17) is the largest δ such that the series converges whenever $|x - x_0| < \delta$. The radius of convergence of any power series can be determined from its coefficients by Cauchy's formula

$$(19) \qquad \frac{1}{R} = \limsup_{n \to \infty} \sqrt[n]{|p_n|} = \lim_{n \to \infty} \{\sup_{k > n} \sqrt[k]{|p_k|}\}$$

The series diverges for all x with $|x - x_0| > R$. The *interval of convergence* of (17) is the interval $(x_0 - R, x_0 + R)$.

For functions of a complex variable, the radius of convergence of the series (17) is still determined by Eq. (19). The series is convergent in the *circle of convergence* $|x - x_0| < R$ and divergent if $|x - x_0| > R$; it defines a single-valued analytic (or *holomorphic*) complex function inside its circle of convergence. When $R = \infty$, the power series (17) defines an analytic function for all x, real or complex; such functions are called *entire* functions.

Example 8.   The substitution $\xi = C - x$ reduces the DE

$$(20) \qquad u'' + \frac{A}{C - x} u' + \frac{B}{(C - x)^2} u = 0, \qquad C > 0$$

to *Euler's homogeneous* DE

$$(20') \qquad \frac{d^2 u}{d\xi^2} - \frac{A}{\xi} \frac{du}{d\xi} + \frac{B}{\xi^2} u = 0$$

already discussed in Ch. 3, §2.

To solve (20), try the function $u = \xi^\nu = (C - x)^\nu$. This satisfies (20) if and only if ν is a root of the *indicial equation* of (20'),

$$\nu(\nu - 1) - A\nu + B = 0$$

When $B < 0$, this indicial equation has one positive root and one negative root $-\mu$. Hence, (20) has two linearly independent real solutions, given by the binomial series

$$(21) \quad \left(1 - \frac{x}{C}\right)^\nu = 1 - \nu\left(\frac{x}{C}\right) + \frac{\nu(\nu - 1)}{2}\left(\frac{x}{C}\right)^2 - \cdots, \qquad |x| < C$$

---

† Courant and John, pp. 542–544, 555; Widder, pp. 303–306 and 318–320. For a more complete discussion, see K. Knopp, *Theory and Application of Infinite Series*, Dover, 1956.

and a like series with ν replaced by $-\mu$. When ν is a nonnegative integer, a polynomial solution is obtained. Otherwise, the radius of convergence of the series is C, the same as that of the power series expansions of the coefficient-functions

$$p(x) = \frac{A}{(C - x)} = \left(\frac{A}{C}\right)\left[1 + \sum_{k=1}^{\infty}\left(\frac{x}{C}\right)^k\right]$$

and $q(x) = B/(C - x)^2$ of the DE (20).

6 METHOD OF MAJORANTS

If one keeps in mind the results of §4, one can show quite easily that the formal power series solutions of (4), obtained by the Method of Undetermined Coefficients of §2, have for all choices of a_0 and a_1 radii of convergence at least as large as the smaller of the radii of convergence of the coefficient functions. To prove this, one uses an ingenious method due to Cauchy, the so-called Method of Majorants.

A power series $\Sigma a_k x^k$ is said to be *majorized* by the series $\Sigma A_k x^k$ if and only if $|a_k| \leq A_k$ for all $k = 0, 1, 2, 3, \ldots$. By the Comparison Test, the radius of convergence of $\Sigma a_k x^k$ is then at least as large as that of $\Sigma A_k x^k$, and all A_k are positive or zero. Therefore, we say that the DE

$$(22) \qquad u'' = P(x)u' + Q(x)u, \qquad P(x) = \Sigma P_k x^k, \qquad Q(x) = \Sigma Q_k x^k$$

majorizes the DE (4) if and only if $P_k \geq |p_k|$ and $Q_k \geq |q_k|$, for all k.

In particular, the choice of coefficient-functions

$$(22') \qquad P(x) = \Sigma |p_k| x^k \qquad \text{and} \qquad Q(x) = \Sigma |q_k| x^k$$

in (22) gives a DE that majorizes (4). Moreover, by (19) the coefficient-functions (22') have the same radius of convergence as $p(x)$ and $q(x)$, respectively.

LEMMA 1.   *Let the* DE (22) *majorize the* DE (4), *and let* $\Sigma_{k=0}^{\infty} c_k x^k$ *be the formal power series solution of* (22) *whose first two coefficients are* $|a_0|$ *and* $|a_1|$. *Then* $c_k \geq |a_k|$ *for all* k.

This lemma may be thought of as a generalized comparison test.

Proof.   For the DE (22), the coefficients of formal power series solutions satisfy, by (8) with $p_k = -P_k$, $q_k = -Q_k$:

$$(*) \qquad c_{n+1} = \frac{1}{n(n + 1)}\left[\sum_{k=0}^{n}(n - k)P_k c_{n-k} + \sum_{k=0}^{n} Q_k c_{n-k-1}\right] \qquad n \geq 1$$

Hence, if $c_0 \geq |a_0|$, $c_1 \geq |a_1|$, ..., $c_n \geq |a_n|$, it follows that $c_{n+1} \geq |a_{n+1}|$, as stated. This is because a_{n+1} is given for $n \geq 1$ by the display (8), like (*) above, with each (positive) term replaced by one having at most as great an absolute value. The lemma follows by induction on n.

Now let x_1 be any number whose absolute value $|x_1| = C$ is less than the smaller of the radii of convergence of the two series (7). Then $p_k x_1^k$ and $q_k x_1^k$ are uniformly bounded† in magnitude for all k, by some finite constant M. Hence we have

$$|p_k| \leq MC^{-k}, \qquad |q_k| \leq MC^{-k}, \qquad k = 0,1,2, \ldots$$

This implies that the power series for $p(x)$ and $q(x)$ are both majorized by the geometric series

$$M \sum_{k=0}^{\infty} \left(\frac{x}{C}\right)^k = \frac{MC}{(C - x)}, \qquad \text{for some } M > 0, C > 0$$

This series being majorized in turn by

$$\frac{MC^2}{(C - x)^2} = M \sum_{k=0}^{\infty} (k + 1) \left(\frac{x}{C}\right)^k$$

the DE (4) is majorized by the DE

(23) $$u'' = \frac{MC}{(C - x)} u' + \frac{MC^2}{(C - x)^2} u$$

But, as in Example 8, one solution of this DE is the function

$$\phi(x) = \left[1 - \left(\frac{x}{C}\right)\right]^{-\mu}$$

where $-\mu$ is the negative root of the quadratic indicial equation of (23). Again as in Example 3, this equation is

$$\nu(\nu - 1) - MC\nu - MC^2 = 0, \qquad \text{where } -MC^2 < 0$$

This function $\phi(x)$ has a power series expansion

(24) $$\phi(x) = \left(1 - \frac{x}{C}\right)^{-\mu} = 1 + \frac{\mu x}{C} + \frac{\mu(\mu + 1)x^2}{2C^2} + \cdots$$

convergent for $|x| < C$, as in (21).

---

† This is because, if a series is convergent, its n-th term tends to zero as $n \to \infty$.

Now apply the foregoing lemma. Each solution of (4) is majorized by K times the solution $\phi(x)$ of (23), provided that

(24')
$$K = \max \left\{ |a_0|, \left| \frac{a_1 C}{\mu} \right| \right\}$$

But $K\phi(x)$ has the radius of convergence C. Hence, by the preceding lemma, the radius of convergence of the series (6) is at least $C = |x_1|$. This proves the following result.

THEOREM 2.   *For any choice of a_0 and a_1, the radius of convergence of any power series solution defined by the recursion formula (8) is at least as large as the smaller of the radii of convergence for the series defining the coefficient functions in (4).*

We now recall (§5) that power series can be added, multiplied together, and differentiated term-by-term within their intervals (circles) of convergence. It follows from Theorem 2 that when applied to power series defined by (8), the three equations displayed in §2 between formulas (7) and (8) are identities in the common interval of convergence specified. Hence, the power series defined by (8) are solutions of (4), and we have proved the following local existence theorem.

THEOREM 3.   *Any initial value problem defined by a normal second-order linear homogeneous DE (4) with analytic coefficient functions and initial conditions $f) = a_0, f'(0) = a_1$ has an analytic solution near $x = 0$, given by (8).*

EXERCISES C

1. Let $\Sigma a_k x^k$ have the radius of convergence R. Show that, for any $r < R$, the series is majorized by $\Sigma (m/r^k) x^k$ for some $m > 0$.

2. Using Ex. 1, prove Cauchy's formula (19).

3. Prove that, unless ν is a nonnegative integer, the radius of convergence of the binomial series (21) is C.

4. Using the symbols A, B, C to denote the series $\Sigma a_k x^k$, $\Sigma b_k x^k$, $\Sigma c_k x^k$, and writing $A \ll B$ to express the statement that series A is majorized by series B, prove the following results:
   (a) $A \ll B$ and $B \ll C$ imply $A \ll C$.
   (b) $A \ll B$ and $B \ll A$ imply $A = B$.
   (c) If $A \ll B$, then the derivative series A': $\Sigma k a_k x^k$ and B': $\Sigma k b_k x^k$ satisfy $A' \ll B'$.
   (d) If $A \ll B$ and $C \ll D$, then $A + C \ll B + D$ and $AC \ll BD$.

5. Prove that the radius of convergence of $\Sigma a_k x^k$ is unaffected by term-by-term differentiation or integration.

6. (a) Obtain a recursion relation on the coefficients a_k of power series solutions $\Sigma a_k x^k$ of Pearson's DE $y' = (D + Ex)y/(A + Bx + Cx^2)$, $A \neq 0$.
   (b) What is the radius of convergence of the solution?
   (c) Integrate this solution by quadratures, and compare.

*7. Extend the Method of Majorants of §6 to prove the convergence of the power series solutions of the inhomogeneous DE $u'' + p(x)u' + q(x)u = r(x)$, when the functions

p, q, r are all analytic. [HINT: Show that the DE is majorized by setting $p(x) = -MC/(C - x)$, $q(x) = -MC^2/(C - x)^2$, $r(x) = M/(C - x)$, for some finite $M > 0$, $C > 0$.]

*8. Let the coefficients of $u = \Sigma a_k x^k$ satisfy a recursion relation of the form $a_{k+1}/a_k = P(k)/Q(k + 1)$, where P and Q are polynomials without common factors and $Q(0) \neq 0$. Show also that u must satisfy a DE of the form

$$Q\left(x\frac{d}{dx}\right)[u] = xP\left(x\frac{d}{dx}\right)[u]$$

and conversely.

*7   SINE AND COSINE FUNCTIONS

To illustrate the fact that properties of solutions of DEs can often be derived from the DEs themselves, we will now study the *trigonometric* DE

(25) $y'' + y = 0$

The general solution of this DE is $y = a \cos x + b \sin x$, where a and b are arbitrary constants, and the functions $\cos x$ and $\sin x$ are defined geometrically.

We will pretend that we do not know this, and deduce properties of the trigonometric functions $\sin x$ and $\cos x$ from general theoretical principles, assuming only the trigonometric DE (25) and the initial conditions that they satisfy. In this spirit, we *define* the functions $C(x)$ and $S(x)$ as the solutions of this DE that satisfy the initial conditions $C(0) = 1$, $C'(0) = 0$, and $S(0) = 0$, and $S'(0) = 1$, respectively. Applying the Method of Undetermined Coefficients to (25) with these initial conditions, we get easily the familiar power series expansions

(25') $C(x) = 1 - \dfrac{x^2}{2!} + \dfrac{x^4}{4!} - \cdots$, $S(x) = x - \dfrac{x^3}{3!} + \dfrac{x^5}{5!} - \cdots$

whose convergence for all x follows by the Ratio Test.

Differentiation of (25) gives the DE $y''' + y' = 0$. Therefore, the function $C'(x)$ is also a solution of the DE (25); moreover, since the function satisfies the initial conditions $C'(0) = 0$ and $C''(0) = -C(0) = -1$, it follows from the Uniqueness Theorem (Ch. 2, Theorem 1) that $C'(x) = -S(x)$. This proves the differentiation rule for the cosine function. A similar computation gives $S'(x) = C(x)$.

The Wronskian of the functions $C(x)$ and $S(x)$ can be computed from these two formulas; it is $W(C,S;x) = C(x)^2 + S(x)^2$. From Theorem 3 of Ch. 2, $W(C,S;x) \equiv C(0)^2 + S(0)^2 = 1$ follows. This proves the familiar trigonometric formula $\cos^2 x + \sin^2 x = 1$.

Again, by Theorem 2 of Ch. 2, every solution of the DE (25) is a linear combination of the functions S and C. We now use this fact to derive the addition formula for the sine function:

$$\sin(a + x) = \cos a \sin x + \sin a \cos x$$

First, by the chain rule for differentiating composite functions, the function $S(a + x)$ is also a solution of the DE (25). Therefore (Ch. 2, Theorem 2), this function must be a linear combination of $S(x)$ and $C(x)$:

(26) $$S(a + x) = AS(x) + BC(x)$$

Furthermore, if we write $f(x) = S(a + x)$, then $f(0) = S(a)$ and $f'(0) = C(a)$. But if we differentiate the right side of (26) and set $x = 0$, we find that $f(0) = B$ and $f'(0) = A$, whence $S(a + x) = C(a)S(x) + S(a)C(x)$. This proves the addition formula for the sine function. The addition formula for $C(x)$

$$C(a + x) = C(a)C(x) - S(a)S(x)$$

can be derived similarly.

Finally, the fact that the functions S and C are periodic can be proved from the addition formulas. Define $\pi/4$ as the least positive x such that $S(\pi/4) = 1/\sqrt{2}$. Since $S' = C = \sqrt{1 - S^2} \geq 1/\sqrt{2}$ on any interval $[0, b]$ where $S(x) \leq 1/\sqrt{2}$, we see that $S(x)$ is increasing and satisfies $S(x) \geq x/\sqrt{2}$ there. Hence, $S(x) \leq 1/\sqrt{2}$ on $[0, 1]$ is impossible, which shows that $\pi/4$ exists in $[0, 1]$. Moreover, $\pi/4 = S^{-1}(1/\sqrt{2})$, where S^{-1} is the inverse function of S. Since the derivative of S^{-1} is given by $1/S'(x) = 1/\sqrt{1 - S^2}$, this makes

$$\frac{\pi}{4} = \int_0^{1/\sqrt{2}} \frac{dt}{\sqrt{1 - t^2}}$$

Moreover, C cannot change sign until $S^2 = 1 - C^2 = 1$. Hence, $\cos(\pi/4) = 1/\sqrt{2} = \sin(\pi/4)$. Consequently, by the addition formulas proved above

$$\sin\left(\frac{\pi}{4} + x\right) = \frac{1}{\sqrt{2}}(\sin x + \cos x) = \cos\left(\frac{\pi}{4} - x\right)$$

In particular, $\sin(\pi/2) = (2/\sqrt{2})/\sqrt{2} = 1$ and, therefore, $\cos(\pi/2) = 0$. Using the addition formulas again, we get the formulas $\sin(\pi/2 + x) = \sin(\pi/2 - x)$, $\cos(\pi/2 + x) = -\cos(\pi/2 - x)$, $\sin(\pi + x) = -\sin x$, $\cos(\pi + x) = -\cos x$ and, finally, the periodicity relations $\cos(2\pi + x) = \cos x$, $\sin(2\pi + x) = \sin x$.

*8   BESSEL FUNCTIONS

The Bessel functions of integral order n and half-integer order $n + \frac{1}{2}$ are among the most important functions of mathematical physics (see Exercises D below). In §3, we defined $J_n(x)$ as Example 3, and derived a basic recursion formula (13) expressing $J_n(x)$ algebraically in terms of $J_0(x)$ and its derivatives. In this section, we shall derive many other useful facts and formulas involving Bessel functions from the results proved in §3. We emphasize that *all* of these formulas can be derived from their defining DEs (10), the fact that $J_n(x)$ is *analytic* at 0, and the choice of leading coefficient in formula (12).

Specifically, one can prove all the properties of the Bessel functions of integral order from (10) and the recursion relations (13). For example, one can obtain such useful formulas as

$$\int x J_0 \, dx = x J_1, \qquad \int x J_1 \, dx = -x J_0 + \int J_0 \, dx$$

$$\int x J_0^2 \, dx = \frac{x^2}{2} (J_0^2 + J_1^2) = \frac{x^2}{2} (J_0^2 + J_0'^2)$$

More generally, we can obtain useful expressions representing, in closed form, integrals of arbitrary polynomial functions times Bessel functions and products of Bessel functions. The basic formulas are (13) and

(27a) $$\int x^k J_1 \, dx = -x^k J_0 + k \int x^{k-1} J_0 \, dx$$

(27b) $$\int x^k J_0 \, dx = x^k J_1 - (k - 1) \int x^{k-1} J_1 \, dx$$

(27c) $$2 \int x^k J_0 J_1 \, dx = -x^k J_0^2 + k \int x^{k-1} J_0^2 \, dx$$

(27d) $$\int x^k (J_0^2 - J_1^2) \, dx = x^k J_0 J_1 - (k - 1) \int x^{k-1} J_0 J_1 \, dx$$

(27e) $$\int x^k [(k + 1) J_0^2 + (k - 1) J_1^2] \, dx = x^{k+1} (J_0^2 + J_1^2)$$

Equation (27a) follows from $J_1 = -J_0'$, integrating by parts. To derive formula (27b), note that since $(x J_0')' = -x J_0$, it follows that

$$(x^k J_1)' = -[x^{k-1} (x J_0')] = -(k - 1) x^{k-1} J_0' + x^k J_0$$

To derive (27c)–(27e), differentiate $x^k J_0^2$, $x^k J_0 J_0'$, and $x^{k+1} (J_0^2 + J_0'^2)$, respectively, and use (10) to eliminate J_0''. The integral $\int J_0 \, dx$ cannot be reduced further, and so it has been calculated (by numerical integration) and tabulated.†

Important *qualitative* information can also be obtained from a study of the DE (10) which the Bessel functions satisfy. Substituting $u = v/\sqrt{x}$ into (10), we obtain the equivalent DE

(28) $$v'' + \left[1 - \frac{4n^2 - 1}{4x^2} \right] v = 0$$

---

† G. N. Watson, *Bessel Functions*, Chapter 8. Cambridge University Press 1926; A. N. Lowan et al., *J. Math and Phys.* 22 (1943).

The oscillatory behavior of nontrivial solutions of the Bessel DE (10), for large x, can now be shown, using the Sturm Comparison Theorem (Ch. 2, §4). When applied to (28), this result shows that, for large x, the distance between successive zeros of $J_0(x)$ is inferior to π by a small quantity (at most $\pi/8x^2$), while that between successive zeros x_i and x_{i+1} of $J_n(x)$ exceeds π by about $n^2\pi/2x_i^2$ if $n \geq 1$. Also, since $J_1 = -J_0'$, there is a zero of J_1 between any two successive zeros of J_0.

Setting $\sigma(x) = v^2 + v'^2 = x(J_n^2 + J_n'^2) + J_n J_n' + J_n^2/4x$, it also follows from (28) that $\sigma'(x) = (4n^2 - 1)vv'/2x^2$. Since $|2vv'| \leq v^2 + v'^2 = \sigma(x)$, there follows

(28')
$$|\sigma'(x)| \leq \frac{K_n\sigma(x)}{x^2}, \qquad K_n = \left|n^2 - \frac{1}{4}\right|, \qquad \sigma(x) > 0$$

Using the Comparison Theorem of Ch. 1, §11, we get from (28')

$$Ae^{-K_n/x} \leq \sigma(x) = x(J_n^2 + J_u'^2) + J_n J_n' + \frac{J_n^2}{4x} \leq Ae^{K_n/x}$$

where $K_n = |n^2 - \frac{1}{4}|$. For large x, therefore, $\sigma(x)$ must approach a constant A. Clearly, $1/\sqrt{x}$ is the asymptotic *amplitude* of the oscillations of the Bessel functions for large x, since J_j' vanishes at maxima and minima of $J_n(x)$, so that $\sigma(x) = xJ_n^2(x)$ there. Much more precise asymptotic results about the oscillations of Bessel functions will be proved in Ch. 10.

The General Solution.   The general solution of the Bessel DE of zero order is, setting $W(x) = e^{-\int dx/x} = 1/x$ in formula (13) of Ch. 2,

(29)
$$Z_0(x) = J_0(x)\left[A + B\int \frac{dx}{xJ_0^2(x)}\right]$$

But a straightforward computation with power series gives

$$\frac{1}{J_0^2(x)} = 1 + \frac{x^2}{2} + \frac{5x^4}{32} \cdots$$

From this formula, substituting back into (19) and integrating the resulting series term-by-term, we see that the general solution $Z_0(x)$ of the Bessel DE of zero order is

(30)
$$Z_0(x) = J_0(x)\left[A + B\left(\ln x + \frac{x^2}{4} + \frac{5x^4}{128} + \cdots\right)\right]$$

It follows that every solution not a constant multiple of $J_0(x)$ becomes logarithmically infinite as $x \downarrow 0$, since $B \neq 0$. For further information, see Ch. 9, §7.

Generating Functions.   Given a sequence $\{a_n\}$ of constants $a_0, a_1, a_2, \ldots$, the power series

$$g(x) = a_0 + a_1 x + a_2 x^2 + \cdots = \sum_{n=0}^{\infty} a_n x^n$$

is called its *generating function.* When the series on the right converges in an interval, this defines a function $g(x)$ there; otherwise, the infinite series is just a formal power series. In many cases, useful information can be obtained about a sequence $\{a_n\}$ by studying its generating function.

Likewise, given a sequence of functions $F_n(r)$, the function defined by the power series $\Sigma t^n F_n(r)$ is also called the "generating function" of the sequence. Thus, the generating function of the sequence of Legendre polynomials is

(*)
$$\sum_{n=0}^{\infty} r^n P_n(\theta) = (1 - 2r \cos \theta + r^2)^{-1/2}.$$

The same phrase is used when the sum is taken over all integers; for example, the Bessel functions of integral order have the generating function

(31)
$$\sum_{-\infty}^{\infty} t^n J_n(r) = e^{r(t-t^{-1})/2}$$

See Ex. D13.

EXERCISES D

1. Define $E(x)$ as in §1, by the DE $E' = E$ and the initial condition $E(0) = 1$. Prove in turn, justifying your arguments by referring to theorems, that
   (a) $E(x) = \Sigma_{k=0}^{\infty} x^k/(k!)$ (b) $E(a + x) = E(a)E(x)$
   (c) $E(-x) = 1/E(x)$ [*Suggestion:* Show that for any a, $E(a + x)/E(a)$ satisfies the conditions defining $E(x)$.]

2. Define sinh x and cosh x as the solutions of the DE $u'' = u$ that satisfy the initial conditions $u(0) = 0$, $u'(0) = 1$ and $u(0) = 1$, $u'(0) = 0$, respectively. Show that sinh x has only one real zero and cosh x has no real zeros. Relate this to the Sturm Comparison Theorem.

3. Using methods like those of §7, establish the following formulas (cf. Ex. 4):
   (a) $\cosh^2 x - \sinh^2 x = 1$ (b) $\cosh(-x) = \cosh x$
   (c) $\sinh(-x) = -\sinh x$ (d) $\sinh(x + y) = \sinh x \cosh y + \cosh x \sinh y$

4. (a) Show that sinh x + cosh x satisfies the conditions used to define $E(x)$ in Ex. 1.
   *(b) Using this result, and the formulas of Ex. 3, show $\sinh^{-1}(x) = \ln(x + \sqrt{x^2 + 1})$.

5. Prove formulas (27a) and (27b) in detail, expanding on the remarks in the text.

6. Prove formulas (27c)–(27e) similarly.

Establish the identities for Bessel functions of integral order in Exs. 7–10.

7. $J_n(-x) = (-1)^n J_n(x)$ *8. $\left(\dfrac{1}{x}\dfrac{d}{dx}\right)^k (x^n J_n(x)) = x^{n-k} J_{n-k}(x)$

*9. $J_{n-1}(x) + J_{n+1}(x) = 2nx^{-1} J_n(x)$

*10. $J_{n-1}(x) - J_{n+1}(x) = 2J_n'(x)$

11. In polar coordinates, $\nabla^2 u = u_{rr} + r^{-1} u_r + r^{-2} u_{\theta\theta}$.

   (a) Show that $J_n(r) \begin{Bmatrix} \cos \\ \sin \end{Bmatrix} n\theta$ satisfies $\nabla^2 u + u = 0$.

   *(b) Show that conversely, if $J(r) \begin{Bmatrix} \cos \\ \sin \end{Bmatrix} n\theta$ satisfies $\nabla^2 u + u = 0$ and is bounded near $r = 0$, then $J(r)$ is a constant multiple of $J_n(r)$.

12.   Show that $(\sin r)/\sqrt{r} = \sqrt{r}\, j_0(r)$ is a constant multiple of $J_{1/2}(r)$. (Cf. Ex. A.)

*13. (a) Show that the real and imaginary parts $\cos(r \sin \theta)$ and $\sin(r \sin \theta)$ of $e^{iy} = e^{ir\sin\theta}$ satisfy $\nabla^2 u + u = 0$.

   (b) Show that $e^{iy} = e^{r(t-t-1)/2}$, $t = e^{i\theta}$, $y = r \sin \theta$.

   (c) Show that the functions $F_n(r)$ in the Laurent series expansion

$$e^{r(t-t-1)/2} = \sum_{-\infty}^{\infty} t^n F_n(r)$$

   satisfy the Bessel equation of order n. [Use (a) and (b).]

   (d) Comparing the coefficients of $t^n r^n$, prove the identity $e^{r(t-t-1)/2} = \sum_{-\infty}^{\infty} t^n J^n(r)$ where $J_{-n}(x) = J_n(-x) = (-)^n J_n(x)$.

*14. (a) Show that Kummer's confluent hypergeometric function

$$M(a,b;x) = 1 + \frac{a}{b} x + \frac{a(a+1)}{b(b+1)} \frac{x^2}{2!}$$
$$+ \cdots + \frac{a(a+1)\cdots(a+n)}{b(b+1)\cdots(b+n)} \frac{x^n}{n!} + \cdots$$

   is a solution of the DE $xF'' + (b - x)F' - aF = 0$.

   (b) Show that the preceding function is an entire function.

*15. (a) Show that $u(r,\theta,\alpha) = \cos[r\cos(\theta - \alpha)]$ satisfies $\nabla^2 u + u = 0$ for all α, and hence so does its average $\dfrac{1}{2\pi}\displaystyle\int_0^{2\pi} u(r,\theta;\alpha)\, d\alpha = U(r)$

   (b) Prove that $J_0(r) = \dfrac{1}{2\pi}\displaystyle\int_{-x}^{x} \cos[r\cos\alpha]\, d\alpha = \dfrac{1}{\pi}\displaystyle\int_x^0 \cos[r\cos\alpha]\, d\alpha$

9   FIRST-ORDER NONLINEAR DEs

The Method of Undetermined Coefficients and the Method of Majorants can also be applied to any normal analytic first-order DE. For any function $F(x,y)$

analytic near (0,0), consider the DE

$$(32) \qquad \frac{dy}{dx} = F(x,y) = \sum_{j=0}^{\infty} \sum_{k=0}^{\infty} b_{jk} x^j y^k$$

The DE $y' = 1 + y^2$ of Example 7 is one of the simplest *nonlinear* such DEs; we refer the reader back to §4 for a preliminary discussion of how to solve this particular DE.

In this section we will explain how to solve a general DE of the form (32) by the same method. Namely, we substitute into the DE (32) the formal power series

$$(33) \qquad y = f(x) = a_1 x + a_2 x^2 + a_3 x^3 + \cdots$$

assuming that we are looking for the solution of (32) satisfying the initial condition $y(0) = 0$, which we can always do by a translation of coordinates. Accordingly, setting

$$(33') \qquad y' = a_1 + 2a_2 x + 3a_3 x^2 + \cdots$$

and substituting into (32), we obtain successively

$$(34) \qquad a_1 = b_{00}, \qquad 2a_2 = b_{10} + b_{01}a_1$$
$$3a_3 = b_{01}a_2 + b_{20} + b_{11}a_1 + b_{02}a_1^2$$

and so on. The expression on the right side of each of these equations is a polynomial with positive integral coefficients. Equations (34) can be solved recursively, giving the formulas

$$a_1 = b_{00}, \qquad a_2 = \frac{(b_{10} + b_{00}b_{01})}{2}$$

$$a_3 = \frac{(b_{20} + b_{11}b_{00} + b_{02}b_{00}^2)}{3} + \frac{(b_{10}b_{01} + b_{00}b_{01}^2)}{6}$$

and so on. When we substitute the series (33) for y into the series (32) for $F(x,y)$, the coefficient of x^h is a sum of products of factors b_{jk} (with $j + k \leq h$) times polynomials of degree k in the a_i obtained by raising the series (33) to the kth power. The coefficient of x^h on the left side of (32) is, however, $(h + 1)a_{h+1}$, by (33'). Equating coefficients of like powers of x, we have, therefore,

$$(34') \quad (h + 1)a_{h+1} = q_h(b_{00}, \ldots, b_{0h}; b_{10}, \ldots, b_{1,h-1}; \ldots; b_{h0}; a_1, \ldots, a_h),$$

where the coefficients of q_h are positive integers. Substituting for $a_1, \ldots, a_h$ already available formulas, we obtain

$$(34'') \qquad a_{h+1} = p_h(b_{00}, \ldots, b_{0h}; b_{10}, \ldots, b_{1,h-1}; \ldots; b_{h0})$$

The *polynomial* functions p_h have *positive, rational* numbers as coefficients. They are the same, no matter which function $F(x,y)$ is used in (32).

Solving the resulting equation at $x = 0$ for $a_{n+1} = y^{(n+1)}/(n + 1)!$, we get the following result.

THEOREM 4. *There exists a power series (33) which formally satisfies any analytic first-order DE (32). The coefficients of this formal power series are polynomial functions of the b_{hk} with positive rational coefficients.*

The preceding formulas can also be obtained in another way. Let $y = f(x)$ be the graph in the (x,y)-plane of any solution of the DE $y' = F(x,y)$, and let $u(x,y)$ be any analytic function in a domain containing this graph. Differentiating with respect to x along the graph, we get the formula

$$\frac{du}{dx} = \frac{\partial u}{\partial x} + \left(\frac{\partial u}{\partial y}\right)\left(\frac{dy}{dx}\right) = \frac{\partial u}{\partial x} + F\frac{\partial u}{\partial y}$$

This formula can be differentiated repeatedly, giving the operator identity

(35)
$$\frac{d^n}{dx^n} = \left(\frac{\partial}{\partial x} + F(x,y)\frac{\partial}{\partial y}\right)^n$$

Applying this identity to the function $F(x,y) = y'$, we get, in succession,

$$y'' = F_x + FF_y, \qquad y'' = F_{xx} + 2FF_{xy} + F^2F_{yy} + F_xF_y + FF_y^2$$

and so on. The general formula is

(35′)
$$y^{(n+1)} = \frac{d^{n+1}y}{dx^{n+1}} = \left(\frac{\partial}{\partial x} + F\frac{\partial}{\partial y}\right)^n [F(x,y)]$$

The right side of (35′), evaluated at $x = y = 0$, is a polynomial in the variables b_{jk} with positive integers as coefficients. This is because the operations used in evaluating this expression are addition, multiplication, and differentiation.

EXERCISES E

In Exs. 1–7, calculate the first four nonzero terms of the power series expansion of the solutions of the DE indicated, for the initial value $y(0) = 0$.

1. $y' = x + y$
2. $y' = 1 + x^2y^2$
3. $y' = g(x) = \Sigma b_k x^k$
4. $y' = g(y) = \Sigma b_k y^k$
5. $y' = xy^2 + yx^2 + 1$
6. $y' = 1 + y^2$
7. $y' = \cos\sqrt{y} + 1$
8. Calculate explicitly the polynomial $p_3 = a_4$ of Theorem 4.
9. Compute the first five polynomials p_n of Theorem 4 when $F(x,y) = b(x)y + c(x)$ in (30).

10. Apply the same question for the Riccati DE $y' + y^2 = b(x)y + c(x)$.

11. Show that the DE $y' = x^2 + y^2$ has a solution of the form $\sum_1^\infty a_k x^{4k-1}$ with all $a_k > 0$. For $a_1 = 1$, compute a_2, a_3, a_4.

12. (a) From the DE $y' = 1 + y^2$, prove that the coefficient a_{n+1} in the expansion tan $x = \sum a_k x^k$, $a_0 = 0$, satisfies $(n + 1)a_{n+1} = \sum_{k=1}^{n-1} a_k a_{n-1}$. [HINT: Differentiate y^2 using the binomial expansion of $(uv)^{(n)}$.]

   (b) Compute the first five nonzero coefficients, and compare with those obtained by solving for y recursively from $x = y - y^3/3 + y^5/5 - \dots$, the series for $x =$ arctan y.

13. Show that if $y' = F(x,y)$, where $F \in \mathcal{C}^3$, then

$$y^{iv} = F_{xxx} + 3y'F_{xxy} + 3y'^2 F_{xyy} + y'^3 F_{yyy} + 3y'' F_{xy} + 3y'y'' F_{yy} + y'' F_y$$

14. For the DE $y' = 2y/x$ and the initial condition $y(1) = 1$, calculate the first four terms of the Taylor series of the solution.

10   RADIUS OF CONVERGENCE

The DE $y' = 1 + y^2$ of Example 7 shows that the radius of convergence of power series solutions of a *nonlinear* DE $y' = F(x,y)$ can be much less than that of the function $F(x,y)$. For, the radius of convergence of the solution tan $(x + c)$ of the DE $y' = 1 + y^2$ which satisfies $y(0) = \tan c = \gamma$ is only $\pi/2 - c$, the distance to the nearest singular point of the solution. This can be made arbitrarily small by making γ large enough, even though the radius of convergence of $F(x,y) = 1 + y^2$ is infinite.

The preceding situation is typical of nonlinear DEs and shows that we cannot hope to establish an existence theorem for nonlinear DEs as strong as Theorem 3. The contrast with the situation for nonlinear DEs is further illustrated by the Riccati equation of Ch. 2, §5, (*).

Example 5.   The Riccati equation

(36) $dv/dx = -v^2 - p(x)v - q(x)$

is satisfied by the ratio $v(x) = u'(x)/u(x)$, if $u(x)$ is any nontrivial solution of the second-order linear DE (4). Conversely, if $v(x)$ is any solution of the Riccati DE (36), then the function $u = \exp(\int v(x)\, dx)$ satisfies the linear DE (4).

Now, let p, q be analytic and a, c given constants. Using Theorem 2 to find a solution u of (4) satisfying the initial conditions $u(a) = 1$ and $u'(a) = c$, we obtain an analytic solution $v = u/'u$ of the Riccati equation (36) which satisfies $v(a) = c$. This gives a local existence theorem for the nonlinear Riccati DE (36).

However, this solution becomes infinite when $u = 0$; hence the radius of convergence of its power series can be made arbitrarily small by choosing c sufficiently large. Also, the location of the singular points of solutions of (36) is variable, depending on the zeros of u.

The Riccati equation also serves to illustrate Theorem 4. A simple computation gives, for $v = \Sigma_1^\infty a_k x^k$, the formula

$$v^2 = a_1^2 x^2 + 2a_1 a_2 x^3 + (a_2^2 + 2a_1 a_3)x^4 + \cdots + \sum_{i=1}^{h-1} a_i a_{h-i} x^h + \cdots$$

Substituting back into (36), we get the recursion relation

$$(h+1)a_{h+1} = -\sum_{i=1}^{h-1} a_i a_{h-i} - \sum_{i=1}^{h} a_i p_{h-i} - q_h$$

This is, of course, just the special case of formula (34′) corresponding to setting $F(x,y) = -y^2 - \Sigma_0^\infty p_k x^k y - \Sigma_0^\infty q_k x^k$.

We shall now return to the general case. Let F be any function of x and y analytic in some neighborhood of $(0,0)$. This means that F can be expanded into a double power series

$$(37) \qquad F(x,y) = b_{00} + (b_{10}x + b_{01}y) + (b_{20}x^2 + b_{11}xy + b_{02}y^2) \cdots$$

where b_{jk} are given real numbers, and the series is convergent for sufficiently small x and y. We shall show that the series (33) referred to in Theorem 4, and defined by formulas (34)-(34′)-(34″) has a positive radius of convergence.

Analytic Functions of Two Variables. To prove this, we shall need a few facts about analytic functions of two variables and the convergence of double power series like (37). The terms of any absolutely convergent series can be rearranged in any order without destroying the convergence of the series or changing the value of the sum. If we substitute into a double power series like (37) convergent near $(0,0)$ any power series $y = \Sigma_{k=0}^\alpha a_k x^k$ having a positive radius of convergence, we will obtain an analytic function $F(x, f(x))$ which itself has a positive radius of convergence.

Finally, let the double power series (37) be convergent at (H,K), where $H > 0$ and $K > 0$. Then the terms of the series $\Sigma b_{jk} H^5 K^k$ are bounded in magnitude by some finite constant $M = \max |b_{jk}| H^j K^k$. This gives the bound

$$|b_{jk}| \leq M/H^j K^k, \qquad M < +\infty,$$

to the terms of the series (37). Comparing with the double geometric series mentioned in §6

$$(38) \qquad G(x,y) = \frac{M}{\left(1 - \dfrac{x}{H}\right)\left(1 - \dfrac{y}{K}\right)} = \sum_{j=0}^{\infty} \sum_{k=0}^{\infty} \left(\frac{M}{H^j K^k}\right) x^j y^k.$$

and applying the Comparison Test, we see that the series (37) is absolutely convergent in the open rectangle $|x| < H$, $|y| < K$, and can be differentiated there, term-by-term, any number of times.

The preceding remarks have the following immediate consequence.

COROLLARY. *If the power series* (33) *of Theorem* 4 *has a positive radius of convergence, then the function which it defines is an analytic solution of the* DE (30) *for the initial condition* $y(0) = 0$.

11* METHOD OF MAJORANTS, II

We will now complete the proof of an existence theorem for analytic (normal) first-order DEs by showing that the series (33) has a positive radius of convergence. This is again shown by the Method of Majorants, which we now extend to functions of two variables.

Consider the power series

$$y: a_1 x - a_2 x^2 + a_3 x^3 + \cdots$$

$$F: b_{00} - b_{10} x + b_{01} y + b_{20} x^2 + b_{11} xy + \cdots$$

as infinite arrays of real or complex numbers, irrespective of any questions of convergence. Such power series can be added, subtracted, and multiplied using Cauchy's product formula (18′) algebraically as infinite polynomials, without ever being evaluated as functions of x and y. Such expressions are called *formal power series*. It is possible to substitute one formal power series into another; after rearranging terms, another formal power series is obtained. Two formal power series are considered identical when all their coefficients coincide.

Comparing F with the formal power series

$$G: c_{00} + c_{10} x + c_{01} y + c_{20} x^2 + c_{11} xy + \cdots,$$

one says that the formal power series G *majorizes* the formal power series F when

$$|b_{jk}| \le c_{jk} \qquad \text{for all } j, k = 0, 1, 2, \cdots.$$

In symbols, one writes

(39) $$F \ll G.$$

This implies that all coefficients c_{jk} are nonnegative.

The following lemma is immediate.

LEMMA 1. *Let* F, G, H *be any three formal power series. Then* $F \ll G$ *and* $G \ll F$ *imply* $F = G$, *and* $F \ll G$ *and* $G \ll H$ *imply* $F \ll H$.

It is not true, however, that $F \ll F$ if F has any negative coefficient.

The relation (39) of majorization between formal power series is useful in estimating the radius of convergence of such series. From the Comparison Test for convergence, we obtain the following directly.

LEMMA 2.   *If F and G are formal power series and if $F \ll G$, then F converges absolutely at any point (x,y) if G converges at $(|x|, |y|)$.*

The crucial result for the proof of convergence of the formal power series (33), obtained from Theorem 4, is the following.

LEMMA 3.   *Let $F \ll G$, and let f and g be the formal power series (without constant terms) obtained by solving $y' = F(x,y)$ and $y' = G(x,y)$ formally, as in Theorem 4, for the initial condition $y(0) = 0$. Then g majorizes f (that is, $f \ll g$).*

Proof.   The polynomials p_h in Theorem 4 have nonnegative coefficients. It follows that

$$|a_{h+1}| = |p_h(b_{00}, b_{10}, b_{01}, \ldots)| \leq p_h(c_{00}, c_{10}, c_{01}, \ldots)$$

for all h; hence, the absolute value of each coefficient a_h is less than or equal to the corresponding coefficient of the formal power series g, q.e.d.

It is now a straightforward matter to prove our main result.

THEOREM 5.   *Let $F(x,y)$ be analytic in the closed rectangle $|x| \leq H$, $|y| \leq K$, where H and K are positive. Then the formal power series solution (33) of the DE (32) has a positive radius of convergence.*

Proof.   The power series for the function F is convergent at (H,K); as in §9, it follows that for some finite $M = \max |b_{jk}H^jK^k|$,

$$|b_{jk}H^jK^k| \leq M, \qquad \text{whence} \qquad |b_{jk}| \leq \frac{M}{H^jK^k}.$$

That is, the formal power series F is majorized by the double geometric series (38):

$$G(x,y) = M \left/ \left[\left(1 - \frac{x}{H} \right) \left(1 - \frac{y}{K} \right) \right] = \sum_{j,k=0}^{\infty} \frac{M}{H^jK^k} x^j y^k. \right.$$

This series is the product of two geometric series, each absolutely convergent if $|x| < H$, $|y| < K$. Therefore, it is also absolutely convergent in this rectangular domain, and defines an analytic function there.

Furthermore, the DE $y' = G(x,y)$ can be solved in closed form by separation of variables. The solution satisfying $y(0) = 0$ is

(40) $$y = K[1 - \sqrt{1 + (2MH/K) \ln (1 - (x/H))}] \,,$$

where the principal values of the logarithmic and square root functions are taken, corresponding to the usual expansions of the functions $\sqrt{1 + t}$ and $\ln (1 + t)$ in power series with center at $t = 0$. The radius of convergence is given by the equation $(2MH/K) \ln [1 - x/H] = -1$, or

(41) $$R = H[1 - e^{-K/2MH}]$$

since the binomial series for the radicand in (40) converges so long as $(2MH/K)|\ln [1 - (x/H)]| < 1$. This completes the proof.

By Theorem 6 of Ch. 1, whose hypotheses are satisfied since F is continuously differentiable, the solution satisfying the initial condition $f(0)$ is unique. This proves the following result.

COROLLARY. *Every solution of the analytic DE (30) is analytic. The solution satisfying the initial condition $f(0) = 0$ is unique, and given by the power series (33).*

12* COMPLEX SOLUTIONS

Up to now, we have been assuming tacitly that all variables x, y, u, etc. referred to were real. However, the discussion in this chapter also applies, with very minor changes to *complex* power series. In particular (see Theorems 2–3), any solution of the DE (4) obtained by power series methods defined a complex analytic function within a circle of convergence $|z| < C$ whose radius is the smaller of the radii of convergence, of the series in (5). For instance, from the DE $w'' + w = 0$, we obtain the complex sine and cosine functions $\sin z = z - z^3/3! + z^5/5! - \cdots$, and $\cos z = 1 - z^2/2! + z^4/4! - \cdots$, where as usual $w = u + iv$ and $z = x + iy$ refer to dependent and independent complex variables.

Similarly, let

(42) $$dw/dz = F(z,w) = \sum_{j,k} b_{jk} z^j w^k$$

be any analytic first-order DE, whose right side is a complex analytic function. (For F to be analytic, it is sufficient for the function F to be differentiable, since differentiability implies analyticity in any complex domain.†) Then all the formulas of §9 remain valid; so do the lemmas concerning majorants of §10.

For complex z, w, the domain D: $|z| \leq H$, $|w| \leq K$ of Theorem 5 is not a

---

† Ahlfors, pp. 24–25; Hille, pp. 72, 196.

rectangle, but the four-dimensional product of two discs. Though this domain is harder to visualize than a rectangle, it has the advantage that Cauchy's integral formulas hold on it: the constant M in (41) is given explicitly by‡

$$M = \sup |F(z,w)|.$$

These remarks cover the extension of Theorem 5 to complex DEs.

Dependence on Initial Value. We now consider the dependence of the solutions of real or complex analytic first-order DEs (32) or (42) on their initial values. We will prove that this dependence is analytic; it will follow that the solution curves of any real, normal first-order DE form a *normal curve family*.

THEOREM 6. *Let $f(x,c)$ be the solution of the first-order analytic DE $y' = F(x,y)$ which satisfies the initial condition $f(a) = c$. Then $f(x, c)$ is an analytic function of the independent variable x and the initial value c.*

Proof. By a translation of coordinates, we can reduce to the case $a = 0$, and consider $f(x,c)$ in the neighborhood of $(0,0)$. But clearly for small fixed c, the function $\eta(x, c) = f(x, c) - c$ is the solution of

$$(43) \qquad d\eta/dx = F_1(x, \eta) = F(x, c + \eta) = \Sigma b_{jk}x^j(c + \eta)^k = \Sigma\beta_{ji}x^{jnk},$$

satisfying the initial condition $\eta(0) = 0$. Here each $\beta_{jk} = \beta_{jk}(c)$ is an analytic function of c, the coefficients of whose power series expansion in c are *positive multiples* $\binom{k}{h} b_{jh}$ of some b_{jh}. By Theorems 4 and 5, the solution $\eta(x, c)$ of (43) is $\Sigma_{k=0}^{\infty} \alpha_n x^n$, where each coefficient $\alpha_n = \alpha_n(c)$ is a polynomial in the β_{jk} with *positive rational coefficients*. Hence, the doubly infinite formal power series

$$\eta(x, c) = \Sigma\alpha_n(c)x^n = \Sigma\gamma_{mn}c^m x^n$$

obtained has coefficients γ_{mn} which are polynomials in the b_{jh} (because each b_{jh} only affects the β_{jk} with $k \leq h$) with *positive* coefficients. This series formally satisfies (43).

As in the proof of Theorem 5, this series is majorized by the power series solution without constant term of the DE

$$dz/dx = G(x, c + z) = M/[(1 - x/H)(1 - (c + z)/K)].$$

Integrating, we get

$$(44) \qquad \left(1 - \frac{c}{K}\right)z - \frac{z^2}{2K} = -MH \ln\left(1 - \frac{x}{H}\right).$$

---

‡ See, for example, Picard, Vol. 2, p. 259.

Elementary algebraic manipulation now gives

$$[(z + c) - K]^2 = (c - K)^2 + 2MHK \ln\left(1 - \frac{x}{H}\right).$$

The branch of the function $z(x, c)$ which makes $z(0, 0) = 0$ is given by the formula

$$z = K - c - \left\{(c - K)^2 + 2MHK \ln\left(1 - \frac{x}{H}\right)\right\}^{1/2}.$$

The resulting function (as one sees by expanding the functions $\sqrt{1 + u}$ and $\ln[1 - (x/H)]$ into power series) is analytic for x and c sufficiently small. That is, the power series solution of (44), $g(x, c) = \Sigma\delta_{mn}c^m x^n$ (in which all $\delta_{mn} \geq 0$ as shown above) is also convergent. But this *majorizes* $f(x, c)$, whose power series expansion is therefore also convergent, completing the proof.

EXERCISES F

For the power series expansion of each function defined in Exs. 1–4, determine the domain of convergence:

1. $J_0(x + y)$.

2. $\sqrt{(1 - x)(1 - y)}$.

3. $1/[1 - (x^2 + y^2)]$.

4. $1/[1 + (x^2 + y^2)]$.

5. Show that the solution of $y' = G(x, y)$, where G is the geometric series (38), is the function (40).

6. Find the value of $\Sigma_{j,k=0}^{\infty} x^j y^k/(H^j K^k)$.

7. Find the value of $\Sigma_{j,k=0}^{\infty} jkx^j y^k/(H^j K^k)$.

In Exs. 8–11, establish the properties of the relation "$\ll$" [$P = P(x, y)$, $Q = Q(x, y)$, $p = p(x)$, and $q = q(x)$ are formal power series].

8. If $F \ll G$ and $P \ll Q$, then $F + P \ll G + Q$.

9. If $F \ll G$ and $P \ll Q$, then $FP \ll GQ$.

10. If $F \ll G$ and $p \ll q$, then $F[x, p(x)] \ll G[x, q(x)]$.

11. If $F \ll G$, then $\partial F/\partial x \ll \partial G/\partial x$ (interpret the derivatives formally). Is the converse true?

12. Obtain even and odd power series solutions of the DE $w'' + iw = 0$, and interpret the solutions.

13. Obtain an even power series solution of the DE $B'' + z^{-1}B' + iB = 0$, and show that it defines an entire function.

14. Do the results of Exs. 8–11 hold for complex power series? Justify your answer.

CHAPTER 5
PLANE AUTONOMOUS SYSTEMS

1 AUTONOMOUS SYSTEMS

This and the next three chapters will be concerned with *systems* of first-order ordinary DEs in *normal* form. By this is meant a set of equations

$$\frac{dx_1}{dt} = X_1(x_1, \ldots, x_n; t)$$

(1)

$$\cdots \cdots$$

$$\frac{dx_n}{dt} = X_n(x_1, \ldots, x_n; t)$$

The X_i are given functions of the $n + 1$ real variables $x_1, \ldots, x_n, t$. We want to find *solutions* of (1), that is, sets of n functions $x_1(t), \ldots, x_n(t)$ of class $\mathcal{C}^1$ which satisfy (1). We shall assume the functions X_i to be continuous and real-valued in a given region R of the $(n + 1)$-dimensional space of the independent variables $x_1, x_2, \ldots, x_n, t$.

The simplicity of the concept of a first-order normal system becomes apparent when (1) is written in vector notation. A *vector* is an n-tuple $\mathbf{x} = (x_1, \ldots, x_n)$ of real (or complex) numbers. Thus, the functions $X_i(x_1, \ldots, x_n; t) = X_i(\mathbf{x}, t)$, in (1) define $\mathbf{X} = (X_1, \ldots, X_n)$ as a *vector-valued* function of the vector variable $\mathbf{x}$ and the real variable t. Even more simply, we can define a *vector field* as a vector-valued function $\mathbf{X}(\mathbf{x})$ of the vector variable $\mathbf{x}$ (ranging over a suitably defined domain in IR^n), visualizing this as attaching a small arrow $\mathbf{X}(\mathbf{x})$ to each point $\mathbf{x}$.

In vector notation, the system (1) assumes the very concise form

(2)
$$\frac{d\mathbf{x}}{dt} = \mathbf{X}(\mathbf{x}, t)$$

A *solution* of (2) is a vector-valued function $\mathbf{x}(t)$ of a real (scalar) variable t, such that $\mathbf{x}'(t) = \mathbf{X}(\mathbf{x}(t), t)$. The analogy between (2) and the normal first-order DE $y' = F(x, y)$ studied in Ch. 1 is obvious; the only difference is that the dependent variable in (2) is a vector and not a number (or "scalar"). Hence, one can call (2) a *normal first-order vector* DE.

131

A solution of a normal system (1) or (2), defined by the functions $x_1(t)$, ..., $x_n(t)$, can be visualized as a curve in the $(n + 1)$-dimensional region R. When $n = 1$, this specializes to the concept of a solution curve defined in Ch. 1; when $n = 2$, it is a curve in (x_1, x_2, t)-space. For this reason, the curve in R defined by any solution of (1) is called a *solution curve* of (1).

Chapter 6 will contain proofs of existence, uniqueness, continuity, and differentiability theorems for solutions of first-order systems (1). In the present chapter, attention will be confined to *autonomous* first-order systems. By definition, these are systems of the form

$$(3) \qquad \frac{dx_i}{dt} = X_i(x_1, \ldots, x_n), \qquad i = 1, \ldots, n$$

The characteristic property of autonomous systems is the fact that the functions X_i do not depend on the independent variable t. When this variable is thought of as representing time, autonomous systems are thus *time-independent* or *stationary*.

In vector notation, the autonomous system (3) reduces to

$$(3') \qquad \frac{d\mathbf{x}}{dt} = \mathbf{X}(\mathbf{x})$$

To every autonomous system (3) there thus corresponds a unique vector field $\mathbf{X}(\mathbf{x})$ in Euclidean n-space, and conversely. Throughout this chapter, we will consider only vector fields that are of class $\mathcal{C}^1$, and hence satisfy a Lipschitz condition in every compact domain. As will be shown in Chapter 6, this implies that one and only one solution $\mathbf{x}(t, \mathbf{c})$ of the autonomous system (3) satisfies the initial condition $\mathbf{x}(0) = \mathbf{c}$, and that this solution depends continuously on $\mathbf{c}$.

When $n = 3$, the autonomous system (3) can be imagined as representing the steady flow of a fluid in space: at each point $\mathbf{x}$ in a region of space, the vector $\mathbf{X}(\mathbf{x})$ expresses the velocity of the fluid at that point in magnitude and direction. The flow is called *steady* because its velocity depends only on position and does not vary with time. The solution $\mathbf{x}(t, \mathbf{c})$ of the autonomous system (3) for the initial "value" $\mathbf{c}$ then has a simple physical interpretation: it is the *trajectory* (*path, orbit*, or *streamline*) of a moving fluid particle, whose position (initially at $\mathbf{c}$) is given as a function of the time t.

When the preceding path $\mathbf{x}(t, \mathbf{c})$ is considered as a set of points (that is, as a geometric curve), without reference to its parametric representation, it is also called a *solution curve* of the autonomous system (3), or of the associated vector field $\mathbf{X}(\mathbf{x})$. If $\mathbf{x}(t, \mathbf{c})$ is a solution of the autonomous system (3), then so is $\mathbf{x}(t + a, \mathbf{c})$ for any constant a; this can also be interpreted as the path of a particle that passed through the point $\mathbf{c}$ at time $t = a$.

Plane Autonomous Systems.   This chapter will be largely concerned with the case $n = 2$ of (3). In this case, we can omit subscripts and rewrite (3) in simpler notation as

$$\frac{dx}{dt} = X(x,y), \qquad \frac{dy}{dt} = Y(x,y)$$

The connection of such a "plane autonomous system" with the first-order DE $y' = X(x,y)/Y(x,y)$ will be discussed in §2 below.

Example 1.   The solutions of the autonomous system

(4)
$$\frac{dx}{dt} = mx, \qquad \frac{dy}{dt} = ny$$

are evidently $x = c_1 e^{mt}$, $y = c_2 e^{nt}$ where c_1, c_2 are arbitrary constants. The corresponding solution curves are the loci $y^m = kx^n$. Figure 5.1 depicts sample curves for the case $m = 2$, $n = 3$.

Note that the solution curves ("trajectories") of any autonomous system are endowed with a natural sense or *orientation*, the direction of increasing t. This is indicated in drawings of solution curves by marking on them arrowheads pointing in this direction. See Fig. 5.1, which depicts sample (oriented) solution curves of the system (4).

In Fig. 5.1, the origin $(0,0)$ is evidently a very special point: integral curves emanate from it both horizontally and vertically. This is possible only because the vector field $(X,Y) = (mx,ny)$ reduces there to the null vector $\mathbf{0} = (0,0)$, whose direction is indeterminate. Such points are of particular importance for the study of autonomous systems; they are called critical points.

DEFINITION.   A point $\mathbf{x} = (x_1, \ldots, x_n)$ where all the functions X_i are equal to zero is called a *critical point* of the autonomous system (3) and of the associated vector field $\mathbf{X}(\mathbf{x})$.

If $\mathbf{x} = \mathbf{c}$ is a critical point of (3), then the functions $x_i(t) \equiv c_i$ define a trivial solution $\mathbf{x}(t) = \mathbf{c}$ of (3), which describes not a curve but just a point. In the

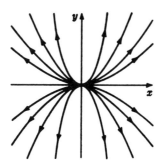

Figure 5.1   Integral curves of $\dot{x} = 2x$, $\dot{y} = 3\dot{y}$.

terminology of hydrodynamics, $\mathbf{c}$ is called a *stagnation point* of the velocity field $\mathbf{X}(\mathbf{x})$.

Every normal system (1) of $(n - 1)$ first-order DEs

$$dx_i/dt = X_i(x_1, \ldots, x_{n-1}; t), \qquad i = 1, \ldots, n - 1$$

is equivalent to an autonomous system in n variables. To see this, introduce an additional variable $x_n = t$, and rewrite (1) as

$$dx_i/dt = X_i(x_1, \ldots, x_n), \qquad i = 1, \ldots, n - 1, \qquad dx_n/dt = 1$$

Evidently, the system so constructed has no critical points.

2   PLANE AUTONOMOUS SYSTEMS

When $n = 2$ in (3), it is convenient to write (3) without subscripts as

(5)
$$\frac{dx}{dt} = X(x,y), \qquad \frac{dy}{dt} = Y(x,y)$$

We then speak of a *plane* autonomous system. The plane autonomous system (5) is evidently equivalent to the first-order DE

(5')
$$\frac{dy}{dx} = \frac{Y(x,y)}{X(x,y)}$$

wherever $X(x,y) \neq 0$. The main advantage of the parametric form (5) is that points $X(x,y) = 0$ of vertical tangency of the solutions of the DE (5') are no longer singular points of the corresponding plane autonomous system (5). Likewise, the solution curves of (5) are just the integral curves of the quasilinear DE

(5'')
$$Y(x,y) = X(x,y)y'$$

and the two have the same critical points.

The advantage of the parametric viewpoint is apparent in the following example, already discussed in Ch. 1, §2.

Example 2.   Consider the autonomous system

(6)
$$\frac{dx}{dt} = -y, \qquad \frac{dy}{dt} = x$$

whose solutions are the function-pairs $x = r \cos(t + c)$, $y = r \sin(t + c)$, where r and c are arbitrary constants. The graphs of these solutions are concentric circles, with center at the origin. The solutions of the corresponding first-order DE

(6')
$$\frac{dy}{dx} = -\frac{x}{y}$$

are the functions $y = \pm \sqrt{r^2 - x^2}$, which are defined only for $|x| < |r|$. Whereas the function $-x/y$ is undefined where $y = 0$, the functions $X(x,y) = -y$ and $Y(x,y) = x$ in the system (6) are defined throughout the plane. This gives the system (6) an obvious advantage over the DE (6').

Referring to the definition of Ch. 1, §12, we see that the circles $x^2 + y^2 = r^2$ form a regular† curve family in the "punctured" xy-plane, the critical point of (6) at the origin $(0,0)$ being deleted. In Ch. 6, §11, it will be shown that this is true of plane autonomous systems in general.

Plane autonomous systems have the following interesting relation to level curves (cf. Ch. 1, §5).

   THEOREM 1. *For any continuously differentiable function $V(x,y)$, each integral curve of the plane autonomous system*

(7)
$$\frac{dx}{dt} = \frac{\partial V}{\partial y}(x,y), \qquad \frac{dy}{dt} = -\frac{\partial V}{\partial x}(x,y)$$

lies on some level curve $V(x,y) = \text{constant}$.

The proof is immediate: along any solution curve we have

$$\frac{dV}{dt} = \frac{\partial V}{\partial x} \cdot \frac{dx}{dt} + \frac{\partial V}{\partial y} \cdot \frac{dy}{dt} = \frac{\partial V}{\partial x} \cdot \frac{\partial V}{\partial y} - \frac{\partial V}{\partial y} \cdot \frac{\partial V}{\partial x} = 0$$

and so $V[x(t),y(t)] = \text{constant}$. Observe that the associated steady flow is divergence-free or *area conserving*, because

$$\text{div}\left(\frac{\partial V}{\partial y}, \frac{-\partial V}{\partial x}\right) = \frac{\partial^2 V}{\partial x \partial y} - \frac{\partial^2 V}{\partial y \partial x} = 0$$

In fluid mechanics, such a steady flow (7) is called *incompressible,* and V is called its stream function.

The representation (7) also reveals the level curves of V as the solution curves of $dx_j/dt = \partial V/\partial x_j$, $j = 1, \ldots, n$—that is, in vector notation, of $dx/dt = \text{grad } V$.

---

† Note that this regular curve family is not normal in the plane, whereas the graphs of the function $y = \sqrt{r^2 - x^2}$ form a normal curve family in the upper half-plane $y > 0$.

The main advantage of the parametric representation (7) over the normal form

(7')
$$\frac{dy}{dx} = -\frac{\partial V/\partial x}{\partial V/\partial y}$$

considered in Ch. 1, §6, is the following. Whereas the solution curves of (7') terminate wherever $\partial V/\partial y$ vanishes, those of (7) terminate only where the function V has a *critical point* (maximum, minimum, or saddle-point) in the sense that grad $V = \mathbf{0}$. This happens exactly where the autonomous system (7) has critical points.

If we set $V = -(x^2 + y^2)/2$ in Theorem 1, we get the system (6) of Example 2, having circular streamlines. If $\mu(x,y)$ is nonvanishing, then the system $dx/dt = -y\mu$, $dy/dt = x\mu$ also has circles for solution curves, and we can construct a wide variety of autonomous systems having the same solution curves in this way, as has been noted before.

Another illustration of Theorem 1 is obtained by setting

$$V(x,y) = \frac{(x^3 + y^3)}{xy} = \frac{r(\cos^3\theta + \sin^3\theta)}{\cos\theta \sin\theta}$$

and letting $\mu(x,y) = x^2y^2$. Evaluating (7), we get the following example.

Example 3.   The plane autonomous system

(8)
$$\frac{dx}{dt} = x(2y^3 - x^3), \qquad \frac{dy}{dt} = -y(2x^3 - y^3)$$

has as solution curves the curves $x^3 + y^3 - 3cxy = 0$, where c is an arbitrary constant. Each such solution curve is a folium of Descartes, as in Figure 5.2. The coordinate axes are also solution curves. The origin is the only critical point of (8); correspondingly, the folia of Descartes in Figure 5.2 form with the axes a curve family that is regular, except at the origin.

Note that the curves of Figure 5.2 form a family of *similar curves*, all similar to $x^3 + y^3 = 3xy$ under a transformation $x \to kx$, $y \to ky$, $t \to t/c^3$, where k is a constant. The reason is that the DE

$$\frac{dy}{dx} = \frac{-y(2x^3 - y^3)}{x(2y^3 - x^3)} = F\left(\frac{y}{x}\right)$$

is homogeneous of degree zero (see the end of Ch. 1, §7).

3   THE PHASE PLANE, II

We have already defined the "phase plane" in Ch. 2, §7, as a way of visualizing the behavior of solutions of (normal) second-order DEs $\ddot{x} = F(x,\dot{x})$. By

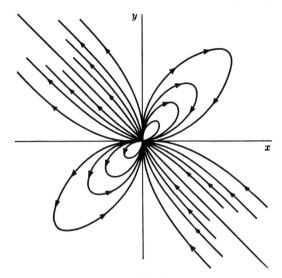

Figure 5.2   Folia of Descartes.

treating $\dot{x} = u$ as a second dependent variable, any such DE is transformed into a (first-order) "plane autonomous system" of the special form $\dot{x} = u$, $\dot{u} = F(x,u)$.

Plane autonomous systems of this special form arise naturally from *dynamical systems* with one degree of freedom. Let a particle be constrained to move on a straight line (or other curve) and let its acceleration $\ddot{x}$ be determined by Newton's Second Law of Motion as a function of its instantaneous position x and velocity $\dot{x}$. Then

$$(9) \qquad\qquad \ddot{x} = F(x,\dot{x})$$

where we have adopted Newton's notation, representing time-derivatives by dots placed over the variable differentiated.

It is usual in dynamics to denote $\dot{x} = dx/dt$ by the letter v, and to call the xv-plane the *phase plane*. Since the variables x and mv are conjugate position and momentum variables, the phase plane is a special instance of the more general concept of phase space in classical dynamics.

Since (9) is time-independent, it is called an *autonomous* second-order DE. In the xv-plane, the second-order autonomous DE (9) is equivalent to the first-order plane autonomous system

$$(9') \qquad\qquad \frac{dx}{dt} = v, \qquad \frac{dv}{dt} = v\frac{dv}{dx} = F(x,v)$$

The integral curves of this autonomous system in the Poincaré phase plane depict graphically the types of motions determined by the DE (9). Note that the solution curves point to the right, to the left, or are vertical according as $\dot{x} > 0$ (upper half-plane), $\dot{x} < 0$ (lower half-plane), or $\dot{x} = 0$ (x-axis). This is because x is increasing, decreasing, or stationary in these three cases, respectively.

Example 4. Consider the *damped linear oscillator* defined by the second-order linear DE with constant coefficients

(10) $$\ddot{x} + p\dot{x} + qx = 0, \qquad p, q \text{ constant}$$

discussed in Ch. 2, §2. The associated autonomous system in the Poincaré phase plane is

(10′) $$\frac{dx}{dt} = v, \qquad \frac{dv}{dt} = -pv - qx$$

The direction field of this system is easily plotted for any p, q.

For example, let $p = 1$, $q = 1$. The resulting DE $\ddot{x} - \dot{x} + x = 0$ describes the free oscillations of a negatively damped particle in an attractive force-field. Sample solution curves in the Poincaré phase plane are sketched in Figure 5.3.

Or again, let $p = 4$, $q = 3$. The resulting DE $\ddot{x} + 4\dot{x} + 3x = 0$ describes heavily damped stable motion; the functions $x = e^{-t}$ and $x = e^{-3t}$ constitute a basis of solutions (the exponential substitution of Ch. 3, §1 gives $\lambda^2 + 4\lambda + 3 = 0$). The graphs of these solutions in the $(x,\dot{x})$-plane are the radii whose slopes -1 and -3 are the roots of this polynomial; they represent solutions of the *first*-order DEs $\dot{x} = -x$ and $\dot{x} = -3x$.

Example 5. The DE of a simple pendulum of length ℓ is

(11) $$\frac{d^2\theta}{dt^2} = -k^2 \sin \theta, \qquad k^2 = \frac{g}{\ell}$$

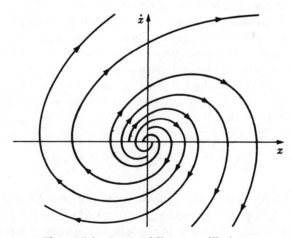

Figure 5.3 Damped linear oscillations.

Here θ is the (counterclockwise) angle made by the pendulum with the vertical. The corresponding plane autonomous system in the phase plane (the $v\theta$-plane) is

$$(11') \qquad\qquad \dot{\theta} = v, \qquad \dot{v} = -k^2 \sin \theta$$

Since the function $\sin \theta$ is periodic, the trajectories form a periodic pattern in the sense that, if $[v(t), \theta(t)]$ is a solution of (11'), so is $[v(t), \theta(t) + 2\pi]$. The case $k^2 = 1$ is sketched in Figure 5.4.

The solutions of (11') correspond to the states of constant energy E: $v^2 - 2 \cos \theta = 2E$ (when $k^2 = 1$). There are two "critical points" in the $v\theta$-plane: the points $(0,0)$ and $(0,\pi)$, corresponding to stable and unstable equilibrium, respectively. Near the "vortex point" $(0,0)$, the pendulum oscillates back and forth; the corresponding trajectories are closed curves, roughly elliptical in shape.

The point $(0,\pi)$ is a saddle-point; the trajectories $v^2 = 2(1 + \cos \theta)$ or $v = \pm 2 \cos(\theta/2)$ that terminate there are called "separatrices," because they separate the closed trajectories from the wavy trajectories $v^2 = 2(E + \cos \theta)$ with $E > 1$, which correspond to whirling the pendulum in circles.

When the amplitude is small, the DE (11) can be approximated well by $\ddot{\theta} + \theta = 0$. In this case, the period of oscillation is independent of the amplitude. For exact solutions, the period increases with the amplitude. We will discuss this phenomenon in §10 below.

EXERCISES A

1. Find and describe geometrically the solution curves of the following vector fields:
   (a) (x, y, z) (b) (ax, by, cz) (c) $(y, -x, 1)$ (d) (y, z, x)

2. Show that the solution curves of the autonomous system

$$\frac{dx}{dt} = e^x - 1, \qquad \frac{dy}{dt} = ye^x$$

are the curves $y = c(e^x - 1)$, and the y-axis $x = 0$.

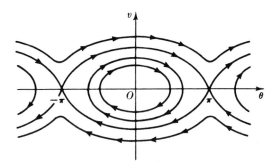

Figure 5.4   Simple pendulum.

3. Show that the functions xyz and $x^2 + y^2$ are integrals of the system

$$\frac{dx}{dt} = xy^2, \qquad \frac{dy}{dt} = -x^2y, \qquad \frac{dz}{dt} = z(x^2 - y^2)$$

Describe the loci xyz = constant, and sketch typical solution curves.

4. (a) Show that the orthogonal trajectories of the level curves of $V = x/(x^2 + y^2)$ are another family of circles. Draw a sketch that displays both families of circles.
   (b) Same question for $V = [(x - 1)^2 + y^2]/[(x + 1)^2 + y^2]$.

The *gradient field* of a scalar function $V(x)$ is defined as the vector field

$$\text{grad } V = (\partial V/\partial x_1, \ldots, \partial V/\partial x_n)$$

The *gradient lines* of V are the solution curves of the autonomous system $dx_i/dt = \partial V/\partial x_i$.

In Exs. 5–7, find the gradient lines of the following functions:

5. $V = xy$ 6. $V = x^2 + y^2 - 2z^2$ 7. $V = \ln [(x - a)^2 + y^2/[(x + a)^2 + y^2]]$

8. Show that a function $\phi(x_1, \ldots, x_n)$ of class $\mathcal{C}^1$ is an integral of the system (3) if and only if it satisfies the partial DE $X_1 \, \partial\phi/\partial x_1 + \cdots + X_n \, \partial\phi/\partial x_n = 0$.

9. Show that if $\partial X/\partial x = \partial \gamma/\partial y$ and $\partial X/\partial y = -\partial \gamma/\partial x$, the plane autonomous system (5) is the real form of a single first-order *complex* analytic DE, and conversely.

*10. Let $e_1, \ldots, e_n$ and $a_1, \ldots, a_n$ be real constants, and let

$$r_j = [(x - a_j)^2 + y^2 + z^2]^{1/2}$$

Show that, if $V = \Sigma e_j/r_j$, the functions

$$\psi = \sum e_j \arccos \alpha_j, \qquad \alpha_j = (x - a_j)/r_j, \qquad \text{and} \qquad \theta = \arctan (z/y)$$

are integrals of $\dot{x} = \partial V/\partial x$, $\dot{y} = \partial V/\partial y$, $\dot{z} = \partial V/\partial z$. Express the integral curves as intersections of the surfaces defined by the preceding equations.

Exercises 11–14 derive some of the main properties of elliptic functions by the methods of Ch. 4, §7, and give an application to Example 5.

11. The *elliptic functions* $u = \text{sn } t$, $v = \text{cn } t$, $w = \text{dn } t$, may be defined as the solutions of the autonomous system

$$(*) \qquad \frac{du}{dt} = vw, \qquad \frac{dv}{dt} = -wu, \qquad \frac{dw}{dt} = k^2 uv$$

having the initial values $u(0) = 0$, $v(0) = w(0) = 1$.
(a) Establish the identities $(\text{sn } t)^2 + (\text{cn } t)^2 = 1$, $k^2 (\text{sn } t)^2 + (\text{dn } t)^2 = 1$.
(b) Using (a), show that the three functions specified are defined and analytic for all real t.
(c) Expand their solutions in power series through terms in t^5.

*12. (a) Show that, in Ex. 11, if $k^2 < 1$, the function $\text{cn } t$ vanishes at

$$t = K = \int_0^1 \frac{dx}{\sqrt{(1 - x^2)\, (1 - k^2 x^2)}}$$

(b) Prove the following addition formulas, valid with $k' = \sqrt{1 - k^2}$: $sn(t + K) = cn\ t/dn\ t$, $cn(t + K) = -k'\ sn\ t/dn\ t$, $dn(t + K) = k'/dn\ t$. [HINT: Show that the vector-valued function v/w, $-k'u/w$, k'/w satisfies (*), and that this vector reduces at $t = K$ to $(1,0,k')$.]

(c) Prove that $sn(-t) = -sn\ t$, $cn(-t) = cn\ t$, $sn(t + 2K) = -sn\ t$, $cn(t + 2K) = -cn\ t$, $dn(t + 2K) = dn\ t$.

(d) Show that $sn\ t$ and $cn\ t$ have infinitely many zeros, and that the zeros of $sn\ t$ separate those of $cn\ t$.

13. (a) From the assumptions of Ex. 11, show that the function $u = sn\ t$ satisfies the second-order DE

(**) $$\ddot{u} + (1 + k^2)u - 2k^2u^3 = 0$$

(b) Infer from (**) that $\dot{u}^2 + (1 + k^2)u^2 - k^2u^4 = $ constant.

(c) Sketch the integral curves of the DE (**) in the phase plane, marking the special curve $(sn\ t, cn\ t, dn\ t)$ and any critical points.

(d) Determine the nature of the other critical points, if any.

14. (a) Show that a one-parameter family of solutions of $\ddot{\theta} + k^2 \sin \theta = 0$ is given by $\sin (\theta/2) = \sin (\alpha/2)sn[k(t - t_0)]$, where α is the amplitude of oscillation.

(b) Show that $\dot{\theta} = 2k \sin (\alpha/2)cn[k(t - t_0)]$.

4 LINEAR AUTONOMOUS SYSTEMS

An autonomous system (3) is called *linear* when all the functions X_i are linear homogeneous functions of the x_k so that

(12) $$\frac{dx_i}{dt} = a_{i1}x_1 + \cdots + a_{in}x_n, \qquad i = 1, \ldots, n$$

Hence, a linear autonomous system is just another name for a (homogeneous) linear system of DEs with constant coefficients. Such a system is determined by the square *matrix* $A = \|a_{ij}\|$ of its coefficients, and its vector field satisfies $\mathbf{X}(\mathbf{x}) = A\mathbf{x}$.

Initial Value Problems. For any autonomous system $\mathbf{x}'(t) = \mathbf{X}(\mathbf{x})$, the "initial value problem" consists in determining, for each $\mathbf{c}$ in the domain of the vector field $\mathbf{X}(\mathbf{x})$, the *solution* $\mathbf{x}(t)$ of the DE that satisfies the "initial condition" $\mathbf{x}(0) = \mathbf{c}$. We will now show how to solve this problem for any linear *plane* autonomous system (i.e., in the case $n = 2$).

Any such system has the form

(13) $$\frac{dx}{dt} = ax + by, \qquad \frac{dy}{dt} = cx + dy$$

where a, b, c, d are constants. The coefficient matrix $A = \begin{pmatrix} a & b \\ c & d \end{pmatrix}$ of constants is nonsingular unless its determinant is $ad - bc = 0$. The origin $(0,0)$ is always a critical point of the system (13). Since the simultaneous linear equations $ax + by = cx + dy = 0$ have no solution except $x = y = 0$ unless A is singular, we see that the origin is the only critical point of the system (13), unless $ad = bc$ (the *degenerate* case $|A| = 0$).

To solve the initial value problem for the system (13), it is convenient to introduce a new concept.

DEFINITION.   The *secular equation* of (13) is

$$(14) \qquad\qquad \ddot{u} - (a + d)\dot{u} + (ad - bc)u = 0$$

THEOREM 2.   *If $(x(t), y(t))$ is any solution of the plane autonomous system* (13), *then $x(t)$ and $y(t)$ are solutions of the secular equation* (14) *of* (13).

Proof.   We shall prove that $x(t)$ is a solution of (14); the proof for $y(t)$ is the same, replacing a with d and b with c. The first equation (13) implies $\dot{x} - ax = by$, which implies $\ddot{x} - a\dot{x} = b\dot{y}$. From the second equation it follows that

$$\ddot{x} - a\dot{x} = bcx + bdy = bcx + d(\dot{x} + ax)$$

Transposing, we see that $x(t)$ satisfies (14).

Conversely, the secular equation (14) can also be used to solve the linear system (13) as follows. First, find a basis of solutions $u(t)$ and $v(t)$ of (14) by the methods of Ch. 2. Then, if $b \neq 0$ in (13), set

$$(15) \qquad\qquad x = bu \qquad \text{and} \qquad y = \dot{u} - au$$

The first equation of (13) will be automatically satisfied, whereas the second will be equivalent to

$$b\dot{y} = cx + dy \qquad \text{or} \qquad b\ddot{u} - ab\dot{u} = bcu + d\dot{u} - au$$

which holds by (14). In the same way, if $b = 0$ but $c \neq 0$ in (13), set $x = u - du$ and $y = cu$; (13) follows similarly. In both cases, a second solution can be constructed from $v(t)$.

In the remaining case that $b = c = 0$, the obvious formula

$$(15') \qquad\qquad x \equiv x_0 e^{at}, \qquad y = y_0 e^{dt}$$

solves (13) for any initial $x(0) = x_0$ and $y(0) = y_0$, as in Example 1.

The preceding recipes are effective computationally. Thus, to solve the initial value problem

$$\dot{x} = x - y, \qquad \dot{y} = x + y, \qquad x(0) = 1, \qquad y(0) = 0$$

we can use (14) to obtain $\ddot{x} - 2\dot{x} + 2x = 0$. Since the roots of the characteristic polynomial $|A - \lambda I| = \lambda^2 - 2\lambda + 2$ are $\lambda = 1 \pm i$, the system has the general solution

$$x = e^t(A \cos t + B \sin t) \qquad y = e^t(A \sin t - B \cos t)$$

Moreover, the initial condition $x(0) = 1$ implies that $A = 1$, while $y(0) = x(0) - \dot{x}(0)$ implies that $B = 0$. The solution of the initial value problem stated is therefore $x = e^t \cos t$, $y = e^t \sin t$.

Systems in n Variables.   Most of the preceding results have straightforward generalizations to (homogeneous) linear autonomous systems (12) in n variables. In vector and matrix notation, the system (12) simplifies to $d\mathbf{x}/dt = A\mathbf{x}$. We can define its "secular equation," for any matrix A, to be

$$(16) \qquad L[u] = c_o u - c_1 u' + \cdots + u^{(n)} = p_A(D)[u] = 0$$

where $p_A(\lambda) = |A - \lambda I| = c_0 + c_1\lambda + \cdots \pm \lambda^n$ is the characteristic polynomial of the matrix A, and $D = d/dt$. We then have the following.

THEOREM 3.   *If $\mathbf{x}(t)$ is any solution of (12), then every component $x_i(t)$ of $\mathbf{x}(t)$ satisfies the secular equation (16) of (12).*

The proof of this result depends on theorems about matrices and so will be deferred until Appendix A.

Eigensolutions.   By an "eigensolution" of the (constant-coefficient) linear autonomous system $\dot{\mathbf{x}} = A\mathbf{x}$ is meant a solution of the form $\mathbf{x}(t) = c(t)\boldsymbol{\phi}$, where $\boldsymbol{\phi}$ is a nonzero vector. Since this implies $\dot{\mathbf{x}} = c'(t)\boldsymbol{\phi} = Ac(t)\boldsymbol{\phi}$, there follows $A\boldsymbol{\phi} = \lambda\boldsymbol{\phi}$ with $\lambda = c'(t)/c(t)$, whence $c(t) = Ke^{\lambda t}$. Conversely, if $\boldsymbol{\phi}$ is any eigenvector of the matrix A with eigenvalue λ, then $\mathbf{x}(t) = e^{\lambda t}\boldsymbol{\phi}$ is evidently an eigensolution, since $\dot{\mathbf{x}} = \lambda\mathbf{x} = A\mathbf{x}$.

Eigensolutions (and generalized eigensolutions) of complex linear constant-coefficient systems $\dot{\mathbf{z}} = C\mathbf{z}$ provide a general canonical basis in the "solution space." But for *real* linear plane autonomous systems, they arise only when $\mathbf{0}$ is a saddle-point. For example, since

$$\begin{pmatrix} -3 & 2 \\ -2 & 2 \end{pmatrix}\begin{pmatrix} 2 \\ 1 \end{pmatrix} = -2\begin{pmatrix} 2 \\ 1 \end{pmatrix} \qquad \text{and} \qquad \begin{pmatrix} -3 & 2 \\ -2 & 2 \end{pmatrix}\begin{pmatrix} 1 \\ 2 \end{pmatrix} = \begin{pmatrix} 1 \\ 2 \end{pmatrix}$$

the vector-valued functions $\boldsymbol{\phi}(t) = e^{-2t}\begin{pmatrix} 2 \\ 1 \end{pmatrix}$ and $\boldsymbol{\psi}(t) = e^t\begin{pmatrix} 1 \\ 2 \end{pmatrix}$ form a basis of

eigensolutions of the system $\mathbf{x}'(t) = \begin{pmatrix} -3 & 2 \\ -2 & 2 \end{pmatrix}\mathbf{x}$. They clearly correspond to the

invariant lines of the linear fractional DE $y' = (-2x + 2y)/(-3x + 2y)$ (cf. Ch. 1, §7).

When the matrix A has a basis of eigenvectors $\boldsymbol{\phi}_j$, there is an especially elegant way to solve the initial value problem $\mathbf{x}'(t) = A\mathbf{x}$. Namely, expand the initial data $\mathbf{x}(0) = \mathbf{c}$ into a linear combination of eigenvectors of A: $\mathbf{c} = \Sigma a_j \boldsymbol{\phi}_j$. Then the solution of the system $\mathbf{x}'(t) = A\mathbf{x}$ for these initial data is

$$(17) \qquad\qquad \mathbf{x}(t) = \sum_{j=1}^{n} a_j e^{\lambda_j t} \boldsymbol{\phi}_j$$

where λ_j is the eigenvalue of $\boldsymbol{\phi}_j$.

5   LINEAR EQUIVALENCE

The secular equation (14) of a *linear* plane autonomous system (13) establishes a clear connection between its solutions and those of an associated (linear) *constant-coefficient* second-order DE. As we shall now show, it also throws light on the rough classification, made in Ch. 2, §7, into "focal," "nodal" and "saddle" points of the critical points of such DEs.

Consider first the case of *focal* points. Anticipating what will soon be proved, we begin by considering system of the special form

$$(18) \qquad\qquad \dot{x} = ax - by, \qquad \dot{y} = bx + ay$$

Substituting into (14), we find that its secular equation is

$$(18') \qquad\qquad \ddot{u} - 2a\dot{u} + (a^2 + b^2)\,u = 0$$

Here $-2a$ is clearly arbitrary, while the discriminant $\Delta = 4a^2 - 4(a^2 + b^2) = -4b^2$ can be any negative number. Hence (cf. Ch. 2, §7, Case A), all secular equations of focal point type can be obtained from linear plane autonomous systems of the special form (18).

In *polar coordinates,* on the other hand, one easily verifies that (18) reduces to

$$(19) \qquad\qquad \dot{r} = \alpha r, \qquad \dot{\theta} = b, \qquad \alpha = -2a$$

Hence the *orbits* (trajectories) of (18) are *equiangular spirals* $\theta = \theta_0 + bt$, $r = r_0 e^{\gamma \theta}$, $\gamma = -2a/b$, except in two *degenerate* cases:

(i) $\qquad\qquad x = x_0 e^{at}, \qquad y = y_0 e^{at} \qquad$ when $\qquad b = 0$

(ii) $\qquad x = r_0 \cos(bt - \beta), \qquad y = r_0 \sin(bt - \beta) \qquad$ when $\qquad a = 0$

In the first case, the origin is said to be a *star point* of (18); in the second, it is said to be a *vortex point* of (18). It should be noted that these two "degenerate" cases (occurring when $q = 0$ resp. $\Delta = 0$) were explicitly omitted in the discussion of Ch. 2, §7.

Clearly the *phase-plane* representation of (18), which is

$$(19') \qquad\qquad \dot{x} = y, \qquad \dot{y} = 2ax - (a^2 + b^2)y$$

is less attractive than (18); cf. Ex. 1. Yet the two are linearly equivalent in the following sense.

DEFINITION. Two first-order (linear homogeneous) autonomous systems, $\mathbf{x}'(t) = A\mathbf{x}$ and $\mathbf{u}'(t) = B\mathbf{u}$ are called *linearly equivalent* when there exists a nonsingular matrix K such that $B = KAK^{-1}$, that is, when

(20) $$u_i = k_{i1}x_1 + \cdots + k_{in} x_n, \qquad i = 1, \cdots, n$$

where as usual K denotes the matrix $[k_{ij}]$.

The reason for this definition is that, if we write $\mathbf{u} = K\mathbf{x}$ and $\mathbf{x} = K^{-1}\mathbf{u}$, then $\mathbf{x}'(t) = A\mathbf{x}$ is transformed into

(21) $$\frac{d\mathbf{u}}{dt} = K\frac{d\mathbf{x}}{dt} = KA\mathbf{x} = (KAK^{-1})\mathbf{u}$$

under the *change of basis* associated with the nonsingular matrix K. Thus, in algebraic language, linearly equivalent linear autonomous systems are associated with *similar*† matrices A and KAK^{-1}.

Therefore, the reduction of linear autonomous systems to a standard simplified (or "canonical") form under linear equivalence amounts to reducing matrices to canonical form under "similarity." We will treat this problem here only for linear plane autonomous systems

(22) $$\frac{dx}{dt} = ax + by, \qquad \frac{dy}{dt} = cx + dy$$

Its solution will throw considerable light onto the classification of critical points of linear *and* nonlinear plane autonomous systems into those of focal, nodal, and saddle-point type.

LEMMA. *Linearly equivalent linear plane autonomous systems have the same secular equation.*

Proof. This result follows immediately from (14′), (21), and general identities of linear algebra. If $B = KAK^{-1}$, then‡

$$|B - \lambda I| = |KAK^{-1} - \lambda I| = |K(A - \lambda I)K^{-1}|$$
$$= |K| \cdot |A - \lambda I| \cdot |K^{-1}| = |A - \lambda I|$$

---

† Birkhoff and MacLane, p. 264.

‡ Birkhoff and MacLane, p. 264. For reduction to diagonal form and Jordan canonical form, see *ibid.*, pp. 294, 354. For the companion matrix form $\begin{pmatrix} 0 & 1 \\ -p & -q \end{pmatrix}$, see *ibid.*, p. 338.

(The lemma is also a corollary of Theorem 3 unless there are two linearly independent equations $\ddot{u} + p\dot{u} + qu = 0$ satisfied by both components of all solutions of $d\mathbf{x}/dt = A\mathbf{x}$.)

THEOREM 4.　*Unless $a = d$ and $b = c = 0$, the linear plane autonomous system (22) is linearly equivalent to the phase plane representation*

$$(23)\qquad \frac{du}{dt} = v, \qquad \frac{dv}{dt} = -qu - pv, \qquad p = -(a + d), \qquad q = (ad - bc)$$

of its secular equation (14).

(22)　*Proof.*　If $b \neq 0$, let $u = x$, $v = ax + by$; that is, let $K = \begin{pmatrix} 1 & 0 \\ a & b \end{pmatrix}$. Then

reduces to

$$(24)\qquad \dot{u} = v, \qquad \dot{v} = a\dot{x} + b\dot{y} = (a^2 + bc)x + (ab + bd)y$$

The last expression in (24) is equal to

$$(a^2 + ad)x + (ab + bd)y - (ad - bc)x = (a + d)v - (ad - bc)u$$

by definition of u, v; hence (22) is equivalent to

$$\dot{u} = v, \qquad \dot{v} = (a + d)v - (ad - bc)u = -pv - qu$$

that is, to (23). This shows, in particular, that

$$KAK^{-1} = \begin{pmatrix} 0 & 1 \\ -q & -p \end{pmatrix}$$

is the *companion matrix* of the secular equation of (22).

This proves Theorem 4 for the case $b \neq 0$. The case $c \neq 0$ can be treated in the same way, letting $u = y$ and $v = cx + dy$.

When $b = c = 0$, let $u = x + y$, $v = ax + dy$; if $a \neq d$, we can set $x = (du - v)/(d - a)$, $y = (au - v)/(a - d)$. By (22) with $b = c = 0$, we have $\dot{u} = \dot{x} + \dot{y} = ax + dy = v$. Similarly,

$$\dot{v} = a\dot{x} + d\dot{y} = a^2x + d^2y$$

Comparing this with the expression

$$(a + d)v - (ad)u = [(a^2 + ad)x + (ad + d^2)y - adx - ady]$$

we also verify (23) in this case.

Exceptional Case.   The case $a = d$, $b = c = 0$ of a scalar matrix,

(25) $$\dot{x} = ax, \qquad \dot{y} = ay$$

is genuinely exceptional. The secular equation of (25) is $\ddot{x} - 2a\dot{x} + a^2 x = 0$, just as it is for the system

(25′) $$\dot{x} = ax, \qquad \dot{y} = x + ay$$

but (25) and (25′) are not linearly equivalent. Every component of every solution of (25) satisfies $\dot{u} = au$, but this is not true of the solution (e^{at}, te^{at}) of (25′).

The exceptional case arises when the characteristic polynomial of the secular equation (14) has equal roots, that is, when its *discriminant* $\Delta = p^2 - 4q = (a - d)^2 + 4bc$ vanishes. This gives us the following corollary of Theorem 4.

COROLLARY.   *Unless the discriminant $(a - d)^2 + 4bc$ vanishes, two linear plane autonomous systems are linearly equivalent if and only if they have the same secular equation.*

Complete Classification.   We now use Theorem 4 to provide a complete classification of linear plane autonomous systems, giving for each type a simple canonical form. This classification is based on a study of the possible root-pairs λ_1, λ_2 of the characteristic equation (14), which we renumber as

(26) $$\lambda^2 - (a + d)\lambda + (ad - bc) = \lambda^2 + p\lambda + q = 0$$

By Theorem 4, unless $a = d$ and $b = c = 0$, it suffices to display one autonomous system for each such root pair. We now enumerate the different possibilities, which depend largely on the sign of the *discriminant*

(26′) $$\Delta = p^2 - 4q = (a - d)^2 + 4bc$$

of the characteristic equation (26). We begin with the case $\Delta \neq 0$, $q \neq 0$ of distinct nonzero roots $\lambda_1 \neq \lambda_2$.

A. Focal Points.   Suppose $\Delta < 0$, so that the characteristic equation (26) has distinct *complex* roots $\lambda_j = \mu \pm i\nu$ ($\nu \neq 0$). This is the case $q = (\mu^2 + \nu^2) > 0$ and $0 \leq p^2 = 4\mu^2 < 4q$ of a harmonic oscillator. We choose the canonical form (see the Corollary of Theorem 4 and Exs. B1–B3)

(26a) $$dx/dt = \mu x - \nu y, \qquad dy/dt = \nu x + \mu y$$

for (26), whose solutions are the equiangular *spirals* $r = \rho e^{\mu t}$, $\theta = \nu t + \tau$ in polar coordinates, where $\rho \geq 0$ and τ are arbitrary constants. When $p > 0$, the spirals approach the origin (*stable* focal point); when $p < 0$, they diverge from it (*unsta-*

ble focal point); when $p = 0$, they are closed curves representing periodic oscillations (neutrally stable *vortex* points). See Figures 5.5a and 5.5b.

B. Nodal Points.   Suppose that $\Delta > 0$ and $q > 0$, so that the roots $\lambda = \mu_1$, μ_2 of the characteristic equation (26) are real, distinct, and of the same sign. We choose, as the linearly equivalent canonical form,

$$(26b) \qquad dx/dt = \mu_1 x, \qquad dy/dt = \mu_2 y, \qquad 0 < |\mu_1| < |\mu_2|$$

whose general solution is $(ae^{\mu_1 t}, be^{\mu_2 t})$. The system is stable when μ_1 and μ_2 are negative and unstable when they are positive (the two subcases are related by the transformation $t \to -t$ of time reversal). Geometrically, the integral curves $y = cx^m$, $m = \mu_2/\mu_1$, look like a sheaf of parabolas, tangent at the origin, as in Figure 5.5c.

C. Saddle-Points.   Suppose that $\Delta > 0$ but $q < 0$, so that the roots of the characteristic equation (26) are real and of opposite sign. We again have the canonical form (26b). But since μ_1 and μ_2 have opposite signs, the integral curves $x^m y = c$, $m = -\mu_2/\mu_1 > 0$ look like a family of similar hyperbolas having given asymptotes, as in Figure 5.5d. A saddle-point is always *unstable*.

There remain various degenerate cases and subcases, in which $\Delta = 0$ or $q = 0$. The simplest such case is the exceptional case (25), in which $\Delta = 0$, $q = a^2 > 0$. The integral curves consist of the straight lines through the origin, and the configuration formed by them is called a *star*, as in Figure 5.7e. In the nonexceptional subcase, we have the canonical form of (25′)

$$(26c) \qquad\qquad dx/dt = ax, \qquad dy/dt = x + ay,$$

whose integral curves have the appearnce of Figure 5.5f. Such a point is also called a *nodal* point, and it is stable or unstable according as $a < 0$ or $a > 0$.

The case $q = 0$, $\Delta \neq 0$ corresponds to the phase plane representation of the second-order DE $\ddot{x} + p\dot{x} = 0$, $p \neq 0$. This corresponds to a rowboat "coasting" on a lake, with no wind and its oars shipped. The boat comes to rest at a finite distance, in infinite time. The integral curves form a family of parallel straight lines $\dot{x} + px = $ constant, as in Figure 5.5g. The origin is a stable (but not strictly stable) critical point if $p > 0$, unstable if $p < 0$.

Finally, the case $q = \Delta = 0$ reduces to $\dot{x} = \dot{y} = 0$ in the exceptional case (25) and to the phase plane representation of $\ddot{x} = 0$ otherwise. The former case is (neutrally) stable; the latter case is unstable.

EXERCISES B

1. Show that the secular equation of the system

$$\dot{x} = \alpha x - \beta y, \qquad \dot{y} = \beta x + \alpha y \qquad (\alpha, \beta \text{ real})$$

has the complex roots $\lambda = \alpha \pm i\beta$.

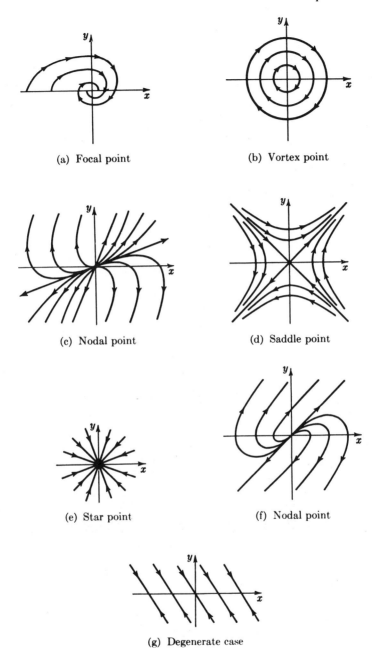

(a) Focal point

(b) Vortex point

(c) Nodal point

(d) Saddle point

(e) Star point

(f) Nodal point

(g) Degenerate case

Figure 5.5

2. Show that the solution curves of the system

$$\dot{x} = ax, \qquad \dot{y} = by, \qquad a \neq 0, \qquad \text{are} \qquad y = C|x|^\gamma, \qquad \gamma = \frac{b}{a}$$

3. Solve the following initial value problems:

(a) $\dfrac{d}{dt}\begin{pmatrix} x \\ y \end{pmatrix} = \begin{pmatrix} 2 & -1 \\ 3 & -2 \end{pmatrix}\begin{pmatrix} x \\ y \end{pmatrix},$ $[x(0),y(0)] = (1,0)$

(b) $\dfrac{d}{dt}\begin{pmatrix} x \\ y \end{pmatrix} = \begin{pmatrix} 1 & -2 \\ 2 & 1 \end{pmatrix}\begin{pmatrix} x \\ y \end{pmatrix},$ $x(0) = 1, \qquad y(0) = 0$

4. (a) Solve the initial value problem for the DE

$$\frac{d}{dt}\begin{pmatrix} x \\ y \end{pmatrix} = \begin{pmatrix} 1 & 2 \\ 2 & 1 \end{pmatrix}\begin{pmatrix} x \\ y \end{pmatrix}, \qquad \text{for general} \cdot \qquad \begin{pmatrix} x(0) \\ y(0) \end{pmatrix} = \begin{pmatrix} a \\ b \end{pmatrix}$$

(b) Show that the trajectories are hyperbolas or straight lines through the origin.

5. (a) Show that the characteristic equation of the system

(*) $\dot{x} = \mu x - \nu \dot{y}, \qquad \dot{y} = \nu x + \mu y \qquad (\mu, \nu \text{ real})$

has the complex roots $\lambda = \mu \pm i\nu$.

(b) Infer that any linear plane autonomous system (13) with discriminant $\Delta < 0$ is linearly equivalent to (*), for some μ, ν.

(c) Show that, in polar coordinates, the system (*) defines the flow $r \to e^{\mu t}r$, $\theta \to \theta + \nu t$.

(d) Show that (*) is the real form of the first-order complex linear DE $\dot{z} = \lambda z$, $z = x + iy$.

6. Consider the system $dx/dt = \alpha x$, $dy/dt = \beta y$ for $\alpha \neq 0$.

(a) Show that its solution curves are $y = C|x|^p$, $p = \beta/\alpha$.

(b) Prove that any linear plane autonomous system (13) with positive discriminant is linearly equivalent to a DE of the foregoing form under a suitable change of basis.

*(c) Show that, in the punctured plane $x^2 + y^2 > 0$, the system $dx/dt = \alpha x$, $dy/dt = \beta t$ and $dx/dt = k\alpha x$, $dy/dt = k\beta y$ are equivalent if $k \neq 0$, but that they are not equivalent on any domain that contains the origin unless $k = 1$.

7. Show that any linear plane autonomous system (13) with zero discriminant is equivalent to $dx/dt = ax$, $dy/dt = ay$, or to $dx/dt = ax$, $dy/dt = x + ay$. Describe the associated flows geometrically.

8. (a) Show that, if $ad \neq bc$, the linear fractional DE

$$\frac{dy}{dx} = \frac{cx + dy + f}{ax + by + e}$$

is equivalent by an (affine) transformation to one of the canonical forms of Exs. 5–7.

(b) Derive a set of canonical forms for the exceptional case $ad = bc$.

6   EQUIVALENCE UNDER DIFFEOMORPHISMS

In the preceding sections, we have analyzed properties of linear autonomous systems that are preserved under linear transformations; in the rest of this chapter we will study properties that are preserved under the far more general class of continuously differentiable transformations.

Such transformations often enable one to greatly simplify the form of a DE or system of DEs. For instance, consider the system

(27)
$$\frac{dx}{dt} = \frac{3x^4 - 12x^2y^2 + y^4}{(\dot{x}^2 + y^2)^{3/2}}$$
$$\frac{dy}{dt} = \frac{6x^3y - 10xy^3}{(x^2 + y^2 3/2}$$

In polar coordinates $r = (x^2 + y^2)^{1/2}$, $\theta = \arctan(y/x)$ with inverse functions $x = r\cos\theta$, $y = r\sin\theta$, this system reduces to

(27′)
$$\frac{dr}{dt} = 3r\cos 3\theta, \qquad \frac{d\theta}{dt} = \sin 3\theta$$

In this form, one sees at a glance that the rays $\theta = n\pi/3$ are integral curves, for $n = 0, \ldots, 5$. Other integral curves are sketched in Figure 5.6.

We shall study below how far we can simplify *linear* autonomous systems by such coordinate transformations. Our study will be based on a general concept of *equivalence* under diffeomorphism, which we now define precisely. Let

(28) $u_i = f_i(x_1, \ldots, x_n), \qquad i = 1, \ldots, n$

be continuously differentiable functions with inverse functions

(28′) $x_j = g_j(u_1, \ldots, u_n), \qquad j = 1, \ldots, n$

so that $\mathbf{f}(\mathbf{g}(\mathbf{u})) = \mathbf{u}$ and $\mathbf{g}(\mathbf{f}(\mathbf{x})) = \mathbf{x}$. For such inverse functions to exist locally and be continuously differentiable, it is necessary and sufficient (by the Implicit

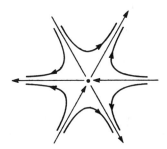

Figure 5.6   Solution curves at (18).

Function Theorem†) that the Jacobian of (28) be nonvanishing: that $|\partial f_i/\partial x_j| \neq 0$.

If $x(t)$ is any solution of the autonomous system

(29)
$$\frac{dx_j}{dt} = X_j(x_1, \ldots, x_n), \qquad j = 1, \ldots, n$$

then the functions

(30)
$$u_i(t) = f_i[x_1(t), \ldots, x_n(t)], \qquad i = 1, \ldots, n$$

satisfy the autonomous system

(31)
$$\frac{du_i}{dt} = U_i(u), \qquad U_i = \sum_{j=1}^{n} \frac{\partial f_i}{\partial x_j} \frac{dx_j}{dt} = \sum_{j=1}^{n} \frac{\partial f_i}{\partial x_j} (g(u)) X_j(g(u))$$

and conversely. In this sense, the autonomous systems (29) and (31) are *equivalent*. We formalize the preceding discussion in a definition.

DEFINITION.   Let $dx/dt = \mathbf{X}(\mathbf{x})$ and $du/dt = \mathbf{U}(\mathbf{u})$ be autonomous systems, defined in regions R and R' of n-dimensional space, respectively. The two systems are *equivalent* if and only if there exists a one–one transformation $\mathbf{u} = \mathbf{f}(\mathbf{x})$ of coordinates, of class $\mathcal{C}^1$ and with nonvanishing Jacobian, which maps R onto R' and carries $dx/dt = \mathbf{X}(\mathbf{x})$ into $du/dt = U(u)$.

Under these circumstances, the inverse transformation is also of class $\mathcal{C}^1$ with nonvanishing Jacobian. Note that the relation of equivalence is symmetric, reflexive, and transitive: it is an equivalence relation.‡ It follows from the preceding discussion that equivalent autonomous systems have solution curves obtainable from each other by a coordinate transformation. If the systems (29) and (31) are equivalent under the change of coordinates (30), and if $V(\mathbf{u})$ is an integral of (31), then $V[\mathbf{f}(\mathbf{x})]$ is an integral of (29).

However, two autonomous systems may have the same solution curves without being equivalent. Thus, the solution curves of

(32)
$$\dot{x} = (x^2 + y^2)y, \qquad \dot{y} = -(x^2 + y^2)x$$

are concentric circles, as in Example 2 ($\dot{x} = -y, \dot{y} = x$). Yet the two systems are not equivalent: all solutions of $\dot{x} = -y, \dot{y} = x$ are periodic with the same period

---

† See Ch. 1, §5. The Jacobian of (28) is the determinant of the square matrix $\|\partial f_i/\partial x_j\|$ of first partial derivatives. See Widder, p. 28 ff. Transformations with the properties stated are called "diffeomorphisms."

‡ Birkhoff and Mac Lane, p. 34.

2π, whereas the periods of the solutions of (32) vary like $1/r^2$ with distance from the origin.

7 STABILITY

The concepts of stability and strict stability, already defined for linear DEs with constant coefficients in Ch. 2, §3, apply to the critical points of any autonomous system. Loosely speaking, a critical point P is stable when the solution curves originating near P stay uniformly near it at all later times; P is strictly stable if, in addition, each such individual solution curve gets and stays arbitrarily near P as t increases without limit. In vector notation, the precise definitions are as follows.

DEFINITION. Let a be a critical point of the autonomous system $\mathbf{x}'(t) = \mathbf{X}(\mathbf{x})$, so that $\mathbf{X}(\mathbf{a}) = \mathbf{0}$. The critical point a is called:

(i) *stable* when, given $\epsilon > 0$, there exists a $\delta > 0$ so small that, if $|\mathbf{x}(0) - \mathbf{a}| < \delta$, then $|\mathbf{x}(t) - \mathbf{a}| < \epsilon$ for all $t > 0$

(ii) *attractive* when, for some $\delta > 0$,

$$(33) \qquad |\mathbf{x}(0) - \mathbf{a}| < \delta \qquad \text{implies} \qquad \lim_{t \to \infty} |x(t) - a| = 0$$

(iii) *strictly stable* when it is stable and attractive. A stable critical point which is not attractive is called *neutrally* stable; a critical point which is not stable is called *unstable*.

Evidently, the preceding definitions are invariant under diffeomorphisms (§5); thus, they describe important qualitative distinctions between the kinds of critical points. For first-order autonomous DEs, being "attractive" has a simple interpretation.

THEOREM 5. *The critical point* 0 *of the one-dimensional autonomous* DE $x'(t) = X(x)$ *is attractive if and only if* (ii'). *For some* $\delta > 0$, $0 < |x| < \delta$ *implies* $xX(x) < 0$. *In this case, the* DE *is strictly stable.*

Explanation. In other words, in the first-order case (ii) is equivalent to (ii') while either condition implies (iii).

Proof. If $X(x_1) = 0$ for some x_1 with $0 < |x_1| < \delta$, then $x(t) = x_1$ is a solution, violating (ii') above. In the same way, if $x_1 X(x_1) > 0$ [that is, if x_1 and $X(x_1)$ have the same sign], then the solution with initial value $x(0) = (x_1 + \delta \operatorname{sgn} x_1)/2$ could never cross x_1; hence it would also violate (ii). Therefore, condition (ii') is necessary for being "attractive." It is sufficient since, if $0 < x(0) < \delta$, then $\sup_{[\delta_1, x(0)]} X(x) = -a_1 < 0$ for all $\delta_1 \in [0, x(0)]$. Hence, we have $0 < x(t) < \delta_1$ for all $t \geq x(0)/a_1$, proving (30); we omit the details. A similar argument covers the case $-\delta < x(0) < 0$.

In both cases, stability is immediate; indeed, stability follows if 0 is a limit point of x with $xX(x) < 0$.

Attractiveness also implies strict stability for *linear* autonomous systems in n dimensions, that is, for linear DEs with constant coefficients. We now prove this important result, which goes back to Lagrange.

THEOREM 5'.   *The critical point* **0** *of the constant-coefficient linear autonomous system* $\mathbf{x}'(t) = A\mathbf{x}$ *is attractive if and only if every eigenvalue of A has a negative real part. In this case, the system is also strictly stable.*

Proof.   If some eigenvalue λ of A has a nonnegative real part, then (12) has a solution of the form $\mathbf{x}(t) = e^{\lambda t}\mathbf{f}$ (an "eigensolution" or "normal mode"), where the initial eigenvector $\mathbf{x}(0) = \mathbf{f}$ can have arbitrarily small length. Conversely, note that by Theorem 3, every component $x_i(t)$ of every solution $\mathbf{x}(t)$ of (12) satisfies the secular equation $P_A(D)x_i(t) = 0$, where the roots of the polynomial equation $P_A(\lambda) = 0$ are just the eigenvalues λ_j of A. From Theorem 4 of Ch. 3, we know that every $x_i(t) \mapsto 0$ if these λ_i all have negative real parts, which implies (33). This proves the first statement of Theorem 5'.

To prove the second statement, we extend the concept of solution basis to vector DEs with constant coefficients. If $\mathbf{x}(0) = \mathbf{0}$ and $\mathbf{x}'(t) = A\mathbf{x}$, then by repeated differentiations, we have $\mathbf{x}^{(n)}(0) = 0$ for all n. Hence, in Theorem 2, every $x_i^{(k)}(0) = 0$ and, by the crucial Lemma of Ch. 3, §4, every $x_i(t) \equiv 0$. It follows from this that the vector form $\mathbf{x}(t) = A\mathbf{x}$ of (12) can have, at most n linearly independent solutions (it will be proved in Ch. 6 that it has exactly n of them). Calling them $\mathbf{u}^1(t), \ldots, \mathbf{u}^n(t)$, we see that the general solution of (12) is $\mathbf{x}(t) = \Sigma c_l \mathbf{u}^l(t) \to 0$, where $|\mathbf{x}(t)| < \epsilon$ for all sufficiently large t, *uniformly,* provided only that $\Sigma c_l < \delta$, some sufficiently small number. The stability condition (i) above follows.

It follows from Ch. 3, §5, that the conditions for strict stability of the second-order system $dx/dt = ax + by$, $dy/dt = cx + dy$, are $p = -(a + d) > 0$, $q = ad - bc > 0$ or equivalently $a + d < 0$, $ad > bc$.

Caution.   One should not conclude from Theorems 5 and 5' that attractiveness implies strict stability for all autonomous systems. Indeed, we now construct an attractive critical point of a nonlinear plane autonomous system that is unstable.† Figure 5.7 depicts sample solution curves.

Example 6.   Let D_1 be the lower half-plane $y \leq 0$; let D_2 be the locus $x^2 + y^2 \leq 2|x|$, consisting of the discs $(x \pm 1)^2 + y^2 \leq 1$; let D_3 be the half-strip $|x| \leq 2, y > 0$, exterior to D_2; let D_4 be the locus $|x| > 2, y > 2$, all as

---

† The authors are indebted to Dr. Thomas Brown for constructing Example 6.

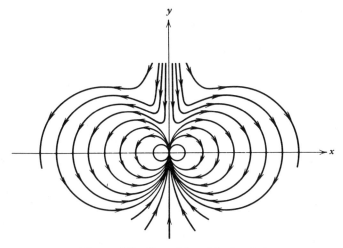

Figure 5.7   Unstable critical point.

depicted in Figure 5.7. The system

$$\dot{x} = \begin{cases} 2xy & \text{on} \quad D_1 \cup D_2 \cup D_3 \\ 2xy/[3 - (4|x|)] & \text{on} \quad D_4 \end{cases}$$

$$\dot{y} = \begin{cases} y^2 - x^2 & \text{on} \quad D_1 \cup D_2 \\ 4|x| - y^2 - 3x^2 & \text{on} \quad D_3 \\ (4|x| - y^2 - 3x^2)/[3 - 4/|x|] & \text{on} \quad D_4 \end{cases}$$

is unstable, yet $\lim_{t \to \infty} |\mathbf{x}(t)| = \mathbf{0}$ for all orbits $\mathbf{x}(t)$.

Dynamical Systems.   If the DEs for an autonomous dynamical system are written in normal form, as

(34) $$\frac{d^2 q_i}{dt^2} = F_i\left(\mathbf{q}, \frac{d\mathbf{q}}{dt}\right) = F_i\,(\mathbf{q}, \mathbf{p}), \qquad i = 1, \dots, m$$

and conjugate velocity variables $p_i = dq_i/dt$ are introduced, then (34) defines an autonomous system of first-order DEs

(34') $$\frac{dq_i}{dt} = p_i, \qquad \frac{dp_i}{dt} = F_i(\mathbf{q}, \mathbf{p})$$

in an associated $2m$-dimensional *phase space*. A given point $(\mathbf{q}, \mathbf{p})$ of phase space is a *critical* point for the system (34') if and only if $\mathbf{p} = \mathbf{0}$ and $\mathbf{F}(\mathbf{q}, \mathbf{0}) = \mathbf{0}$, so that $\mathbf{q}$ is an *equilibrium* point of the dynamical system (34).

The dynamical system (34) is called *conservative* when $F_i(\mathbf{q}, \dot{\mathbf{q}}) = -\partial V/\partial q_i$ for a suitable *potential energy* function $V(\mathbf{q})$. For any conservative system, the total energy function $E(\mathbf{q}, \mathbf{p}) = (\Sigma\, m_i p_i^2/2) + V(\mathbf{q})$ is an integral of the system (34). Moreover, the point $(\mathbf{a}, 0)$ is a critical point of (34') if and only if the gradient $\nabla V(\mathbf{a}) = \mathbf{0}$, so that the potential energy function has a stationary value.† The point is neutrally stable if the potential energy has a (strict) local minimum at $\mathbf{q} = \mathbf{a}$; it is never strictly stable.

Thus, consider the simple pendulum of Example 5, §3, with $k = 1$. For $c = -2$, the "solution curve" reduces to a set of isolated critical points $v = 0$, $\theta = \pm 2n\pi$. If $|c| < 2$, then $|\theta - 2n\pi| < \theta_0$, where $\theta_0 < \pi$ is the smallest positive angle such that $\cos \theta_0 = -c/2$. Therefore, the solution curves for $-2 < c < 2$ are closed curves (loops) surrounding the origin or any one of the critical points for $\theta = \pm 2n\pi$. As $c \to -2$, these loops tend to the origin. Consequently, the origin and its translates $\theta = \pm 2n\pi$, $v = 0$ are neutrally *stable*. For $c = 2$, we have the separatrix curve defined by $v = \pm 2 \cos (\theta/2)$. From the first of equations (11'), it is seen that the direction of the motion is from $-\pi$ to π for $v > 0$ and from π to $-\pi$ for $v < 0$. Therefore, the critical points $v = 0$, $\theta = \pm(2n + 1)\pi$ are *unstable*. These unstable critical points occur when the pendulum is balanced vertically above the point of support.

EXERCISES C

1. For the following DEs, determine the stability and type of the solution curves in the phase plane, sketching typical curves in each case:
   (a) $\ddot{u} + u = 0$ (b) $\ddot{u} + \dot{u} + u = 0$ (c) $\ddot{u} - \dot{u} + u = 0$
   (d) $\ddot{u} + \dot{u} - u = 0$ (c) $\ddot{u} + 2\dot{u} + u = 0$ (f) $\ddot{u} + 4\dot{u} + u = 0$.

2. For which of the following is $x = 0$ a stable critical point:

$$\dot{x} = x^2, \qquad \dot{x} = -x^2, \qquad \dot{x} = x^3, \qquad \dot{x} = -x^3$$

3. Determine conditions on the coefficients a, b, c, d of (13) necessary and sufficient for neutral stability.

4. Show that $\dot{x} = X(x)$ has a strictly stable critical point at $x = 0$ if $X(0) = 0$ and $X'(0) < 0$. [HINT: x^2 is a Liapunov function.]

5. Show that the plane autonomous system $\dot{x} = x - y$, $\dot{y} = 4x^2 + 2y^2 - 6$ has critical points at $(1, 1)$ and $(-1, -1)$, both of them unstable.

6. Show that the system $dx/dt = \ln (1 + x + 2y)$, $dy/dt = (x/2) - y + (x^2/2)$ has an unstable critical point at the origin.

*7. Is the system (*) of Ex. B5 strictly stable, neutrally stable, or unstable at $(0, 0, 0)$?

---

† Courant and John, Vol. 2, p. 326. Points where the value of a function is stationary are also often called "critical points."

8. Show that the autonomous system

$$\frac{dx}{dt} = \left(\frac{1}{2}\right)e^{2x} \sin 3y + 2 \sin x \cos x + e^{z} - 1$$

$$\frac{dy}{dt} = \sin (2x + 3y), \qquad \frac{dz}{dt} = \tan (2x + z)$$

has an unstable critical point at $x = y = z = 0$.

9. Show that the DE $\ddot{x} + x \sin (1/x) = 0$ is neutrally stable at $x = 0$.

*10. Let $S = S(\mathbf{x})$ be a function of class $\mathcal{C}^1$ in some neighborhood of the origin, such that $S(\mathbf{0}) = 0$, while $S(\mathbf{x}) > 0$ if $0 < |\mathbf{x}| < \epsilon$ for some $\epsilon > 0$.
   (a) Prove that the system $d\mathbf{x}/dt = \mathbf{X}(\mathbf{x})$ is *unstable* at the origin if $\Sigma X_i \partial S/\partial x_i > 0$ there and is strictly stable if $\Sigma X_i \partial S/\partial x_i < 0$ there.
   (b) Derive from (a) a stability criterion for the autonomous nth order DE

$$d^n x/dt^n = \Phi(x, dx/dt, \ldots, d^{n-1}x/dt^{n-1})$$

11. Show that the integral curves in Example 6 are the semicircles

$$x^2 + y^2 = \pm 2ax \qquad \text{in} \qquad D_1 \cup D_2$$

and of the form $x(x^2 - 2|x| + y^2) = $ constant in $D_3 \cup D_4$.

8   METHOD OF LIAPUNOV

In studying a critical point of an autonomous system (3), we can assume without loss of generality that it is at the origin. For $\mathbf{X}(\mathbf{x})$ of class $\mathcal{C}^2$, we can therefore rewrite the system as

$$x_i'(t) = \sum_{j=1}^{n} a_{ij}x_j + R_i(\mathbf{x}), \qquad i = 1, \ldots, n$$

where the R_i are infinitesimals of the second order. This suggests that the behavior of solutions near the critical point will be like that of solutions of the associated *linearized* system (12). We now show, at least for $n = 2$, that this is indeed true as regards *strict stability*.

THEOREM 6.   *If the critical point* $(0, 0)$ *of the linear plane autonomous system* (12) *is strictly stable, then so is that of the perturbed system*

(35) $\dot{x} = ax + by + \xi(x,y), \qquad \dot{y} = cx + dy + \eta(x,y)$

provided† that $|\xi(x,y)| + |\eta(x,y)| = O(x^2 + y^2)$.

---

† The symbol $O(x^2 + y^2)$ stands for a function bounded by $M(x^2 + y^2)$ for some constant M and all sufficiently small x,y.

Idea of Proof.   The proof is based on a simple geometrical idea due to Liapunov. Let $E(x,y)$ be any function having a strict local minimum at the origin. For a small positive C, the level curves $E(x,y) = E(0, 0) + C$ constitute a family of small concentric closed loops, roughly elliptical in shape, enclosing the origin. Now, examine the direction of the vector field defined by (35) on these small loops. Intuition suggests that the critical point will be strictly stable whenever, for all small enough loops, the vector field points inward. For this implies that any trajectory which once crosses a loop is forever trapped inside it, because, to get outside, the trajectory would have to cross the loop in an outward direction. At such a crossing point, the vector could not point inward, giving a contradiction.

To make the preceding intuitive argument precise, we define a *Liapunov function* for a critical point $\mathbf{a}$ of an autonomous system $\mathbf{x}'(t) = \mathbf{X}(\mathbf{x})$ to be a function $E(\mathbf{x})$ that assumes its minimum value at $\mathbf{a}$ and satisfies

$$\sum X_i(\mathbf{x}) \, \partial E/\partial x_i < 0 \qquad \text{for·all} \qquad \mathbf{x} \neq \mathbf{a}$$

Since $\dot{E} = dE/dt = \Sigma \, x_i'(t) \, \partial E/\partial x = \Sigma X_i(\mathbf{x}) \, \partial E/\partial x_i$ for any solution $\mathbf{x}(t)$, this implies that E is decreasing along any trajectory $\mathbf{x}(t)$.

To construct a Liapunov function for the critical point $(0, 0)$ under the hypotheses of Theorem 6, we consider, in turn, each of the three canonical forms (26a) through (26c) derived in §5. Since the definition of stability is invariant under linear transformations of coordinates, it suffices to consider these three cases.

For the linear terms in (27), the necessary calculations are simple. The Liapunov function can be taken as a positive definite quadratic function $E = \alpha x^2 + 2\beta xy + \gamma y^2$, $\alpha > 0$, $\alpha\gamma > \beta^2$, whose level curves are a family of concentric coaxial ellipses. We will show, by considering cases separately, that $\dot{E} \leq -kE$ for some positive constant k.

In (26a), the Liapunov function $E = x^2 + y^2$ satisfies

$$\dot{E} = 2(x\dot{x} + y\dot{y}) = 2\mu E$$

In the strictly stable case, $\mu < 0$. In (26b), the same Liapunov function satisfies $\dot{E} = 2(x\dot{x} + y\dot{y}) = 2\mu_1 x^2 + 2\mu_2 y^2 \leq 2\mu_1 E$, in the strictly stable case $0 > \mu_1 \geq \mu_2$. (By allowing equality, we also take care of the exceptional case of a star point.) In (26c), the Liapunov function $E = x^2 + a^2 y^2$ satisfies, in the strictly stable case $a < 0$

$$\dot{E} = 2(x\dot{x} + a^2 y\dot{y}) = 2(ax^2 + a^2 xy + a^3 y^2) = aE + a(x + ay)^2 \leq aE$$

Hence, in the three possible cases of strict linear stability, we have $\dot{E} = 2\mu E$, $\dot{E} \leq 2\mu_1 E$, or $\dot{E} \leq aE$, where the coefficient on the right side is negative. It follows that $\dot{E} \leq -kE$ for some $k > 0$, in every case. Since the quadratic function $E(x, y)$ is positive definite, we conclude that $E(x, y) \geq K(x^2 + y^2)$ for some constant $K > 0$.

We now consider the nonlinear system (27). Since $E_x = \partial E/\partial x$ and $E_y = \partial E/\partial y$ are linear functions of x and y, we have

$$|E_x\xi + E_y\eta| = O(|x| + |y|)(x^2 + y^2)$$

Hence, for some $\epsilon > 0$, $E(x, y) \leq \epsilon$ implies $|E_x\xi + E_y\eta| \leq kE(x, y)/2$.

Now let $[x(t), y(t)]$ be a trajectory of (27) such that $E(x(t_0), y(t_0)) \leq \epsilon$. Along this trajectory, for $t \geq t_0$, $E(t) = E[x(t), y(t)]$ will satisfy

$$\dot{E}(t) \leq -kE + |E_x\xi + E_y\eta| \leq -kE(t)/2$$

By Theorem 7 of Ch. 1, it follows that $E(t) \leq E(t_0)\exp[-k(t-t_0)/2]$. Hence, $E(t)$ approaches zero exponentially. Since $E(x, y) \geq K(x^2 + y^2)$, it follows at once that the trajectory tends to the origin.

Using similar arguments, we can prove the following classic generalization of Theorem 6, which we state without proof.

POINCARÉ-LIAPUNOV THEOREM.   *If the critical point **0** of the linear autonomous system $\mathbf{x}'(t) = A\mathbf{x}$ is strictly stable, then so is that of the perturbed system*

$$x_i'(t) = \sum a_{ij}x_j + \xi(\mathbf{x})$$

provided that $|\xi(\mathbf{x})| = O(|\mathbf{x}|^2)$.

9   UNDAMPED NONLINEAR OSCILLATIONS

The classification made in §5 covers linear oscillators near equilibrium points, which correspond to critical points in the phase plane. We will now study the nonlinear oscillations of a particle with one degree of freedom, about a position of stable equilibrium. The case of undamped (i.e., frictionless) oscillations will be treated first. This case is described by the second-order DE

(36) $$\ddot{x} + q(x) = 0$$

which can be imagined as describing the motion of a particle in a conservative force field. This DE is equivalent (for $v = \dot{x}$) to the first-order quasilinear DE $v\, dv/dx + q(x) = 0$, or, in the phase plane, to the plane autonomous system

(37) $$dx/dt = v, \qquad dv/dt = -q(x)$$

By a translation of coordinates, we can move any position of equilibrium to $x = 0$; hence we can let $q(0) = 0$. If equilibrium is stable, then the "restoring force" $q(x)$ must act in a direction opposite to the displacement x, at least for small displacements. Therefore we assume that $xq(x) > 0$ for $x \neq 0$ sufficiently small. This will make the system (37) have an isolated critical point at $x = v = 0$, as in the case of the simple pendulum (Example 5, §3).

The key to the analysis of systems (36) is the *potential energy* integral

(38)
$$V(x) = \int_0^x q(\xi)\, d\xi$$

Since $xq(x) > 0$, the function $V(x)$ is increasing when x is positive and decreasing when x is negative; it has a local minimum $V(0) = 0$ at $x = 0$. Differentiating the *total energy*

(39)
$$E(x,v) = \frac{v^2}{2} + V(x)$$

with respect to t, we get $\dot{E} = \dot{x}[\ddot{x} + q(x)] = 0$; hence $E(x, v)$ is a constant along any trajectory in the phase plane. Therefore $E(x, v)$ is an *integral* of the system (37), and of its first-order equivalent $dv/dx = -q(x)/v$. Although the energy integral $E(x,v)$ has a strict local minimum at $(0, 0)$, the following result is of more interest to us now.

THEOREM 7.   *If $q \in \mathcal{C}^1$ and if $xq(x) > 0$ for small nonzero x, then the critical point $(0, 0)$ of the system (36) is a vortex point.*†

Proof.   For any given positive constant E, the locus $v^2/2 + V(x) = E$ is an integral curve, where $V(0) = 0$ and $V(x)$ increases with $|x|$, on both sides of $x = 0$. These curves are symmetric under reflection $(x, v) \mapsto (x, -v)$ in the x-axis; they slope down with slope $-q(x)/v$ in the first and third quadrants and up in the second and fourth quadrants. For any given small value of E, the function $E - V(x)$ has a maximum at $x = 0$ and decreases monotonically on both sides, crossing zero at points $x = -B$ and $x = A$, where B and A are small and positive. Hence each locus $v^2 = 2[E - V(x)]$ is a simple closed curve, symmetric about the x-axis.

As the energy parameter E decreases, so does $|v| = \sqrt{2[E - V(x)]}$; thus, the simple closed curves defined by the trajectories of (32) shrink monotonically toward the origin as $E \downarrow 0$. In fact, consider the new coordinates (u, v), defined by $u = \pm\sqrt{2V(x)}$, according as x is positive or negative. The transformation $(x, v) \to (u, v)$ is of class $\mathcal{C}^1$ with a nonvanishing Jacobian near $(0, 0)$, if $q'(0)$ exists and is positive. Hence the integral curves of (37) resemble a distorted family of circles $u^2 + v^2 = 2E$.

10 SOFT AND HARD SPRINGS

The most familiar special case of (36) is the undamped linear oscillator

(40) $\ddot{x} + qx = 0, \qquad q = k^2 > 0$

---

† As in the linear case, a critical point of a plane autonomous system is called a *vortex point* when nearby solution curves are concentric simple closed curves.

for which $q(x) = k^2 x$. The general solution of (40) is the function $x = A \cos [k(t - t_0)]$, representing an *oscillation* of *amplitude A, frequency $k/2\pi$* (period $2\pi/k$), and *phase t_0*.

In other cases, the DE (36) can be imagined as determining the motion of a unit mass, attached to an *elastic spring* that opposes a displacement x by a force $q(x)$, independent of the velocity $\dot{x}$. The ratio $h(x) = q(x)/x$ is called the *stiffness* of the spring; it is bounded for bounded x if $q \in \mathcal{C}^1$ and $q(0) = 0$. The case (40) of a linear spring is the case of constant stiffness (Hooke's Law). For linear springs, the formulas of the last paragraph show that the frequency $f = k/2\pi$ is proportional to the square root of the stiffness k^2 and is independent of the amplitude. We will now show that, for *nonlinear* springs, the frequency f still increases with the stiffness but is amplitude-dependent in general.

Indeed, the force law (36) implies that $\dot{x} = v = \sqrt{2[E - V(x)]}$. Hence, if the limits of oscillation [i.e., the smallest negative and positive roots of the equation $V(x) = E$] are $x = -B$ and $x = A$, the period T of the complete oscillation is

(41)
$$T = 2 \int_{-B}^{A} \frac{dx}{\sqrt{2[E - V(x)]}}$$

The integral (41) is improper, but it converges, provided that $q(x)$ does not vanish at $-B$ or A; hence it converges for all sufficiently small amplitudes if $q \in \mathcal{C}^1$ in the stable case $h(0) > 0$.

We now compare the periods T and T_1 of the oscillation of two springs, having stiffness $h(x)$ and $h_1(x) \geq h(x)$, and the same limits of oscillation $-B$ and A. By (39), $E = \int_0^A q(u)\, du = V(A)$; hence, $E - V(x) = \int_x^A q(u)\, du$. From the stiffness inequality $h_1(x) \geq h(x)$ assumed, therefore, we obtain,

$$E - V(x) = \int_x^A q(u)\, du \leq \int_x^A q_1(u)\, du = E - V_1(x), \qquad 0 \leq x \leq A$$

Reversing the sign of x, we get the same inequality for $-B \leq x \leq 0$. Substituting into (41), we obtain $T \geq T_1$. We thus get the following comparison theorem.

THEOREM 8.   *For any two oscillations having the same span $[-B, A]$, the period becomes shorter and the frequency greater as the stiffness $q(x)x$ increases in (36).*

Springs for which $h(x) = h(-x)$ are called symmetric; this makes $q(-x) = -q(x)$ and $V(-x) = V(x)$, so that $B = A$ in the preceding formulas: symmetric springs oscillate symmetrically about their equilibrium position. Hence, for symmetric springs, the phrase "span $[-B, A]$" in Theorem 8 can be replaced by "amplitude A."

For any symmetric spring, $h'(0) = 0$; if $h''(0)$ is positive, so that $h(x)$ increases with $|x|$, the spring is said to be "hard"; if $h''(0)$ is negative, so that $h(x)$ decreases as $|x|$ increases, it is said to be "soft." Thus the simple pendulum of Example 5, §3, acts as a "soft" spring. We now show that the period of oscillation is amplitude-dependent, at least for symmetric hard and soft springs.

THEOREM 9. *The period of a hard symmetric spring decreases, whereas the period of a soft symmetric spring increases as the amplitude of oscillation increases.*

Proof.   The period is given by (41); it suffices to compare the periods of quarter-oscillations, say from 0 to A and from 0 to A_1, with $A_1 > A$. We write $A_1 = cA$, with $c > 1$. To study the period p_1 of the quarter-oscillation from 0 to $cA = A_1$, we let $x = cy$ in (36). The equivalent DE for y is

$$(42) \qquad\qquad d^2y/dt^2 + yh(cy) = 0$$

where $h(x) = q(x)/x$. The oscillation of amplitude cA for (36) corresponds to the oscillation of amplitude A for (42); and, since the independent variable t is unchanged, the periods of oscillation are the same for both. Therefore, it suffices to compare the quarter periods p and p_1 for amplitude A for the two springs (36) and (42), respectively. Using Theorem 8, we find that, for $y > 0$, if $yh(cy) \geq q(y) = yh(y)$, that is, if $h(cy) \geq h(y)$ for $c > 1$ (hard spring), then we have $p_1 \leq p$, and so the period decreases as the amplitude increases. Soft springs can be treated similarly.

EXERCISES D

1. (a) Show that the integral curves of $\ddot{x} - x + x^3 = 0$ in the phase plane are the curves $v^2 - x^2 + x^4/2 = C$.
    (b) Sketch these curves.
    (c) Show that the autonomous system defining these curves has a saddle point at $(0, 0)$ and vortex points at $(\pm 1, 0)$.

2. Duffing's equation without forcing term is $\ddot{x} + qx + rx^3 = 0$. Show that, for oscillations of small but finite half-amplitude L, the period T is

$$T = 4\sqrt{2} \int_0^{\pi/2} \frac{d\theta}{\sqrt{2q + rL^2(1 + \sin^2\theta)}}$$

Verify Theorem 9 in this special case as a corollary.

3. (a) Draw sample trajectories of the DE $\ddot{x} = 2x^3$ in the phase plane, including $\dot{x} = \pm x^2$.
    (b) Show that $x = 0$ is the only solution of this DE defined for all $t \in (-\infty, \infty)$.

*4. Show that if $q(0) = 0$ and $q'(0) < 0$ in (36), the origin is a saddle-point in the phase plane.

5. Discuss the dependence on the sign of the constant μ, of the critical point at the origin of the system

$$u = -v + \mu u^3, \qquad v = u + \mu v^3$$

6. (a) Show that the trajectories of $\ddot{x} + q(x) = 0$ in the phase plane are convex closed curves if $q(x)$ is an increasing function with $q(0) = 0$.
    (b) Is the converse true?

7. Show that, if $\dot{x} = ax + by$, $\dot{y} = cx + dy$ is unstable at the origin, and if

$$X(x, y) = ax + by + O(x^2 + y^2), \qquad Y(x, y) = cx + dy + O(x^2 + y^2)$$

where $ad \neq bc$ and $\Delta \neq 0$, then the system $\dot{x} = X(x, y)$, $\dot{y} = Y(x, y)$ is also unstable there.

8. The equation of a falling stone in air satisfies approximately the DE

$$\ddot{x} = g - \frac{g\dot{x}|\dot{x}|}{v^2}$$

where the constant v is the "terminal velocity." Sketch the integral curves of this DE in the phase plane, and interpret them physically.

9. Show that the plane autonomous system $\dot{x} = y - x^3$, $\dot{y} = -x^3$ is stable, though its linearization is unstable. [HINT: Show that $x^4 + 2y^2$ is a Liapunov function.]

*10. Show that for the analytic plane autonomous system

$$\dot{x} = 2x^3y, \qquad \dot{y} = x^2y^2 - x^4 - y^{11}$$

the origin is an unstable critical point that is asymptotically stable. [HINT: Study Example 6. To prove instability, show that the ellipse $4y^2 = x - x^2$ cannot be crossed from the left in the first quadrant.]

11 DAMPED NONLINEAR OSCILLATIONS

The equation of motion for a particle of mass m having an equilibrium point at $x = 0$, in the presence of a restoring force $mq(x)$ and a friction force equal to $f(x, v) = mvp(x, v)$, is

(43) $\quad \ddot{x} + p(x, \dot{x})\dot{x} + q(x) = 0, \qquad q(x) = xh(x), \qquad p \in \mathcal{C}^2, \qquad h \in \mathcal{C}^1$

When $h(0)$ is positive, the equilibrium point is called *statically stable,* because the restoring force tends to restore equilibrium under static conditions (when $v = 0$). The conservative system (36) obtained from any statically stable system (43) by omitting the friction term $\dot{x}p(x, \dot{x})$, is neutrally stable by Theorem 7.

The differential equation (43) has a very simple interpretation in the phase plane, as

(44) $$\frac{dx}{dt} = v, \qquad \frac{dv}{dt} = -vp(x, v) - q(x)$$

The critical points of the system (44) are all on the x-axis, where $\dot{x} = v = 0$; they are the equilibrium points $(x, 0)$ where $q(x) = 0$ in (43). Since in (43), $q(0) = 0h(0) = 0$ the origin is always a critical point of (44); unless $h(0)$ changes sign, there is no other.

We shall consider only the case $h(x) > 0$ of static stability in the large, which is the case of greatest interest for applications. For simplicity, we will also assume that $p(0, 0) \neq 0$.

Under these assumptions, the origin is the only critical point of (44). Moreover the direction field, whose slope is

$$\frac{dv}{dx} = \tau(x, v) = -p(x, v) - \frac{q(x)}{v}$$

points to the right in the upper half-plane, where $\dot{x} = v > 0$, and to the left in the lower half-plane. On the x-axis, the solution curves have finite curvature $q(x)$; they cut it vertically downward on the positive x-axis and vertically upward on the negative x-axis. Thus, the solution curves have a general clockwise orientation.

Conversely, any continuous oriented direction field with the properties specified represents a DE of the form (43) in the phase plane. From this it is clear that the behavior of the solutions of the DEs of the form (43) can be extremely varied in the large (see §13). However, the local possibilities are limited.

THEOREM 10.   *If p and h are of class $\mathcal{C}^l$ in (43), and if p (0, 0) and $h(0)$ are positive, then the origin is a strictly stable critical point of (44).*

Proof.   Under the hypotheses of Theorem 10, we have that $p > 0$ near the critical point (0, 0), and we can write (44) as

$$dx/dt = v, \qquad dv/dt = -h(0)x - p_0 v + O(x^2 + y^2), \qquad p_0 = p(0, 0)$$

An easy computation shows that the linearization of the system (44) has the secular equation $\lambda^2 + p_0\lambda + h(0) = 0$. Since this quadratic polynomial has positive coefficients, it is of stable type (Ch. 3, §5). Hence, by Theorem 6, the origin is a strictly stable critical point of (43) when the *damping* factor $p(0, 0)$ is *positive*. The equilibrium point $x = 0$ of (43) is then said to be *dynamically stable*: the solution curves tend to the origin in the vicinity of the origin.

When $p(0, 0)$ is negative, the system is said to be *negatively damped,* and the equilibrium point to be *dynamically unstable.* Since (43) can be rewritten as

(45)
$$\frac{d^2x}{d(-t)^2} + p\left(x, \frac{-dx}{d(-t)}\right)\frac{-dx}{d(-t)} + xh(x) = 0$$

we see that the substitutions $t \to -t$, $x \to x$, $v \to -v$, of time reversal, reverse the sign of $p(x, \dot{x})$ but do not affect (43) otherwise. Hence, if $p(0, 0) < 0$, all solution curves of (44) spiral outward near the origin.

*12   LIMIT CYCLES

We now come to a major difference between nonlinear oscillations and linear oscillations. When a linear oscillator is negatively damped, the amplitude of oscillation always increases exponentially without limit. In contrast the amplitude of oscillation of a negatively damped, statically stable, nonlinear oscillator

commonly tends to a finite limit. The limiting periodic oscillation of finite amplitude so approached is called a *limit cycle*.

The simplest DE that gives rise to a limit cycle is the *Rayleigh equation*,

$$(46) \qquad \ddot{x} - \mu(1 - \dot{x}^2)\dot{x} + x = 0, \qquad \mu > 0$$

The characteristic feature of this DE is the fact that the damping is negative for
small $\dot{x}$ and positive for large $\dot{x}$. Hence it tends to increase the amplitude of small
oscillations and to decrease the amplitude of large oscillations. Between these
two types of motions, there is an oscillation of contant amplitude, a limit cycle.

If we differentiate the Rayleigh equation (46) and set $y = \dot{x}\sqrt{3}$, we obtain
the *van der Pol equation*

$$(47) \qquad \ddot{y} - \mu(1 - y^2)\dot{y} + y = 0, \qquad \mu > 0$$

This DE arises in the study of vacuum tubes. The sign of the damping term
depends on the magnitude of the displacement y. The remarks about the Rayleigh equation made above apply also to the van der Pol equation.

As stated in §11, negatively damped nonlinear oscillators can give rise to a
great variety of qualitatively different solution curve configurations in the phase
plane. For any particular DE of the form (43), such as the Rayleigh or van der
Pol equation with given μ, one can usually determine the qualitative behavior of
solutions by integrating the DE

$$v \, dv + [vp(x,v) + q(x)] \, dx = 0$$

graphically (Ch. 1, §8). More accurate results can be had by use of numerical
integration. With modern computing machines, using the techniques to be
described in Ch. 8, it is a routine operation to obtain such a family of solution
curves. Figure 5.8 depicts sample integral curves for the van der Pol equation
with $\mu = 0.1$, $\mu = 1$, and $\mu = 10$ so obtained.

Liénard Equation.   General criteria are also available which determine the
qualitative behavior of the oscillations directly from that of the coefficient-functions. Such criteria are especially useful for DEs depending on parameters,
because graphical integration then becomes very tedious. They are available for
DEs of the form

$$\ddot{x} + f(x)\dot{x} + q(x) = 0 \qquad \text{(Liénard equation)} \qquad (48)$$

The van der Pol equation is a Liénard equation; moreover, it is *symmetric* in the
sense that $-x(t)$ is a solution if $x(t)$ is a solution. This holds whenever $q(-x) = -q(x)$ is odd and $f(-x) = f(x)$ is even.

One can prove the *existence* of limit cycles for a wide class of Liénard equations; we can even prove that *every* nontrivial solution is either a limit cycle, or
a spiral that tends toward a limit cycle as $t \to +\infty$. This is true if: (i) $xq(x) > 0$

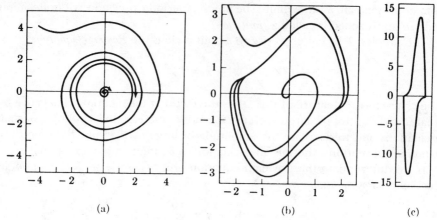

Figure 5.8   Van der Pol equation.

for $x \neq 0$; (ii) $f(x)$ in (48) is negative in an interval $a < x < b$ containing the origin and positive outside this interval, and

$$(49) \qquad \int_0^\infty f(x)\, dx = \int_{-\infty}^0 f(x)\, dx = +\infty$$

We sketch the proof.†

In the xv-plane, solution curves satisfy

$$(50) \qquad \frac{dv}{dx} + f(x) + \frac{q(x)}{v} = 0 \qquad \text{if} \qquad v \neq 0$$

It follows, since $xq(x) > 0$, that they can cross the x-axis only downward if $x > 0$, and upward if $x < 0$. Also by (50), between successive crossings of the x-axis, $v(x)$ is a bounded single-valued function, decreasing in magnitude if $x > b$ in the upper half-plane, and if $x < a$ in the lower half-plane.

Now, consider the *Liénard function*

$$(51) \qquad E(x, v) = \tfrac{1}{2}[v + F(x)]^2 + U(x) \qquad F(x) = \int_0^x f(x)\, dx$$

$$U(x) = \int_0^x q(x)\, dx \geq 0$$

A straightforward calculation gives $dE/dt = -q(x)F(x)$, where, by (49), $|F(x)|$ becomes positively infinite as $|x| \to \infty$. For sufficiently large $|x|$, since $xq(x) > 0$, dE/dt is identically negative. Since the set of all (x, v) for which $E(x, v) \leq E_0$,

---

† For a complete proof, see Lefschetz, p. 267, or Stoker, Appendices III and IV.

any finite constant, is contained in the bounded strip $|v + F(x)| \leq \sqrt{2E_0}$, it also follows that solution curves can stay in one half-plane ($v > 0$ or $v < 0$) for only a finite distance (and time); if nontrivial, they must cut the x-axis infinitely often.

Let $x_0, x_1, x_2, \ldots$ be successive zero-crossings; we can assume that $x_{2n} > 0$ and $x_{2n+1} < 0$ without loss of generality. If $x_2 = x_0$, then (by uniqueness) $x_{2n} = x_0$ and $x_{2n+1} = x_1$ for all $n > 0$; the solution curve is a limit cycle. Likewise, if $x_2 > x_0$, then we find that $x_{2n} > x_{2n-2}$ and $x_{2n+1} < x_{2n-1}$ for all $n > 0$ and the solution curve spirals outward as t increases. Similarly, if $x_2 < x_0$, then the solution curve must spiral inward.

Finally, since $f(0) < 0$, Theorem 6 applies if $f, g \in \mathcal{C}^1$: solution curves near the origin must spiral outward. Also, we know that

$$E(x_{2n}, 0) - E(x_{2n-1}, 0) = - \int_{x_{2n-1}}^{x_{2n}} \frac{q(x)F(x)}{r(x)}\, dx$$

and the definite integral is negative for sufficiently large oscillations since $xq(x) > 0$ and $F(\pm\infty) = \pm\infty$, by (49). Hence, every solution curve sufficiently far from the origin must spiral inward. Therefore, every oscillation of sufficiently large initial amplitude must spiral inward toward a limit cycle of maximum amplitude. Similarly, every oscillation of sufficiently small initial amplitude must spiral outward to a smallest limit cycle.

For the Rayleigh and van der Pol equations, these limit cycles are the same. Therefore, every nontrivial solution tends to a unique limit circle, which is stable. The preceding result holds under much more general conditions. We quote one set of such conditions without proof.

LEVINSON–SMITH THEOREM.   *In* (48), *let $q(x) = xh(x)$, where $h(x) > 0$ and let $f(x)$ be negative in an interval (a, b) containing the origin and positive outside this interval. Let $q(-x) = -q(x)$, $f(-x) = f(x)$, and let* (49) *hold. Then* (48) *has a unique stable limit cycle in the phase plane, toward which every nontrivial integral curve tends.*

EXERCISES E

1. Show that any DE

$$\ddot{x} + (px^2 - q)\dot{x} + rx = 0$$

   where q and r are positive constants, can be reduced to the van der Pol DE by a change of dependent and independent variables.

2. (a) Show that the autonomous plane system

$$\dot{u} = u - v - u^3 - uv^2, \qquad \dot{v} = u + v - v^3 - u^2v$$

   has a unique critical point, which is unstable, and a unique limit cycle.

   (b) Discuss the stability of the related system

$$\dot{u} = -u - v + u^3 + uv^2, \qquad \dot{v} = u - v + v^3 + u^2v$$

   with special reference to oscillations of very small and very large amplitude.

3. In the DE $\ddot{x} + q(x) = 0$, let $V(x) = \int_0^x q(u)\,du$, and let q be a continuous function satisfying a Lipschitz condition. Show that, if $V(x_1) = V(x_2)$ and $V(x) < V(x_1)$ for $x_1 < x < x_2$, the equivalent autonomous system (44) has a periodic solution passing through the points $(x_1, 0)$ and $(x_2, 0)$.

4. Show that the plane autonomous system

$$\dot{r} = \frac{(r^3 - r)}{100}, \qquad \dot{\theta} = 1 \qquad \text{(polar coordinates)}$$

has just one limit cycle.

5. Discuss the limit cycles of the system

$$\dot{r} = \left(\frac{r}{100}\right) \sin\left(\frac{1}{r}\right), \qquad \dot{\theta} = 1$$

6. Prove in detail that, for $\mu = 1$ and the initial condition $x(0) = 10$, $\dot{x}(0) = 0$, the amplitude of successive oscillations decreases in the Rayleigh DE (46).

7. Answer the same question for the van der Pol DE, if $y(0) = 10$, $\dot{y}(0) = 0$.

8. Sketch the integral curves of the van der Pol DE in the phase plane for $\mu = 100$. [HINT: Most of the time, the integral curves are "relaxation oscillations," near $\dot{y} = 0$ or $y = 1$.]

9. Do the same for the Rayleigh DE.

ADDITIONAL EXERCISES

1. Locate the critical points of the DE $\dot{x} = x(1 - x)(a - x)$, and discuss how their stability or instability varies with a.

2. Do the same for $\dot{x} = x(1 - x)(x - a)$.

3. Show that, in the complex domain, every system $\dot{x} = ax + by$, $\dot{y} = cx + dy$ is linearly equivalent to either $\dot{x} = \lambda x$, $\dot{y} = \mu y$ or $\dot{x} = \lambda x$, $\dot{y} = \lambda y + x$, for suitable λ, μ.

4. Show that, in the *punctured* plane $x^2 + y^2 > 0$, two linear systems (13) are equivalent, provided that their discriminants Δ, Δ' are not zero, and they both have either (a) stable focal points, (b) unstable focal points, (c) vortex points, (d) stable nodal points, (e) unstable nodal points, or (f) saddle-points.

5. Consider the linear autonomous system

$$\dot{x} = a_1 x + a_2 y + a_3 z, \qquad \dot{y} = b_1 x + b_2 y + b_3 z, \qquad \dot{z} = c_1 x + c_2 y + c_3 z$$

(a) Show that the x-component of any solution satisfies

$$\frac{d^3 x}{dt^3} = p_1 \ddot{x} - p_2 \dot{x} + p_3 x, \qquad \text{where} \qquad p_1 = a_1 + b_2 + c_3$$

$$(\sigma) \qquad p_2 = a_1 b_2 - a_2 b_1 + b_2 c_3 - b_3 c_2 + c_3 a_1 - c_1 a_3,$$

$$p_3 = \begin{vmatrix} a_1 & a_2 & a_3 \\ b_1 & b_2 & b_3 \\ c_1 & c_2 & c_3 \end{vmatrix}$$

(b) Show that the *secular equation* (σ) is invariant under any nonsingular linear transformation of the variables x, y, z.

(c) Conversely, show that, if the polynomial $\lambda^3 = p_1\lambda^2 - p_2\lambda + p_3$ has distinct real roots λ_1, λ_2, λ_3, then the given DE is linearly equivalent to $\xi_j = \lambda_j\xi_j (j = 1,2,3)$.

*(d) Work out a set of real canonical forms for the given DE, with respect to linear equivalence, in the general case.

6. Let $\nabla U = \mathbf{X}$, $U = U(x, \sqrt{y^2 + z^2})$ be any axially symmetric gradient field of class $\mathcal{C}^1$. Show that the DE $\dot{x} = \nabla U$ admits as an integral the "stream function" $V = \int r[(\partial U/\partial x)\, dr + (\partial U/\partial r)\, dx]$, the integral being independent of the path.

7. Sketch the integral curves in the phase plane for

   (a) $\ddot{x} + x + \dfrac{A}{x - a} = 0$ (b) $\ddot{x} + \dot{x}/|\dot{x}| = 0$ (Coulomb friction)

   (c) $\ddot{x} + \dot{x}|\dot{x}| + k \sin x = 0$

8. Let $q(x)$ be an increasing function, with $q(0) = 0$ and $q(-L) = -q(L)$; let $Q(x) = [q(x) - q(-x)]/2$. Show that the period of oscillation of half-amplitude L for $\ddot{x} + Q(x) = 0$ is less than that for $\ddot{x} + q(x) = 0$ unless $q(-x) = -q(x)$ for all $x \in [0, L]$.

9. Show that the DEs $\dot{x} = 2x + \sin x$ and $\dot{x} = 2x$ are equivalent on $(-\infty, +\infty)$ but that $\dot{x} = x + x^3$ is not equivalent to $\dot{x} = x$. [HINT: Consider the escape time.]

10. Show that, if $a_1 < a_2 < \cdots < a_n$ and $b_1 < b_2 < \cdots < b_n$, the DEs $\dot{x} = \Pi(x - a_i)$ and $\dot{x} = \Pi(x - b_i)$ are equivalent.

EXISTENCE
AND UNIQUENESS
THEOREMS

1 INTRODUCTION

In the earlier chapters of this book, we have proved a number of theorems establishing the *existence* of solutions of DEs and the *well-posedness* of initial value problems, but always under special hypotheses. In this chapter, we shall study these questions systematically, in the general context of normal systems of first-order DEs, of the form

$$\frac{dx_1}{dt} = X_1(x_1, \ldots, x_n; t)$$

(1)

$$\cdots$$

$$\frac{dx_n}{dt} = X_n(x_1, \ldots, x_n; t)$$

For the most part, we shall restrict attention to the existence and uniqueness of such solutions. But in the later sections, we shall consider more sophisticated questions, such as the *analyticity* of solutions and their dependence on the initial value vector $\mathbf{c} = (c_1, \ldots, c_n)$. We shall prove that, as one might expect, this dependence is differentiable and shall derive *perturbation* formulas that express the relevant partial derivatives explicitly.

The theorems proved in this chapter will include as special cases all the existence, uniqueness, and continuity theorems proved in earlier chapters. In Chapters 7 and 8 to follow, we shall describe and analyze algorithms for effectively *computing* the solutions whose existence is established in this chapter. Those who are willing to assume plausible results, and who are mainly interested in applications, may wish to skip to Chapter 8.

We will continue to make the assumptions of Ch. 5, §1: that the X_i are continuous, real-valued functions of the independent variables $x_1, x_2, \ldots, x_n, t$ in some region $\mathcal{R}$ of interest in $(x_1, \ldots, x_n, t)$-space. We shall also use the vector notation introduced there, rewriting (1) as

(2) $\qquad dx/dt = \mathbf{X}(\mathbf{x}, t) \qquad \text{or} \qquad \mathbf{x}'(t) = \mathbf{X}(\mathbf{x}, t)$

The curve $\mathbf{x}(t)$ in $\mathcal{R}$ defined by any solution of (1) will be called a *solution curve* of the system (1).

Note that we can trivially inflate *any* normal system (1) of n first-order DEs to a normal *autonomous* system of $n + 1$ DEs by the simple device of writing $t = x_{n+1}$. This gives the equivalent system

$$\frac{dx_j}{dt} = X_j(x_1, \ldots, x_n; x_{n+1}), \qquad j = 1, \ldots, n + 1$$

where X_{n+1} is the function 1. However, this does not help to prove the theorems of major interest.

As in Chapter 5, §1, one can view any autonomous system $\mathbf{x}'(t) = \mathbf{X}(\mathbf{x})$ as defining a steady flow in the appropriate region $\mathcal{R}$ of x-space. Although this does not help to prove the theorems of major interest, it does make it easier to visualize their meaning.

Thus, a continuously differentiable function $U(\mathbf{x})$ is called an *invariant* of the autonomous system $\mathbf{x}'(t) = \mathbf{X}(\mathbf{x})$ when $U(\mathbf{x}(t))$ is constant for every solution $\mathbf{x}(t)$ of this vector DE—i.e., when $\Sigma\, X_j(\mathbf{x})\, \partial U/\partial x_j = 0$. This means that each solution curve of (2′) stays on a single level surface $U = \text{const.}$, and thus generalizes the concept of "integral" defined in Chapter 1.

Example 1.   Consider the autonomous system

$$\frac{dx}{dt} = xz, \qquad \frac{dy}{dt} = -yz, \qquad \frac{dz}{dt} = x^2 - y^2$$

It is easy to verify that, when $x = x(t)$ and $y = y(t)$ are solutions, the two functions

$$V(x, y, z) = xy \qquad \text{and} \qquad W(x, y, z) = x^2 + y^2 - z^2$$

satisfy $dV/dt = x\, dy/dt + y\, dx/dt = 0$ and $dW/dt = 0$. Therefore, V and W are integrals of the system. The intersection of two surfaces $V = c_1$ and $W = c_2$ is a solution curve of the system. Thus, every solution curve lies on the intersection of a hyperbolic cylinder $xy = c_1$ and a hyperboloid (or cone) $x^2 + y^2 - z^2 = c_2$.

As a familiar special case, consider also the plane autonomous system $dx/dt = N(x,y)$, $dy/dt = -M(x,y)$ associated with the DE $M(x,y) + N(x,y)y' = 0$, The function $U(x,y)$ is an integral of this system, associated as in Ch. 1, §5, with the integrating factor $\mu(x,y)$, if and only if $\partial U/\partial x = \mu M$ and $\partial U/\partial y = \mu N$.

First-order normal systems (1) provide a standard form to which all normal ordinary DEs and normal systems of DEs can be reduced. For example, one can reduce the solution of a normal nth-order DE to the solution of a system of n first-order normal DEs as follows. Let $u(t)$ be any solution of the given nth order

DE,

$$\frac{d^n u}{dt^n} = F\left(u, \frac{du}{dt}, \frac{d^2 u}{dt^2}, \ldots, \frac{d^{n-1} u}{dt^{n-1}}; t\right)$$

Then the n functions $x_1(t) = u$, $x_2(t) = du/dt$, ..., $x_n(t) = d^{n-1}u/dt^{n-1}$ satisfy the normal first-order system

$$\frac{dx_k}{dt} = x_{k+1}, \qquad k = 1, \ldots, n-1; \qquad \frac{dx_n}{dt} = F(x_1, x_2, \ldots, x_n; t)$$

Conversely, given any solution of the preceding first-order system, the first component $x_1(t)$ will have the other components $x_2, \ldots, x_n$ as its derivatives of orders $1, \ldots, n-1$. Hence, substituting back, $x_1(t)$ will satisfy the given nth-order equation.

In the present chapter, we shall use this standard form to develop a unified theory for the existence and uniqueness of solutions of DEs and systems of DEs of all orders.

2   LIPSCHITZ CONDITION

In order to make use of vector notation for systems of DEs, we recall a few facts about vectors in n-dimensional Euclidean spaces. Addition of two vectors and multiplication of vectors by scalars are defined component-wise, as in the plane and in space. The *length* of a vector $\mathbf{x} = (x_1, x_2, \ldots, x_n)$ is defined as

$$|\mathbf{x}| = (x_1^2 + \cdots + x_n^2)^{1/2}$$

Length satisfies the triangle inequality

$$|\mathbf{x} + \mathbf{y}| \leq |\mathbf{x}| + |\mathbf{y}|$$

The dot product or *inner product* of two vectors is defined as

$$\mathbf{x} \cdot \mathbf{y} = x_1 y_1 + \cdots + x_n y_n$$

and satisfies the Schwarz inequality $|\mathbf{x} \cdot \mathbf{y}| \leq |\mathbf{x}| \cdot |\mathbf{y}|$.

We shall integrate, differentiate, and take limits of vector functions $\mathbf{x}(t)$ of a scalar (real) variable t. All these operations can be carried out component by component, as in vector addition.

For example, the derivative of a vector function

$$\mathbf{x}(t) = (x_1(t), x_2(t), \ldots, x_n(t))$$

is the vector function $\mathbf{x}'(t) = (x_1'(t), x_2'(t), \ldots, x_n'(t))$. The integral $\int_a^b \mathbf{x}(t)\,dt$ is the vector with components $\int_a^b x_1\, t)\,dt, \int_a^b x_2(t)\,dt, \ldots, \int_a^b x_n(t)\,dt$. We shall often make use of the fundamental inequality†

$$(3) \qquad \left| \int_a^b \mathbf{x}(t)\,dt \right| \leq \int_a^b |\mathbf{x}(t)|\,dt$$

A vector field $\mathbf{X}(\mathbf{x})$ is said to be *continuous* when each component X_i of $\mathbf{X}$ is a continuous function of the n variables $x_1, \ldots, x_n$, the components of the vector independent variable $\mathbf{x}$. This is equivalent to the following statement: The vector field $\mathbf{X}(\mathbf{x})$ is continuous at the point (vector) $\mathbf{c}$ whenever, given $\epsilon > 0$, there exists $\delta > 0$ such that, if $|\mathbf{x} - \mathbf{c}| < \delta$, then $|\mathbf{X}(\mathbf{x}) - \mathbf{X}(\mathbf{c})| < \epsilon$. We leave it as an exercise to verify that these definitions are equivalent.

There is no such thing as "the" derivative of a vector field $\mathbf{X}(\mathbf{x})$, but only a "Jacobian matrix" of *partial* derivatives $\partial X_i/\partial x_j$ relative to the different components $x_1, \ldots, x_n$.

The reader who is not accustomed to working with functions of vectors should note the differences between the following types of functions: vector-valued functions of a scalar variable, such as $\mathbf{x}(t) = (x_1(t), x_2(t), \ldots, x_n(t))$; scalar-valued functions of a vector variable, such as $|\mathbf{x}| = \sqrt{x_1^2 + \cdots + x_n^2}$; vector-valued functions of a vector variable such as

$$\mathbf{X}(\mathbf{x}) = (X_1(x_1, x_2, \ldots, x_n), X_2(x_1, \ldots, x_n), \ldots, X_m(x_1, \ldots, x_n))$$

vector-valued function of a vector variable $\mathbf{x}$ and a parameter t, such as $\mathbf{X}(\mathbf{x}, t)$.

A vector-valued function $\mathbf{X}(\mathbf{x})$ of a vector variable is said to be of class $\mathcal{C}^n$ in a given region when each of the component functions $X_i(x_1, \ldots, x_n)$ is of class $\mathcal{C}^n$ there. One can easily extend the definition of a Lipschitz condition to vector-valued functions; as we shall see, this provides a simple sufficient condition for the uniqueness and existence of solutions for normal systems.

DEFINITION. A family of vector fields $\mathbf{X}(\mathbf{x}, t)$ satisfies a *Lipschitz condition* in a region $\mathcal{R}$ of $(\mathbf{x}, t)$-space if and only if, for some Lipschitz constant L,

$$(4) \qquad |\mathbf{X}(\mathbf{x}, t) - \mathbf{X}(\mathbf{y}, t)| \leq L|\mathbf{x} - \mathbf{y}| \qquad \text{if} \qquad (\mathbf{x}, t) \in \mathcal{R} \qquad (\mathbf{y}, t) \in \mathcal{R}$$

Note that both terms on the left side of (4) involve the same value of t.

---

† This inequality is the continuous analog of the triangle inequality

$$|\mathbf{x}^{(1)} + \mathbf{x}^{(2)} + \cdots + \mathbf{x}^{(n)}\| \leq |\mathbf{x}^{(1)}| + |\mathbf{x}^{(2)}| + \cdots + |\mathbf{x}^{(n)}|.$$

It can be obtained from this inequality by recalling the definition of the integral $\int_a^b \mathbf{x}(t)\,dt$ as a limit of Riemann sums, using the triangle inequality for each of the Riemann sums, and passing to the limit on both sides.

LEMMA.   *If* $\mathbf{X}(\mathbf{x}, t)$ *is of class* $\mathcal{C}^1$ *in a bounded closed ("compact") convex†
domain D, then it satisfies a Lipschitz condition there.*

Proof.   Let M be the maximum of all partial derivatives $|\partial X_i/\partial x_j|$ in the closed
domain D. For each component X_i we have, for fixed $\mathbf{x}, \mathbf{y}, t$ and for variable s

$$\frac{d}{ds}\left[X_i(\mathbf{x} + s\mathbf{y}, t)\right] = \sum_{k=1}^{n} \frac{\partial X_i}{\partial x_k}(\mathbf{x} + s\mathbf{y}, t)y_k$$

Hence, by the mean-value theorem applied to the function $X_i(\mathbf{x} + s\mathbf{y}, t)$ of the
variable s on the interval $0 \le s \le 1$, we have

$$X_i(\mathbf{x} + \mathbf{y}, t) - X_i(\mathbf{x}, t) = \sum_{k=1}^{n} \frac{\partial X_i}{\partial x_k}(\mathbf{x} + \sigma_i\mathbf{y}, t)y_k$$

for some σ_i between 0 and 1. Squaring, and applying the Schwarz inequality to
the right side, we obtain

$$|X_i(\mathbf{x} + \mathbf{y}, t) - X_i(\mathbf{x}, t)|^2 \le \left(\sum_{k=1}^{n}\left|\frac{\partial X_i}{\partial x_k}\right|^2\right)\left(\sum_{k=1}^{n}|y_k|^2\right) \le nM^2|\mathbf{y}|^2$$

Consequently, summing over all components i, we have

$$|\mathbf{X}(\mathbf{x} + \mathbf{y}, t) - \mathbf{X}(\mathbf{x}, t)|^2 \le n^2M^2|\mathbf{y}|^2$$

Taking square roots, the Lipschitz condition follows with Lipschitz constant nM.

3 WELL-POSED PROBLEMS

For DEs to be useful in predicting the future behavior of a physical system
from its present state, their solutions must exist, be unique, and depend continuously on their initial values. As stated in Ch. 1, §9, an initial-value problem is
said to be *well-posed* when these conditions are satisfied. We now show that, if $\mathbf{X}$
satisfies a Lipschitz condition, the vector DE (2) defines a well-posed (or "well-
set") initial-value problem.

We begin by proving uniqueness.

THEOREM 1 (UNIQUENESS THEOREM).   *If the vector fields* $\mathbf{X}(\mathbf{x}, t)$ *satisfy a
Lipschitz condition (4) in a domain* $\mathcal{R}$, *there is at most one solution* $\mathbf{x}(t)$ *of the vector
DE (2) that satisfies a given initial condition* $\mathbf{x}(a) = \mathbf{c}$ *in* $\mathcal{R}$.

---

† A set S in n-space is *convex* when the segment joining any two points of the set S lies entirely within
S. This definition applies both to closed domains D and open regions $\mathcal{R}$.

The proof of this theorem parallels that of Theorem 5 of Ch. 1. We show that, if $\mathbf{x}(t)$ and $\mathbf{y}(t)$ are both solutions of (2) and if they are equal for one value of t, say $t = a$, it follows that $\mathbf{x}(t) \equiv \mathbf{y}(t)$ in any domain in which a Lipschitz condition is satisfied.

Consider the square of the n-dimensional distance between the two vectors $\mathbf{x}(t)$ and $\mathbf{y}(t)$. By definition, this is

$$\sigma(t) = \Sigma \, [x_k(t) - y_k(t)]^2 = |\mathbf{x}(t) - \mathbf{y}(t)|^2 \geq 0$$

Differentiating $\sigma(t)$, and using the fact that $\mathbf{x}$ and $\mathbf{y}$ are solutions of the normal system (2), we get

$$\sigma'(t) = 2 \, \Sigma \, [x_k(t) - y_k(t)][X_k(\mathbf{x}(t), t) - X_k(\mathbf{y}(t), t)]$$
$$= 2[\mathbf{x}(t) - \mathbf{y}(t)] \cdot [\mathbf{X}(\mathbf{x}(t), t) - \mathbf{X}(\mathbf{y}(t), t)]$$

By the Schwarz inequality, therefore, we have

$$\sigma'(t) \leq |\sigma'(t)| = 2|(\mathbf{x} - \mathbf{y}) \cdot (\mathbf{X}(\mathbf{x}, t) - \mathbf{X}(\mathbf{y}, t))|$$
$$\leq 2|\mathbf{x} - \mathbf{y}| \cdot |\mathbf{X}(\mathbf{x}, t) - \mathbf{X}(\mathbf{y}, t)| \leq 2L|\mathbf{x} - \mathbf{y}|^2 = 2L\sigma(t)$$

By the result of Lemma 2 of Ch. 1, §10, it follows that if $\mathbf{x}(a) = \mathbf{y}(a)$, that is, if $\sigma(a) = 0$, then $\sigma(t) \equiv 0$ [that is, $|\mathbf{x}(t) - \mathbf{y}(t)|^2 \equiv 0$] for all $t \geq a$.

A similar argument works for $t < a$: replacing t by $-t$, as in proving Theorem 6 of Ch. 1, we obtain

$$\frac{d\sigma}{d(-t)} \leq |\sigma'(t)| \leq 2L\sigma(t)$$

again using the preceding inequality.

We shall prove next that the solutions of a normal first-order system (2) depend continuously on their initial values.

THEOREM 2 (CONTINUITY THEOREM).   *Let $\mathbf{x}(t)$ and $\mathbf{y}(t)$ be any two solutions of the vector DE (2), where $\mathbf{X}(\mathbf{x}, t)$ is continuous and satisfies the Lipschitz condition (4). Then*

(5) $$|\mathbf{x}(a + h) - \mathbf{y}(a + h)| \leq e^{L|h|}|\mathbf{x}(a) - \mathbf{y}(a)|$$

Proof.   Replacing $a + t$ by $a - t$, we can always reduce to the case $h \geq 0$. Consider again $\sigma(t) = |\mathbf{x}(t) - \mathbf{y}(t)|^2$. As in the proof of Theorem 1,

$$\sigma'(t) = 2[\mathbf{x}(t) - \mathbf{y}(t)] \cdot [\mathbf{X}(\mathbf{x}(t), t) - \mathbf{X}(\mathbf{y}(t), t)] \leq 2L|\mathbf{x} - \mathbf{y}|^2 = 2L\sigma(t)$$

Applying Lemma 2 of Ch. 1, §10 to $\sigma(t)$, we get $\sigma(a + h) \leq \sigma(a)e^{2Lh}$. Taking the square root of both sides, we get the desired result.

From Theorem 2 we can easily infer the following important property of the solutions of the DE (2).

COROLLARY. *Let* $\mathbf{x}(t, \mathbf{c})$ *be the solution of the* DE (2) *satisfying the initial condition* $\mathbf{x}(a, \mathbf{c}) = \mathbf{c}$. *Let the hypotheses of Theorem* 2 *be satisfied, and let the functions* $\mathbf{x}(t, \mathbf{c})$ *be defined for* $|\mathbf{c} - \mathbf{c}^0| \leq K$ *and* $|t - a| \leq T$. *Then:*
(a) $\mathbf{x}(t, \mathbf{c})$ *is a continuous function of both variables:*
(b) *if* $\mathbf{c} \to \mathbf{c}^0$, *then* $x(t, \mathbf{c}) \to \mathbf{x}(t, \mathbf{c}^0)$ *uniformly for* $|t - a| \leq T$.

Both properties follow from the inequality (5).

In view of the preceding results, it remains only to prove an *existence* theorem, in order to show that the initial value problem is well-set for normal first-order systems (1). This will be done in Theorems 6–8 later.

EXERCISES A

1. Show that $u = x + y + z$ and $v = x^2 + y^2 + z^2$ are integrals of the linear system $dx/dt = y - z$, $dy/dt = z - x$, $dz/dt = x - y$. Check that the solution curves are circles having the line (t, t, t) as the axis of symmetry.

2. Reduce each of the following DEs to an equivalent first-order system, and determine in which domain or domains (e.g., entire plane, any bounded region, a half-plane, etc.) the resulting system satisfies a Lipschitz condition:
   (a) $d^3x/dt^3 + x^2 = 1$ (b) $d^2x/dt^2 = x^{-1/2}$,   (c) $d^3x/dt^3 = [1 + (d^2x/dt^2)^2]^{1/2}$

3. Reduce the following system to normal form, and determine in which domains a Lipschitz condition is satisfied:

$$\frac{du}{dt} + \frac{dv}{dt} = u^2 + v^2, \qquad \frac{2du}{dt} + \frac{3dv}{dt} = 2uv$$

4. Show that the vector-valued function $(t + be^{at}, -e^{at}/ab)$ satisfies the DE (2) with $\mathbf{X} = [1 - (1/x_2), 1/(x_1 - t)]$, for any nonzero constants a, b.

5. State and prove a uniqueness theorem for the DE $y'' = F(x, y, y')$, with $F \in \mathcal{C}^1$. [HINT: Reduce to a first-order system, and use Theorem 1.]

6. (a) Show that any solution of the linear system $dx/dt = y$, $dy/dt = z$, $dz/dt = x$ satisfies the vector DE $d^3\mathbf{x}/dt^3 = \mathbf{x}$, where $\mathbf{x} = (x, y, z)$.
   (b) Show that every solution of the preceding system can be written $\mathbf{x} = e^t\mathbf{a} + e^{-t/2}[\mathbf{b} \cos \sqrt{3}t/2 + \mathbf{c} \sin \sqrt{3}t/2]$, for suitable constant vectors $\mathbf{a}$, $\mathbf{b}$, and $\mathbf{c}$.
   (c) Express $\mathbf{a}$, $\mathbf{b}$, and $\mathbf{c}$ in terms of $\mathbf{x}(0)$, $\mathbf{x}'(0)$, and $\mathbf{x}''(0)$.

7. Show that the general solution of the system $dx/dt = x^2/y$, $dy/dt = x/2$ is $x = 1/(at + b)^2$, $y = -1/[2a(at + b)]$.

8. Show that the curves defined parametrically as solutions of the system $dx/dt = \partial F/\partial x$, $dy/dt = \partial F/\partial y$, $dz/dt = \partial F/\partial z$ are orthogonal to the surfaces $F(x, y, z) = $ constant. What differentiability condition on F must be assumed to make this system satisfy a Lipschitz condition?

9. (a) Find a system of first-order DEs satisfied by all curves orthogonal to the spheres $x^2 + y^2 + z^2 = 2ax - a^2$.
   (b) By integrating the preceding system, find the orthogonal trajectories in question. Describe the solution curves geometrically.

10. (a) In what sense is the following statment inexact? "The general solution of the DE

$cy'' = (1 + y'^2)^{3/2}$ is the circle $(x - a)^2 + (y - b)^2 = c^2$, where a and b are arbitrary constants."

(b) Correct the preceding statement, distinguishing carefully between explicit, implicit, and multiple-valued functions.

11. (a) Given $\dot{x} = a(t)x + b(t)y$ and $\dot{y} = c(t)x + d(t)y$, prove that, if $b \neq 0$

$$\ddot{x} - \left[(a + d) + \frac{\dot{b}}{b} \right] \dot{x} + \left[(ad - bc) - \dot{a} + \frac{a\dot{b}}{b} \right] x = 0$$

(b) Given that $\ddot{x} + p(t)\dot{x} + q(t)x = r(t)$, prove that, if $q \neq 0$, $v = \dot{x}$ satisfies

$$\ddot{v} + \left[p - \frac{\dot{q}}{q} \right] \dot{v} + \left[p + q - \frac{\dot{q}p}{q} \right] v = \dot{r} - r\dot{q}/q$$

12. For which values of α, β does the function $x^\alpha y^\beta$ satisfy a Lipschitz condition: (a) in the open square $0 < x, y < 1$, (b) in the quadrant $0 < x, y < +\infty$, (c) in the part of the quadrant of (b) exterior to the square of (a)?

13. For each of the following scalar-valued functions of a vector $\mathbf{x}$ and each of the following domains, state whether a Lipschitz condition is satisfied or not:
(a) $x_1 + x_2 + \cdots + x_n$ (b) $x_1 x_2 \cdots x_n$ (c) $y/(x^2 + y^2)$ (d) $|\mathbf{x}|$ in (i)
$|x| < 1$, (ii) $-\infty < x_k < \infty$, (iii) $-\infty < x_1 < \infty$, $|x_k| < 1$, $k \geq 2$.

14. Let $\mathbf{X}(\mathbf{x}, t) = (X_1(\mathbf{x}, t), \ldots, X_n(\mathbf{x}, t))$ be a one-parameter family of vector fields. Show that $\mathbf{X}$ satisfies a Lipschitz condition if and only if each scalar-valued component X_i satisfies a Lipschitz condition, and relate the Lipschitz constant of $\mathbf{X}$ to those of the X_i.

4 CONTINUITY

We shall now prove a much stronger continuity property of the solutions of systems of DEs, namely that the solutions of (2) vary continuously when the function $\mathbf{X}$ varies continuously. Loosely speaking, the solution of a DE depends continuously upon the DE for given initial values.

THEOREM 3. *Let* $\mathbf{x}(t)$ *and* $\mathbf{y}(t)$ *satisfy the DEs*

$$dx/dt = \mathbf{X}(\mathbf{x}, t) \qquad and \qquad dy/dt = \mathbf{Y}(\mathbf{y}, t)$$

respectively, on $a \leq t \leq b$. *Further, let the functions* $\mathbf{X}$ *and* $\mathbf{Y}$ *be defined and continuous in a common domain D, and let*

(6) $$|\mathbf{X}(\mathbf{z}, t) - \mathbf{Y}(\mathbf{z}, t)| \leq \epsilon, \qquad a \leq t \leq b, \qquad \mathbf{z} \in D$$

Finally, let $\mathbf{X}(\mathbf{x}, t)$ *satisfy the Lipschitz condition* (4). *Then*

(7) $$|\mathbf{x}(t) - \mathbf{y}(t)| \leq |\mathbf{x}(a) - \mathbf{y}(a)| e^{L|t-a|} + \frac{\epsilon}{L} [e^{L|t-a|} - 1]$$

The function $\mathbf{Y}$ is not required to satisfy a Lipschitz condition.

Proof. Consider the real-valued function $\sigma(t)$, defined for $a \leq t \leq b$ by

$$\sigma(t) = |\mathbf{x}(t) - \mathbf{y}(t)|^2 = \sum_{k=1}^{n} [x_k(t) - y_k(t)]^2$$

From the last expression we see that σ is differentiable. Its derivative can be written in the form

$$\sigma'(t) = 2[\mathbf{X}(\mathbf{x}(t), t) - \mathbf{Y}(\mathbf{y}(t), t)] \cdot [\mathbf{x}(t) - \mathbf{y}(t)]$$
$$= 2\{[\mathbf{X}(\mathbf{x}(t), t) - \mathbf{X}(\mathbf{y}(t), t)] \cdot [\mathbf{x}(t) - \mathbf{y}(t)]\}$$
$$+ 2\{[\mathbf{X}(\mathbf{y}(t), t) - \mathbf{Y}(\mathbf{y}(t), t)] \cdot [\mathbf{x}(t) - \mathbf{y}(t)]\}$$

We now apply the triangle inequality to the right side, and then the Schwarz inequality to each of the two terms of the last expression. This gives the inequality

$$|\sigma'(t)| \leq 2|\mathbf{X}(\mathbf{x}(t), t) - \mathbf{X}(\mathbf{y}(t), t)| \; |\mathbf{x}(t) - \mathbf{y}(t)|$$
$$+ 2|\mathbf{X}(\mathbf{y}(t), t) - \mathbf{Y}(\mathbf{y}(t), t)| \; |\mathbf{x}(t) - \mathbf{y}(t)|$$

To the first term on the right side we now apply the Lipschitz condition that $\mathbf{X}$ satisfies; to the second term, we apply (6). This gives the following differential inequality for σ:

(8)
$$\sigma'(t) \leq 2L\sigma(t) + 2\epsilon\sqrt{\sigma(t)}$$

The theorem is now an immediate consequence of the following lemma.

LEMMA. *Let $\sigma(t) \geq 0$, $a \leq t \leq b$ be a differentiable function satisfying the differential inequality (8). Then*

(9)
$$\sigma(t) \leq \left[\sqrt{\sigma(a)}\, e^{L(t-a)} + \frac{\epsilon}{L} (e^{L(t-a)} - 1) \right]^2, \qquad a \leq t \leq b$$

Proof. We shall apply Theorem 7 of Ch. 1, §11, on differential inequalities to (8). The right side of (8), the function $F(\sigma, t) = 2L\sigma + 2\epsilon\sqrt{\sigma}$, satisfies a Lipschitz condition in any half plane $\sigma \geq \sigma_0$ that does not include the line $\sigma = 0$. Therefore, Theorem 7 of Ch. 1 applies when $\sigma(a) > 0$. For, if $\sigma(a) > 0$, then the solution of the DE

(9')
$$\frac{du}{dt} = 2\epsilon\sqrt{u} + 2Lu, \qquad u \geq 0$$

which satisfies the initial condition $u(a) = \sigma(a)$, will have a nonnegative derivative, and therefore will remain, for $t > a$, within the half-plane $u \geq \sigma(a)$.

The DE (9') is a Bernoulli DE (Ch. 1, Ex. C7). To find the solution satisfying

$u(a) = \sigma(a)$, make the substitution $v(t) = \sqrt{u(t)}$. (The square root is well-defined because $u(t) \geq \sigma(a) > 0$.) This gives the equivalent DE

$$2vv' = 2\epsilon v + 2Lv^2$$

If $u(a) > 0$, it follows that $u(t) > 0$ for all later t, since the derivative of u is positive. This gives $v(t) > 0$, and so we can divide both sides of this DE by v. The resulting DE is $v' - Lv = \epsilon$, an inhomogeneous linear DE whose solution satisfying the initial condition $v(a) = \sqrt{u(a)}$ is the function

$$\sqrt{u(t)} = v(t) = \sqrt{u(a)}\, e^{L(t-a)} + (\epsilon/L)(e^{L(t-a)} - 1)$$

On applying Theorem 7 of Ch. 1, we obtain the inequality (9).

We must now consider the case $\sigma(a) = 0$ when this theorem does not apply directly. In this case, we consider the solution $u_n(t)$ of the differential equation (9') that satisfies the initial condition $u_n(a) = 1/n$. Since the right side of (9') is positive, $u_n(t)$ is an increasing function of t. We shall prove that $u_n(t) \geq \sigma(t)$. Suppose that at some point $t_1 > a$ we had $u_n(t_1) < \sigma(t_1)$. Then among all numbers t with $a < t < t_1$ such that $u_n(t) \geq \sigma(t)$ there would be a largest, say t_0. Hence, we would have $u_n(t_0) = \sigma(t_0) > 0$ and $u_n(t) < \sigma(t)$ for $t_0 < t \leq t_1$. But this is impossible by what we have already proved, since in the interval $t_0 \leq t \leq t_1$, the functions $u(t)$ and $\sigma(t)$ stay away from 0. Therefore a Lipschitz condition is satisfied for (9'). We infer that

$$\sigma(t) \leq [n^{-1/2}e^{L|t-a|} + (\epsilon/L)(e^{L|t-a|} - 1)]^2$$

for all $n > 0$. Letting $n \to \infty$, we obtain the inequality (9) also in this case.

The following corollary follows immediately from Theorem 3.

COROLLARY. *Let $\mathbf{X}(\mathbf{x}, t; \epsilon)$ be a set of continuous functions of $\mathbf{x}$ and t, defined in the domain D: $|t - a| \leq T, |\mathbf{x} - \mathbf{c}| \leq K$ for all sufficiently small values of a parameter ϵ. Suppose that, as $\epsilon \to 0$, the functions converge uniformly in D to a function $\mathbf{X}(\mathbf{x}, t)$ that satisfies a Lipschitz condition. For each $\epsilon > 0$, let $\mathbf{x}(t, \epsilon)$ be a solution of $d\mathbf{x}/dt = \mathbf{X}(\mathbf{x}, t; \epsilon)$ satisfying the initial condition $\mathbf{x}(a; \epsilon) = \mathbf{c}$. Then the $\mathbf{x}(t; \epsilon)$ converge to the solution of $d\mathbf{x}/dt = \mathbf{X}(\mathbf{x}, t)$ satisfying $\mathbf{x}(a) = \mathbf{c}$, uniformly in any closed sub-interval $|t - a| \leq T_1 < T$ where all functions are defined.*

EXERCISES B

1. Let $\mathbf{X}$ and $\mathbf{Y}$ be as in Theorem 3, and let $\mathbf{x}(a) - \mathbf{y}(a)$. Show that $|\mathbf{x}(t) - \mathbf{y}(t)|/|t - a|$ remains bounded as $t \to a$.

2. Show that if $|\partial F/\partial y| \leq L(x)$, then any two solutions of $u' = F(x, u)$ and $v' = F(x, v)$ satisfy $|u(x) - v(x)| \leq |u(0) - v(0)|e^{\int L(x)dx}$.

For the pairs of DEs in Exs. 3–5, bound the differences on [0, 1] between solutions having the same initial value $y(0) = c$.

3. $y' = e^y$ and $y' = 1 + y + \cdots + \dfrac{y^n}{n!}$.

4. $\dfrac{d^2\theta}{dt^2} = -\theta$ and $\dfrac{d^2\theta}{dt^2} = -\sin\theta$.

5. $y' = \sin xy$ and $y' = xy - \dfrac{x^3y^3}{3!} + \dfrac{x^5y^5}{5!} + \cdots + (-1)^n \dfrac{xy^{2n+1}}{(2n+1)!}$.

6. To what explicit formulas does formula (7) specialize for the system $dx/dt = X(t)$? For the DE $dx/dt = bx + c$?

*7. Show that the conclusion of Theorem 3 holds if only a one-sided Lipschitz condition $(x - y) \cdot (X(x, t) - X(y, t)) \leq L|x - y|^2$ is assumed for X.

8. Let $X(x, t, s)$ be continuous for $|x - c| \leq K$, $|t - a| \leq T$, and $|s - s_0| \leq S$, and let it satisfy $|X(x, t, s) - X(y, t, s)| \leq L|x - y|$. Show that the solution $x(t, s)$ of $x' = X(x, t, s)$ satisfying $x(a, s) = c$ is a continuous function of s.

*5 NORMAL SYSTEMS

Many important mathematical problems have normal systems of DEs of order $m > 1$ as their natural formulation. We now give two examples and show how to reduce every normal system of ordinary DEs to a first-order system.

DEFINITION. A *normal system* or ordinary DE's for the unknown functions $\xi_1(t), \xi_2(t), \ldots, \xi_m(t)$ is any system of the form

$$(10) \qquad \frac{d^{n(k)}\xi_k}{dt^{n(k)}} = F_k\left(\xi_1, \frac{d\xi_1}{dt}, \ldots; \xi_2, \frac{d\xi_2}{dt}, \ldots; \xi_m, \frac{d\xi_m}{dt}, \ldots; t\right)$$

$k = 1, \ldots, m$, in which for each k only derivatives $d^p\xi_j/dt^p$ of any ξ_j of orders $p < n(j)$ occur on the right side.

In other words, the requirement is that the derivative $d^{n(k)}\xi_k/dt^{n(k)}$ of highest order of each ξ_k constitutes the left-hand side of one equation and occurs nowhere else.

THEOREM 4. *Every normal system* (10) *of ordinary DEs is equivalent to a first-order normal system* (1) *(with $n \geq m$).*

Proof. Each function F_k appearing on the right side of (10) is a function of several real variables. To reduce the system (11) to the first-order form (1), it is convenient notationally to rewite $n(i)$ as n_i, and to define new variables $x_1, \ldots, x_n$, where

$$n = n_1 + \ldots + n_m = n(1) + \ldots + n(m)$$

by the formulas

$$x_1 = \xi_1, \; x_2 = \frac{d\xi_1}{dt}, \; x_3 = \frac{d^2\xi_1}{dt^2}, \ldots, \; x_{n_1} = \frac{d^{n_1-1}\xi_1}{dt^{n_1-1}},$$

$$x_{n_1+1} = \xi_2, \; x_{n_1+2} = \frac{d\xi_2}{dt}, \ldots, \; x_{n_1+n_2} = \frac{d^{n_2-1}\xi_2}{dt^{n_1-1}}$$

In terms of these new variables, the system (1) assumes the form

$$\frac{dx_1}{dt} = x_2, \frac{dx_2}{dt} = x_3, \ldots, \frac{dx_{n_1-1}}{dt} = x_{n_1}, \frac{dx_{n_1}}{dt} = F_1(x_1, \ldots, x_n),$$

$$\frac{dx_{n_1+1}}{dt} = x_{n_1+2}, \ldots, \frac{dx_{n_1+n_2}}{dt} = F_2(x_1, \ldots, x_n, t)$$

It is clear that this system satisfies the requirements of the theorem.

The *initial value problem* for the normal system (10) is the problem of finding a solution for which the variables

$$\xi_1, \frac{d\xi_1}{dt}, \ldots, \frac{d^{n_1-1}\xi_1}{dt^{n_1-1}}; \xi_2, \ldots \frac{d^{n_2-1}\xi}{dt^{n_2-1}}, \ldots, \frac{d^{n_m-1}\xi_m}{dt^{n_m-1}}$$

assume given values at $t = a$.

It is easily seen from the proof of Theorem 4 that, if the functions F_k, considered as functions of the vector variables $\mathbf{x} = (x_1, \ldots, x_n)$ and t, satisfy Lipschitz conditions, then so do the functions $X_k(\mathbf{x}, t)$ in the associated first-order systems (1). This gives the following corollary.

COROLLARY. *If the functions F_k of the normal system* (10) *satisfy Lipschitz conditions in a domain D, then the system has at most one solution in D satisfying given initial conditions.*

Example 2 (the n-body problem). Let n mass points with masses m_j attract each other according to an inverse αth power law of attraction. Then, in suitable units, their position coordinates satisfy a normal system of $3n$ second-order differential equations of the form

$$\frac{d^2x_i}{dt^2} = \sum_{j \neq i} \frac{m_j(x_j - x_i)}{r_{ij}^{\alpha+1}}$$

and the same is true for d^2y_i/dt^2 and d^2z_i/dt^2, where

$$r_{ij} = [(x_i - x_j)^2 + (y_i - y_j)^2 + (z_i - z_j)^2]^{1/2} = r_{ji}.$$

Then Theorem 1 asserts that the initial positions $[x_i(0), y_i(0), z_i(0)]$ and velocities $(x_i'(0), y_i'(0), z_i'(0))$ of the mass points *uniquely* determine their subsequent motion (if any motion is possible). That is, the uniqueness theorem asserts the *determinacy of the n-body problem*. This theorem, taken with the continuity theorem, and Theorem 8 to follow, asserts that the n-body problem is well-posed, provided that there are no collisions.

To see this, let $\xi = (\xi_1, \ldots, \xi_{6n})$ be the vector with components defined as follows, for $k = 1, \ldots, n$.

$$\xi_k = x_k, \quad \xi_{n+k} = y_k, \quad \xi_{2n+k} = z_k$$

$$\xi_{3n+k} = x_k', \quad \xi_{4n+k} = y_k', \quad \xi_{5n+k} = z_k'$$

In this notation, the system (10) is equivalent to a first order normal system of the form (1):

$$\frac{d\xi_h}{dt} = F_h(\xi) = \begin{cases} \xi_{h+3n}, & h = 1, \ldots, 3n \\ \sum_j m_j(\xi_{j-3n} - \xi_{h-3n})/r_{hk(j)}^{\alpha+1}, & h = 3n + 1, \ldots, 6n \end{cases}$$

where $k(j)$ is the remainder of j when divided by n, and summation is extended to those $n - 1$ values of j such that $(h - 1)/n$ and $(j - 1)/n$ are distinct and have the same integral part. So long as no $r_{hj} = 0$, that is, so long as there are *no collisions,* a Lipschitz condition is evidently satisfied by the functions F_h. When one or more r_{hj} vanish, however, some of the functions F_h become *singular* (they are undefined) and Theorem 1 is inapplicable.

Example 3.   The *Frenet-Serret formulas*† comprise the following normal system of first-order DEs:

$$\frac{d\alpha}{ds} = \frac{\beta}{R(s)}, \quad \frac{d\beta}{ds} = -\frac{\alpha}{R(s)} + \frac{\gamma}{T(s)}, \quad \frac{d\gamma}{ds} = -\frac{\beta}{T(s)}$$

where α, β, and $\gamma = \alpha \times \beta$ are three-dimensional‡ vectors: the unit tangent, normal, and binormal vectors to a space curve. The curvature $\kappa(s) = 1/R(s)$ and torsion $\tau(s) = 1/T(s)$ are functions of the arc length s; $\alpha = d\mathbf{x}/ds$ is the derivative of vector position with respect to arc length.

If we let $\boldsymbol{\eta}(s)$ be the nine-dimensional vector $(\alpha_1, \alpha_2, \alpha_3, \beta_1, \beta_2, \beta_3, \gamma_1, \gamma_2, \gamma_3)$, the system can be written as the first-order vector DE

$$d\boldsymbol{\eta}/ds = \mathbf{Y}(\boldsymbol{\eta}; s)$$

Here $\mathbf{Y}(\boldsymbol{\eta}; s)$ is obtained by setting

$$Y_h(\boldsymbol{\eta}; s) = \begin{cases} \kappa(s)\eta_{h+3} & h = 1, 2, 3 \\ -\kappa(s)\eta_{h-3} + \tau(s)\eta_{h+3} & h = 4, 5, 6 \\ -\tau(s)\eta_{h-3} & h = 7, 8, 9 \end{cases}$$

If $\kappa(s)$ and $\tau(s)$ are bounded, the vector fields $\mathbf{Y}(\boldsymbol{\eta}; s)$ satisfy a Lipschitz condition (4) with $L = \sup\{|\kappa(s)| + |\tau(s)|\}$; hence, for given initial tangent direction $\alpha(0)$, normal direction $\beta(0)$ perpendicular to $\alpha(0)$, and binormal direction $\gamma(0) = \alpha(0) \times \beta(0)$, there is only one set of directions satisfying the Frenet-Serret formulas. This proves that *a curve with nonvanishing curvature is determined up to a rigid motion by its curvature and torsion.*††

---

† Widder, p. 101.

‡ $\alpha \times \beta$ denotes the cross product of the vectors α and β.

†† This theorem of differential geometry can fail when $\kappa(s)$ is zero, because $\beta = \mathbf{x}''(s)/|\mathbf{x}''(s)|$ is then geometrically undefined, so that the Frenet-Serret formulas do not necessarily hold.

EXERCISES C

1. Find all solutions of the system

$$x\frac{d^2x}{dt^2} - y\frac{d^2y}{dt^2} = 0, \qquad \frac{d^2x}{dt^2} + \frac{d^2y}{dt^2} + x + y = 0$$

2. Show that, if $a_{ihk} = -a_{khi}$, then $\Sigma\, x_i^2$ is an integral of the system

$$\frac{dx_i}{dt} = \sum a_{ihk}x_h x_k$$

3. The *one-body* problem is defined in space by the system

$$\ddot{x} = -xf(r), \qquad \ddot{y} = -yf(r), \qquad \ddot{z} = -zf(r), \qquad r^2 = x^2 + y^2 + z^2$$

   (a) Show that the components $L = y\dot{z} - z\dot{y}$, $M = z\dot{x} - x\dot{z}$, and $N = x\dot{y} - y\dot{x}$ of the angular momentum vector (L, M, N) are integrals of this system.
   (b) Show that any solution of the system lies in a plane $Ax + By + Cz = 0$.
   (c) Construct an energy integral for the system.

4. Let $\alpha = 2$ in the n-body problem (Newton's law of gravitation), and define the potential energy as $V = -\Sigma_{i<j}\, m_i m_j/r_{ij}$.
   (a) Show that the n-body problem is defined by the system

$$\frac{m_i d^2 x_i}{dt^2} = -\frac{\partial V}{\partial x}$$

   (b) Show that the total energy $\Sigma\, m_i \dot{x}_i^2/2 + V(\mathbf{x})$ is an integral of the system.
   (c) Show that the components $\Sigma\, m_i \dot{x}_i$, etc., of linear momentum are integrals.
   (d) Do the same as in (c) for the components $\Sigma\, m_i(y_i\dot{z} - z_i\dot{y}_i)$, etc., of angular momentum.

5. Show that the general solution of the vector DE $d^3\mathbf{x}/dt^3 = d\mathbf{x}/dt$ is $\mathbf{a} + \mathbf{b}e^t + \mathbf{c}e^{-t}$, where $\mathbf{a}$, $\mathbf{b}$, $\mathbf{c}$ are arbitrary vectors.

Exercises 6–9 refer to the Frenet -Serret formulas.

6. Show that, if $\alpha(s)$, $\beta(s)$, $\gamma(s)$ are orthogonal vectors of length one when $s = 0$, this is true for all s, provided they satisfy the Frenet-Serret formulas.

7. Show that if $1/T(s) \equiv 0$, and $d\mathbf{x}/ds = \alpha$, the curve $\mathbf{x}(s)$ lies in a plane. [HINT: Consider the dot product $\gamma \cdot \mathbf{x}$.]

*8. Show that, if $T = kR$ (k constant), the curve $\mathbf{x}(s)$ lies on a cylinder.

*9. Show that, if $R/T + (TR')' = 0$, the curve $\mathbf{x}(s)$ lies on a sphere.

6   EQUIVALENT INTEGRAL EQUATION

We now establish the *existence* of a local solution of any normal first-order system of DEs for arbitrary initial values. To this end, it is convenient to reduce the given initial value problem to an equivalent integral equation. One reason why this restatement of the problem makes it easier to treat is that we do not have to deal with differentiable functions directly, but only with continuous functions and their integrals. Every continuous function has an integral, whereas many continuous functions are not differentiable.

THEOREM 5.   *Let* $\mathbf{X}(\mathbf{x};\, t)$ *be a continuous vector function of the variables* $\mathbf{x}$ *and* t. *Then any solution* $\mathbf{X}(t)$ *of the vector integral equation*

$$(11) \qquad\qquad \mathbf{x}(t) = \mathbf{c} + \int_a^t \mathbf{X}(\mathbf{x}(s),\, s)\, ds$$

is a solution of the vector DE (2) *that satisfies the initial condition* $\mathbf{x}\,(a)\;=\;\mathbf{c}$, *and conversely.*

The vector integral equation (11) is a system of integral equations for r unknown scalar functions $x_1(t), \ldots, x_r(t)$, the components of the vector function $\mathbf{x}(t)$. That is,

$$x_k(t) = c_k + \int_a^t X_k(x_1(s),\, x_2(s),\, \ldots,\, x_r(s),\, s)\, ds, \qquad 1 \le k \le r$$

(In §§6–7, we will deal with r-dimensional vectors.)

Proof.   If $\mathbf{x}(t)$ satisfies the integral equation (11), then $\mathbf{x}(a) = \mathbf{c}$ and, by the Fundamental Theorem of the Calculus, $x_k'(t) = X_k(\mathbf{x}(t);\, t)$ for $k = 1, \ldots, r$, so that $\mathbf{x}(t)$ also satisfies the system (2). Conversely, the Fundamental Theorem of the Calculus shows that $x_k(t) = x_k(a) + \int_a^t x_k'(s)\, ds$ for all continuously differentiable functions $\mathbf{x}(t)$. If $\mathbf{x}(t)$ satisfies the normal system of DEs (2), then $\mathbf{x}(t) = \mathbf{x}(a) + \int_a^t \mathbf{X}(\mathbf{x}(s);\, s)\, ds$; if, in addition, $\mathbf{x}(a) = \mathbf{c}$, the integral equation (11) is obtained, q.e.d.

Example 4.   Consider the DE $dx/dt = e^x$ for the initial condition $x(0) = 0$. Separating variables, we see that this initial-value problem has the (unique) solution $1 - e^{-x} = t$, $x = -\ln(1 - t)$. Theorem 5 shows that it is equivalent to the integral equation $x(t) = \int_0^t e^{x(s)}\, ds$, which therefore has the same (unique) solution. Since the solution is defined only in the interval $-\infty < t < 1$, we see again that only a *local* existence theorem can be proved.

Operator Interpretation.   The problem of finding a solution to the integral equation (11) can be rephrased in terms of operators on vector-valued functions as follows. We define an operator $\mathbf{y} = U[\mathbf{x}] = U\mathbf{x}$, transforming vector-valued functions $\mathbf{x}$ into vector-valued functions $\mathbf{y}$ by the identity

$$(12) \qquad\qquad \mathbf{y}(t) = U[\mathbf{x}(t)] = \mathbf{c} + \int_a^t \mathbf{X}(\mathbf{x}(s),\, s)\, ds$$

If $\mathbf{X}(\mathbf{x},\, t)$ is defined for all $\mathbf{x}$ in the slab $|t - a| \le T$ and is continuous, the domain of this operator can be taken to be the family of continuous vector functions defined in the interval $|t - a| \leq T$ its range consists of all continuously differentiable vector-valued functions defined in this interval, satisfying $\mathbf{y}(a) = \mathbf{c}$. In this case, Theorem 5 has the following corollary.

COROLLARY. *The* DE *(2) has a solution satisfying* $\mathbf{x}(a) = \mathbf{c}$ *if and only if the mapping* U *of* (12) *has a fixpoint in* $\mathcal{C}[a, b]$.

However, if $\mathbf{X}(\mathbf{x}, t)$ is not defined for all $\mathbf{x}$, the domain of the operator U has to be determined with care. This will be done in Theorem 8 below.

7 SUCCESSIVE APPROXIMATION

Picard had the idea of iterating the integral operator U defined by (12), and proving that, for any initial trial function $\mathbf{x}^0$, the successive integral transforms (Picard approximations)

$$\mathbf{x}^0, \qquad \mathbf{x}^1 = U\mathbf{x}_0, \qquad \mathbf{x}^2 = U^2[\mathbf{x}^0] = U[\mathbf{x}^1], \qquad \mathbf{x}^3 = U[\mathbf{x}^2], \ldots$$

converge to a solution. This idea works under various sets of hypotheses; one such set is the following.

THEOREM 6. *Let the vector function* $\mathbf{X}(\mathbf{x}; t)$ *be continuous and satisfy the Lipschitz condition* (4) *on the interval* $|t - a| \leq T$ *for all* $\mathbf{x}, \mathbf{y}$. *Then, for any constant vector* $\mathbf{c}$, *the vector* DE $\mathbf{x}'(t) = \mathbf{X}(\mathbf{x}; t)$ *has a solution defined on the interval* $|t - a| \leq T$, *which satisfies the initial condition* $\mathbf{x}(a) = \mathbf{c}$.

Proof. As remarked at the end of the preceding section, the operator U is defined by (12) for all functions $\mathbf{x}(t)$ continuous for $|t - a| \leq T$. In particular, since $U\mathbf{x}$ is again a continuous function of $|t - a| \leq T$, the function $\mathbf{x}^2 = U^2[\mathbf{x}]$ $= U^2\mathbf{x}$ is well-defined. Similarly, the *iterates* $U^3\mathbf{x}$, $U^4\mathbf{x}$, etc., are well-defined. These iterates will always converge; a typical case is depicted in Figure 6.1.

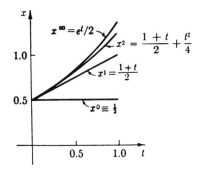

Figure 6.1   Picard approximation for $dx/dt = x$, $x(0) = 1/2$.

LEMMA. *If* $\mathbf{x}^0(t) \equiv \mathbf{c}$, *the sequence of functions defined recursively by* $\mathbf{x}^1 = U[\mathbf{x}^0]$, $\mathbf{x}^2 = U[\mathbf{x}^1] = U^2[\mathbf{x}^0], \ldots, \mathbf{x}^n = U[\mathbf{x}^{n-1}] = U^n[\mathbf{x}^0], \ldots$ *converges uniformly for* $|t - a| \leq T$.

Proof. Let $M = \sup_{|t-a| \leq T} |\mathbf{X}(\mathbf{c}; t)|$; the number M is finite because continuous functions are bounded on a closed interval. Without loss of generality we can assume that $a = 0$ and $t \geq a$, that is, that the interval is $0 \leq t \leq T$; the proof for general a and for $t < a$ can be deduced from this case by the substitutions $t \to t + a$ and $t \to a - t$.

By the basic inequality (3) for vector-valued functions, the function $\mathbf{x}^1(t)$ satisfies the inequality

$$|\mathbf{x}^1(t) - \mathbf{x}^0(t)| = \left| \int_0^t \mathbf{X}(\mathbf{x}^0(s), s) \, ds \right|$$

(13)
$$\leq \int_0^t \left| \mathbf{X}(\mathbf{x}^0, s) \right| ds \leq M \int_0^t ds = Mt$$

Again, by (3), the function $\mathbf{x}^2 = U[\mathbf{x}^1]$ satisfies the inequality

$$|\mathbf{x}^2(t) - \mathbf{x}^1(t)| = \left| \int_0^t [\mathbf{X}(\mathbf{x}^1(s), s) - \mathbf{X}(\mathbf{x}^0(s), s)] \, ds \right|$$

$$\leq \int_0^t |\mathbf{X}(\mathbf{x}^1(s), s) - \mathbf{X}(\mathbf{x}^0(s), s)| \, ds$$

We now use the assumption that the function $\mathbf{X}$ satisfies a Lipschitz condition with Lipschitz constant L. This gives, by (13), the inequality

$$|\mathbf{x}^2(t) - \mathbf{x}^1(t)| \leq \int_0^t |\mathbf{X}(\mathbf{x}^1(s), s) - \mathbf{X}(\mathbf{x}^0(s), s)| \, ds$$

$$\leq \int_0^t L |\mathbf{x}^1(s) - \mathbf{x}^0(s)| \, ds \leq L \int_0^t Ms \, ds = \frac{LMt^2}{2}$$

Similarly, for any $n = 1, 2, 3, \ldots,$

$$|\mathbf{x}^{n+1}(t) - \mathbf{x}^n(t)| \leq \int_0^t |\mathbf{X}(\mathbf{x}^n(s), s) - \mathbf{X}(\mathbf{x}^{n-1}(s), s)| \, ds$$

$$\leq L \int_0^t |\mathbf{x}^n(s) - \mathbf{x}^{n-1}(s)| \, ds$$

We now proceed by induction. Assuming that

$$|\mathbf{x}^n(t) - \mathbf{x}^{n-1}(t)| \leq \frac{(M/L)(Lt)^n}{n!}$$

we infer that

$$(14) \qquad |\mathbf{x}^{n+1}(t) - \mathbf{x}^n(t)| \le L \left(\frac{M}{L} \right) \int_0^t \frac{(Ls)^n}{n!} \, ds = \frac{M(Lt)^{n+1}}{L(n+1)!}$$

Next, we show that the sequence of functions $\mathbf{x}^n(t)$ $(n = 0, 1, 2, \dots)$ is uniformly convergent for $0 \le t \le T$. Indeed, the infinite series

$$\left(\frac{M}{L} \right) \sum_{n=0}^{\infty} \frac{(Lt)^{n+1}}{(n+1)!}$$

of positive terms is convergent to $(M/L)(e^{Lt} - 1)$, and uniformly convergent for $0 \le t \le T$. Hence, by the Comparison Test,† the series $x^0(t) + \sum_{k=0}^{\infty} [\mathbf{x}^{k+1}(t) - \mathbf{x}^k(t)]$ is uniformly convergent for $0 \le t \le T$. The nth partial sum of this series is the function $\mathbf{x}^n(t)$. It follows that the sequence of functions $\mathbf{x}^n(t)$ is uniformly convergent. This completes the proof of the lemma.

To complete the proof of Theorem 6, let $\mathbf{x}^\infty(t)$ denote the limit function of the sequence $\mathbf{x}^n(t)$; it suffices by Theorem 5 to show that $\mathbf{x}^\infty(t)$ is a solution of the integral equation (11). To this end, we consider the limit of the equations $\mathbf{x}^{n+1} = U[\mathbf{x}^n]$, namely the equations

$$\mathbf{x}^{n+1}(t) = \mathbf{c} + \int_0^t \mathbf{X}(\mathbf{x}^n(s), s) \, ds$$

The left side converges uniformly, by the preceding lemma. By the Lipschitz condition, $[\mathbf{X}(\mathbf{x}^m(s), s) - \mathbf{X}(\mathbf{x}^n(s), s)] \le L|\mathbf{x}^m(s) - \mathbf{x}^n(s)|$, and so the integrals on the right side also converge uniformly. It follows that they have a continuous limit $\mathbf{X}(\mathbf{x}^\infty(t); t)$.‡ Passing to the limit, we have

$$\mathbf{x}^\infty(t) = \mathbf{c} + \int_0^t \mathbf{X}(\mathbf{x}^\infty(s), s) \, ds$$

This demonstrates (11) and completes the proof of Theorem 6.

EXERCISES D

In Exs. 1–5, solve the integral equations specified.

1. $u(t) = 1 + \int_0^t s u(s) \, ds$ 2. $u(t) = 1 + \int_0^t s u^2(s) \, ds$

3. $u(t) + e^t = \int_0^t s u(s) \, ds$ 4. $u(t) = 1 - \int_0^t u(s) \tan s \, ds$

5. $u(t) = \int_0^t [u(s) + v(s)] \, ds$ $v(t) = 1 - \int_0^t u(s) \, ds$

---

† Courant and John, p. 535; see also Widder, p. 285.

‡ Courant and John, p. 537; Widder, p. 304.

6. Show that the nth iterate for the solution of $y' = yx$ such that $y(0) = 1$ is the sum of the first $n + 1$ terms of the power series expansion of $e^{x^2/2}$.

For the initial value problems in Exs. 7–10, obtain an expression for the nth function of the sequence of Picard approximations $\mathbf{x}^n = U^n[\mathbf{x}^0]$ to the exact solutions.

7. $dx/dt = x$, $x(0) = 1$ 9. $dx/dt = tx$, $x(0) = 1$

8. $dx/dt = y$, $dy/dt = -4x$ 10. $dx/dt = ty$, $dy/dt = -tx$

   $x(0) = 0$, $y(0) = 1$ $x(0) = 0$, $y(0) = 1$

For the initial value problems of Exs. 11–13, compute the functions x^1, x^2, x^3 of the sequences of Picard approximations.

11. $dx/dt = x^2 + t^2$, $x(0) = 0$

12. $dx/dt = y^2 + t^2$, $dy/dt = x^2 + t^2$, $x(0) = y(0) = 0$

13. $dx/dt = x(1 - 2t)$, $x(0) = 1$

*14. Show that, in Ex. 13, the sequence of Picard approximations converge for all t, but that this is not so in Ex. 11. In Ex. 13, is the convergence uniform?

15. Let $\mathbf{X}(\mathbf{x}, t) = A\mathbf{x}$, where A is a constant matrix. Show that each component of the nth Picard approximation to any solution is a polynomial function of degree at most n.

16. Establish the following inequalities for the sequence of Picard approximations:

$$|\mathbf{x}^n(t) - \mathbf{x}^{n-1}(t)| \leq \frac{M}{L}\left(\frac{(L|t-a|)^n}{n!}\right), \qquad |\mathbf{x}^n(t) - \mathbf{x}^\infty(t)| \leq \frac{M}{L}\sum_{k=n+1}^{\infty}\frac{(L|t-a|)^k}{k!}$$

8   LINEAR SYSTEMS

A first-order system of DEs (1) is said to be *linear* when it is of the form

$$(15) \qquad \frac{dx_i}{dt} = \sum_{j=1}^{n} a_{ij}(t)x_j(t) + b_i(t), \qquad 1 \leq i \leq n$$

In this case, we have $X_i(\mathbf{x}, t) = \sum_{j=1}^{n} a_{ij}(t)x_j + b_i(t)$. In vector notation, the linear system (15) is written in the form

$$(16) \qquad d\mathbf{x}/dt = A(t)\mathbf{x} + \mathbf{b}(t)$$

where $A(t)\mathbf{x}$ stands for the matrix $\|a_{ij}(t)\|$ applied to the vector $\mathbf{x}$, and $\mathbf{b}$ stands for the vector $(b_1, \dots, b_n)$.

When $\mathbf{b}(t) \equiv 0$, the system (16) is said to be *homogeneous*. Otherwise, it is called *inhomogeneous*. The homogeneous system obtained from a given inhomogeneous system (15) by setting the b_j equal to zero is called the *reduced* system associated with (15).

A basic property of a linear system of DEs (16) is that the difference $\mathbf{x} - \mathbf{y}$ of any two solutions of (16) is a solution of the reduced system. It can be immediately verified that any linear combination $a\mathbf{x}(t) + b\mathbf{y}(t)$ of solutions $\mathbf{x}(t)$ and $\mathbf{y}(t)$ of a homogeneous linear system is again a solution.

We shall now establish the existence of solutions of linear systems and describe the set of all solutions.

LEMMA. *Any linear system* (15) *with continuous coefficient functions on a closed interval I satisfies a Lipschitz condition* (4) *with*

$$(17) \qquad\qquad L \leqq \sum_{i,j} \sup_{t \in I} |a_{ij}(t)|$$

Proof. Since $\mathbf{X}(\mathbf{x}, t) - \mathbf{X}(\mathbf{y}, t)$ is the vector sum of n^2 vectors $\mathbf{z}_{ij}$, with ith component $a_j(x_{ij} - y_j)$ and other components zero, repeated use of the triangle inequality gives

$$|\mathbf{X}(\mathbf{x}, t) - \mathbf{X}(\mathbf{y}, t)| \leq \sum_{i,j} |\mathbf{z}_{ij}| \leq \sum_{i,j} |a_{ij}(t)| \cdot |x_j - y_j|$$

$$\leq \sum_{i,j} \sup_{t \in I} |a_{ij}(t)| \cdot |\mathbf{x} - \mathbf{y}|$$

The functions $a_{ij}(t)$, being continuous on a closed interval, are bounded.† Hence, the Lipschitz constant L of (17) is finite. This completes the proof of the lemma.

We can now state the existence theorem for linear systems.

THEOREM 7. *The initial value problem defined by a linear system* (15), *with the $a_{ij}(t)$ and $b_i(t)$ defined and continuous for $|t - a| \leq T$, and the initial condition $\mathbf{x}(a) = \mathbf{c}$, has a unique solution on $|t - a| \leq T$.*

Proof. The preceding lemma shows that such a system satisfies the hypothesis of Theorem 6. This gives the existence of the solution. The uniqueness follows from Theorem 1, again by the preceding lemma.

For *homogeneous* systems, we can construct a basis of solutions, as follows.

COROLLARY 1. *Let $\mathbf{x}^i(t)$ be the solution of a homogeneous linear system $d\mathbf{x}/dt = A(t)\mathbf{x}$ that satisfies the initial condition $x_k^i(a) = 0$, $i \neq k$, $x_i^i(a) = 1$. Then the solution satisfying the initial condition $\mathbf{x}(a) = \mathbf{c} = (c_1, \ldots, c_n)$ is equal to the linear combination $\mathbf{x}(t) = c_1 \mathbf{x}^1(t) + c_2 \mathbf{x}^2(t) + \cdots + c_n \mathbf{x}^n(t)$.*

Proof. The vector-valued function $\mathbf{y}(t) = \mathbf{x}(t) - \sum_{j=1}^{n} c_j \mathbf{x}^j(t)$ is a solution of the linear system, since it is a linear combination of solutions. This function satisfies the initial condition $\mathbf{y}(a) = (0, 0, \ldots, 0)$ because of the way in which the initial conditions for the solutions $\mathbf{x}^i$ have been chosen. Since the identically zero function is also a solution of the linear system, it follows from the uniqueness in Theorem 7 that $\mathbf{y}(t) \equiv 0$, q.e.d.‡

---

† Courant and John, p. 101.

‡ In algebraic terms, Corollary 1 states that the solutions of a homogeneous linear system of dimension n form an n-dimensional vector space of functions. Therefore, any $n + 1$ solutions of such a system are always linearly dependent.

The reduction of an nth order normal DE to a first-order system sketched in in §1, when applied to a linear nth order DE in normal form

$$\frac{d^n u}{dt^n} = p_1(t) \frac{d^{n-1}u}{dt^{n-1}} + p_2(t) \frac{d^{n-2}u}{dt^{n-2}} + \cdots + p_{n-1}(t) \frac{du}{dt} + p_n(t)u$$

transforms the DE into a homogeneous linear system $dx/dt = A(t)\mathbf{x}$, where the matrix $\|a_{ij}(t)\| = A(t)$ is defined as follows: $a_{ij}(t) = 0$ if $1 \le i \le n - 1$ and $j \ne i + 1$; $a_{i,j+1}(t) = 1$ if $1 \le i \le n - 1$; $a_{nj}(t) = p_{n-j+1}(t)$.

We therefore obtain the following result.

COROLLARY 2.   *An nth order DE in normal form, with coefficients $p_j(t)$ continuous for $|t - a| \le T$, has a basis of solutions $u_j(t)$ $(1 \le j \le n)$ satisfying the initial†* *conditions $u_j^{(i)}(0) = \delta_j^{i+1}$, $0 \le i \le n - 1$.*

More results about solutions of linear systems of DEs will be established in Appendix A.

9   LOCAL EXISTENCE THEOREM

In Theorem 6, it was assumed that $\mathbf{X}(\mathbf{x}, t)$ was defined for all $\mathbf{x}$ and satisfied a Lipschitz condition (4) for all $\mathbf{x}$. But often this is not the case. For instance, this assumption does not hold for the DE $dx/dt = e^x$ of Example 4. The ratio

$$\frac{|X(x, t) - X(0, t)|}{|x - 0|} = \frac{(e^x - 1)}{x}$$

is unbounded if the domain of e^x is unrestricted.‡

Correspondingly, the conclusion of Theorem 6 fails for this DE: the solution which takes the value c at $t = 0$ is the function $x(t) = -\ln(e^{-c} - t)$, and this function is defined only in the interval $-\infty < t < e^{-c}$. Hence, there is *no* $\epsilon > 0$ such that the DE $dx/dt = e^x$ has a solution defined on all of $|t| < \epsilon$ for every initial value: the interval of definition of a solution changes with the initial value.

To cover this situation, and also cases where the function $\mathbf{X}$ is defined only in a small region of $(x_1, \ldots, x_n)$-space, we now prove a *local* existence theorem, whose assumptions and conclusions refer only to neighborhoods of a given point.

THEOREM 8.   *Suppose that the function $\mathbf{X}(\mathbf{x}, t)$ in (2) is defined and continuous in the closed domain $|\mathbf{x} - \mathbf{c}| \le K$, $|t - a| \le T$ and satisfies a Lipschitz condition*

---

† δ_j^i is the Kronecker delta function: $\delta_j^i = 0$ if $i \ne j$ and $\delta_j^i = 1$. For the concept of a basis of solutions of nth order DEs, see Ch. 3, §4.

‡ The same complications arise with the DE $y' = 1 + y^2$.

(4) *there. Let M = sup |**X**(**x**, t)| in this domain. Then the* DE (2) *has a unique solution satisfying* **x**(a) = **c** *and defined on the interval* |t − a| ≤ *min* (T, K/M).

Proof. All steps in the proof of Theorem 6 can be carried out, provided we know that the functions **x**ⁿ(t) referred to there take their values within the domain D_1: |**x** − **c**| ≤ K, |t − a| ≤ min (T, K/M), in which **x**(t) is surely defined. In particular, note that since $D_1 \subset D$, the bound M and Lipschitz constant L of Theorem 6 can be used in D_1. Therefore, the proof is a corollary of the following lemma.

LEMMA. *Under the hypotheses of Theorem 8, the operator U defined by* (12) *carries functions* **x**(t) *satisfying the conditions:* (i) **x**(t) *is defined and continuous on* |t − a| ≤ *min* (T, K/M); (ii) **x**(a) = **c**; (iii) |**x**(t) − **c**| ≤ K *on the interval* |t − a| ≤ *min* (T, K/M), *into functions satisfying the same conditions.*

Proof. In (12), suppose that **x**(s) satisfies conditions (i), (ii), (iii). We must show that **y**(t) satisfies the same conditions. Clearly (i) and (ii) are satisfied by **y**(t). By the inequality (3) we have (taking again t ≥ a for simplicity)

$$|\mathbf{y}(t) - \mathbf{c}| = \left| \int_a^t \mathbf{X}(\mathbf{x}(s), s) \, ds \right| \leq \int_a^t |\mathbf{X}(\mathbf{x}(s), s)| \, ds$$

If M is the maximum of **X** and if |t − a| ≤ K/M, this gives

$$|\mathbf{y}(t) - \mathbf{c}| \leq \frac{MK}{M} = K$$

Therefore, (iii) is satisfied and **y**(t) is defined for |t − a| ≤ min (T, K/M), completing the proof.

Using the reduction of §1, taking an nth-order normal DE

(18) $$u^{(n)} = F(u, u', u'', \ldots, u^{(n-1)}, t)$$

into an equivalent first-order normal system (1), we obtain the following.

COROLLARY. *Let the function F (x₁, x₂, . . . , xₙ, t) be continuous in the cylinder* |t − a| ≤ T, |**x** − **c**| ≤ K. *Let* $(x_2^2 + x_3^2 + \cdots + x_n^2 + F^2)^{1/2} \leq M$, *and let F satisfy a Lipschitz condition there. Then, on the interval* |t − a| ≤ *min* (T, K/M), *the* DE (18) *has one and only one solution that satisfies the initial conditions* $u^{(i)}(a) = c_{i+1}$, $0 \leq i \leq n - 1$.

*10   THE PEANO EXISTENCE THEOREM

The existence theorems for normal systems (1) proved so far have assumed that the functions X_i satisfy Lipschitz conditions. We shall now derive an existence theorem, assuming only continuity. As shown in Ch. 1, solutions of such systems need not be uniquely determined by their initial values.

THEOREM 9 (PEANO EXISTENCE THEOREM). *If the function* $\mathbf{X}(\mathbf{x}, t)$ *is continuous for* $|\mathbf{x} - \mathbf{c}| \leq K$, $|t - a| \leq T$, *and if* $|\mathbf{X}(\mathbf{x}, t)| \leq M$ *there, then the vector DE (2)has at least one solution* $x(t)$, *defined for*

$$|t - a| \leq \min{(T, K/M)}$$

satisfying the initial condition $\mathbf{x}(a) = \mathbf{c}$.

Proof. Using an elegant method due to Tonelli, we shall consider the equivalent integral equation (11) of Theorem 5,

$$(19) \qquad\qquad \mathbf{x}(t) = \mathbf{c} + \int_a^t \mathbf{X}(\mathbf{x}(s), s)\, ds$$

and prove that this has a solution. Let $T_1 = \min{(T, K/M)}$. We may assume that $a = 0$ and that the interval is $0 \leq t \leq T_1$. In this interval we construct a sequence of functions $\mathbf{x}^n(t)$ as follows. For $0 \leq t \leq T_1/n$, set $\mathbf{x}^n(t) \equiv \mathbf{c}$. For $T_1/n < t \leq T_1$ define $\mathbf{x}^n(t)$ by the formula

$$(20) \qquad\qquad \mathbf{x}^n(t) = \mathbf{c} + \int_0^{t-T_1/n} \mathbf{X}(\mathbf{x}^n(s), s)\, ds$$

This formula defines the value of $\mathbf{x}^n(t)$ in terms of the previous values of $\mathbf{x}^n(s)$ for $0 \leq s \leq t - T_1/n$.

It follows, as in the lemma of §9, that the functions $\mathbf{x}^n(t)$ are defined for $0 \leq t \leq T_1$. Also, we have

$$|\mathbf{x}^n(t)| \leq |\mathbf{c}| + \int_0^{T_1} M\, ds \leq |\mathbf{c}| + T_1 M$$

Hence, the sequence of functions $|\mathbf{x}^n(t)|$ $(n = 1, 2, \ldots)$ is uniformly bounded. Next, we prove that the sequence $\mathbf{x}^n$ is *equicontinuous* in the following sense.

DEFINITION. A family $\mathcal{F}$ of vector-valued functions $\mathbf{x}(t)$, defined on an interval I: $|t - a| \leq T$, is said to be *equicontinuous* when, given $\epsilon > 0$, a number $\delta > 0$ exists such that

$$|t - s| < \delta \qquad \text{implies} \qquad |\mathbf{x}(t) - \mathbf{x}(s)| < \epsilon$$

for *all* functions $\mathbf{x} \in \mathcal{F}$, provided that $s, t \in I$.

Indeed, using the inequality (3), we have

$$|\mathbf{x}^n(t_2) - \mathbf{x}^n(t_1)| \leq \int_{t_1 - T_1/n}^{t_2 - T_1/n} |\mathbf{X}(\mathbf{x}^n(s), s)| \, ds \leq M|t_2 - t_1|$$

from which it is evident that the $\mathbf{x}^n(t)$ are equicontinuous.

We now apply to the sequence $\mathbf{x}^n$ the Theorem of Arzelà-Ascoli, which is stated below without proof.[†]

ARZELÀ-ASCOLI THEOREM. *Let $\mathbf{x}^n(t)$ $(n = 1, 2, 3, \ldots)$ be a bounded equicontinuous sequence of scalar or vector functions, defined for $a \leq t \leq b$. Then there exists a subsequence $x^{n_i}(t)$ $(i = 1, 2, \ldots)$ that is uniformly convergent in the interval.*

Applying this result to the sequence $\mathbf{x}^n(t)$, we see that it must contain a uniformly convergent subsequence $\mathbf{x}^{n_i}(t)$, converging to a continuous function $\mathbf{x}^\infty(t)$ as $n_i \to \infty$.

It is now easy to verify that this limit function $\mathbf{x}^\infty(t)$ satisfies the integral equation (19). Indeed, (20) can be written in the form

$$(21) \qquad \mathbf{x}^{n_i}(t) = \mathbf{c} + \int_0^t \mathbf{X}(\mathbf{x}^{n_i}(s), s) \, ds - \int_{t - T_1/n_i}^i \mathbf{X}(\mathbf{x}^{n_i}(s), s) \, ds$$

As $n_i \to \infty$, $\int_0^t \mathbf{X}(\mathbf{x}^{n_i}(s), s) \, ds \to \int_0^t \mathbf{X}(\mathbf{x}^\infty(s), s) \, ds$ because $\mathbf{X}(\mathbf{x}, t)$ is uniformly continuous; and the last term of (21) tends to zero, because, by the inequality (3)

$$\left| \int_{t - T_1/n_i}^t \mathbf{X} \, ds \right| \leq \int_{t - T_1/n_i}^t M \, ds = M \frac{T_1}{n_i} \to 0$$

Therefore, taking limits on both sides of (21) as $n_i \to \infty$, we find that $\mathbf{x}^\infty$ satisfies the integral equation (19), q.e.d.

*11 ANALYTIC EQUATIONS

We shall now consider the vector DE (2) under the assumption that $\mathbf{X}(\mathbf{x}, t)$ is an analytic function of all variables $x_1, \ldots, x_n, t$. The essential principle to be established is that *all solutions of analytic DEs are analytic functions.*[‡]

The result is true whether the variables are real or complex; we shall first consider the complex case. To emphasize that we are dealing with complex vari-

---

[†] Rudin, p. 164 ff. The proof given there is for real-valued functions, but the method applies to vector-valued functions.

[‡] This section requires a knowledge of elementary complex function theory such as is found in the books by Hille (Vol. 1) and Ahlfors.

ables, we rewrite the vector DE (2) as

$$(22) \qquad\qquad dz/dt = \mathbf{z}'(t) = \mathbf{Z}(\mathbf{z},t), \qquad t = r + is$$

where $z_j = x_j + iy_j$ and $Z_j = X_j + iY_j$ are complex-valued functions.

We assume that the $Z_j(z_1, \ldots, z_n, t)$ are analytic functions of the variables z_1, $z_2, \ldots, z_n$ and t in the closed cylindrical domain C: $|t - a| \leq T$, $|\mathbf{z} - \mathbf{c}| \leq K$, with maximum M there. By the lemma of §2, this implies that a Lipschitz condition holds in C, for some constant L.

Vector notation can be adapted to complex vectors with the following changes. The length (or norm) of a vector $\mathbf{z} = (z_1, z_2, \ldots, z_n)$ with complex components z_k is defined as

$$|\mathbf{z}| = (z_1 z_1^* + z_2 z_2^* + \cdots + z_n z_n^*)^{1/2}$$

The *Hermitian inner product* of two complex vectors $\mathbf{z}$ and

$$\mathbf{w} = (w_1, w_2, \ldots, w_n)$$

is defined as

$$\mathbf{z} \cdot \mathbf{w} = (z_1 w_1^* + z_2 w_2^* + \cdots + z_n w_n^*)$$

Note that $\mathbf{z} \cdot \mathbf{w} = (\mathbf{w} \cdot \mathbf{z})^*$: the dot product operation is not commutative for complex vectors. (The set of complex n-vectors with the above inner product is called a unitary space.)

Now let γ be any path in the complex t-plane, defined parametrically by the equation $t = t(\sigma) = r(\sigma) + is(\sigma)$, where $r, s \in \mathcal{C}^1$ and σ is a *real* parameter. On the path γ, (22) is equivalent to the system of *real* DEs

$$(22') \qquad\qquad \begin{aligned} x'(\sigma) &= \mathbf{X}(\mathbf{x}, \mathbf{y}, \sigma)r'(\sigma) - \mathbf{Y}(\mathbf{x}, \mathbf{y}, \sigma)s'(\sigma) \\ y'(\sigma) &= \mathbf{X}(\mathbf{x}, \mathbf{y}, \sigma)s'(\sigma) + \mathbf{Y}(\mathbf{x}, \mathbf{y}, \sigma)r'(\sigma) \end{aligned}$$

Theorems 1 through 8 apply to this system, which satisfies a Lipschitz condition.

Using the complex vector notation described before, we can also prove analogs of these theorems directly, since the DE $\mathbf{z}'(\sigma) = \mathbf{Z}(\mathbf{z}, t(\sigma))t'(\sigma)$ is equivalent to (22'), and hence to (22), on the path γ.

The analog of the operator U of formula (12) is the operator W, defined by the line integral

$$(23) \qquad\qquad \mathbf{w}(t) = W[\mathbf{z}(t)] = \mathbf{c} + \int_a^t \mathbf{Z}(\mathbf{z}(\zeta), \zeta) \, d\zeta$$

Since each component function Z_j is analytic, the line integral defining the operator W is independent of the path from 0 to t in the complex t-plane, pro-

vided that this path stays within the domain C where the function $\mathbf{Z}$ is defined.†
By Morera's theorem, the function $\mathbf{w}$ is therefore also analytic, in the sense that
each component $w_j(z)$ is. Thus, the operator W transforms analytic functions
into analytic functions. Moreover, the lemma of §9 still holds, because the integrals in (23) can be taken along straight line segments in the complex ζ-plane.
This gives the following lemma.

LEMMA. *For* $|t - a| \leq min\ (T, K/M)$*, the operator* W *defined by* (23) *takes anal-
ytic complex-valued vector functions* $\mathbf{z}(t)$ *with* $|\mathbf{z}(t) - \mathbf{c}| \leq K$ *into analytic vector func-
tions* $\mathbf{w}(t)$ *with* $|\mathbf{w}(t) - \mathbf{c}| \leq K$.

By repeated applications of this lemma, it follows that the functions $W^n[\mathbf{w}^0]$
$= \mathbf{w}^n(t)$ defined by the Picard process of iterated quadrature, in the domain
$|t - a| \leq min\ (T, K/M)$ of the complex t-plane, are all analytic.
We now apply the following result‡ from function theory.

Weierstrass Convergence Theorem. If a sequence $\{f_n(t)\}$ of complex analytic functions converges uniformly to $f(t)$ in a domain D of the complex t-plane,
the $f(t)$ is analytic in D.
By this theorem, the sequence of functions $\mathbf{w}^n(t)$ converges uniformly for
$|t - a| \leq min\ (T, K/M)$ to an *analytic* solution $\mathbf{w}^\infty(t)$ of the integral equation

$$(24) \qquad \mathbf{z}(t) = \mathbf{c} + \int_a^t \mathbf{Z}[\mathbf{z}(\zeta), \zeta]\ d\zeta = W[\mathbf{z}(t)]$$

and hence of the complex DE (22). Applying the Existence Theorem of §8 for
real DEs to the system (22′), we infer the next theorem.

THEOREM 10. *In Theorem 8, replace the real variables* t, x_j, X_j *with complex
variables* t, z_j, Z_j*. Under the same hypotheses, if the* $Z_j(\mathbf{z}, t)$ *are complex analytic functions, the vector DE* (22) *has a unique complex analytic solution* $\mathbf{z}(t)$ *for given initial
conditions.*

From this result and the uniqueness theorem, again for real DEs, we obtain
the following corollary.

COROLLARY 1. *Let* $\mathbf{Z}(\mathbf{z}, t)$ *be analytic in any simply-connected domain of* $\mathbf{z}$, t-
space, and let $\mathbf{z}\ (t)$ *be any solution of the DE* (22). *Then* $\mathbf{z}(t)$ *is analytic.*††

---

† This is true because the disk $|t - a| \leq T$ where $\mathbf{Z}$ is defined is *simply connected.*

‡ Ahlfors, p. 173. The result contrasts sharply with the case of functions of a real variable. By the
Weierstrass approximation theorem, every continuous function on a real interval $a \leq x \leq b$ is a
uniform limit of polynomial (hence analytic) functions.

†† An alternative proof can be based directly on (22). If the $z_k(t)$ satisfy (22), they are continuously
differentiable. Hence, they are analytic (Ahlfors, pp. 24, 105; Hille, Vol. 1, pp. 72, 88).

Real Analytic DEs. A *real* function $X(\mathbf{x}, t)$ of real variables $x_1, \ldots, x_n$ and t is said to be analytic at $(\mathbf{c}, a)$ when it can be expanded into a power series with real coefficients in the variables $(x_k - c_k)$ and $(t - a)$, convergent in the cylinder $|\mathbf{x} - \mathbf{c}| < \eta$, $|t - a| < \epsilon$, for sufficiently small positive η and ϵ. When $X(\mathbf{x}, t)$ is analytic, its power series is convergent also in the *complex* cylinder $|\mathbf{z} - \mathbf{c}| < \eta$, $|t - a| < \epsilon$ (t complex), and defines a complex-valued analytic function there.

Now, let a normal system of real DEs (1) be given, the $X_j(x, t)$ being analytic. From Theorem 10, it follows that the resulting complex DE $d\mathbf{x}/dt = \mathbf{X}(\mathbf{x}, t)$ has a unique complex analytic solution for given real initial values. On the other hand, it also has a unique (local) real solution by Theorems 1 and 8. Hence the two solutions must coincide, proving Corollary 2.

COROLLARY 2. *If* $\mathbf{X}(\mathbf{x}, t)$ *is an analytic real function of the real variables* $x_1, \ldots, x_n$ *and* t, *then every solution of* (1) *is analytic.*

EXERCISES E

1. (a) Obtain an equivalent first-order system for $d^2x/dt^2 = t^2x$. Find the nth term of the Picard sequence of iterates for the initial values $x(0) = 1$, $x'(0) = 0$.
   (b) Prove that this initial-value problem has one and only one solution on $(-\infty, \infty)$.

2. (a) Obtain an equivalent first-order system for the DE $d^2x/dt^2 = x^2 + t^2$, and find the Lipschitz constant for the resulting system in the domain $|t| \leq A$, $|x| \leq B$, $|x'| \leq C$.
   (b) State and prove a local existence theorem for solutions of this DE, for the initial conditions $x(0) = b$, $x'(0) = c$. Estimate the largest T, U such that a solution is defined on $-U \leq t \leq T$.

3. Show that, if $F(y)$ is continuous for $|y| \leq K$, and $|F(y)| \leq M$, every solution of $y' = F(y)$ can be uniformly approximated arbitrarily closely for $|x| \leq K/M$ by a solution of a DE $y' = P(y)$, where P is a polynomial.

4. Compute the nth Picard approximation to the solution of the complex system $dw/dt = iz$, $dz/dt = w$, which satisfies the initial conditions $w(0) = 1$, $z(0) = i$.

5. In the complex t-plane, determine a domain in which the system $dw/dt = tz^2$, $dz/dt = tw^2$ has an analytic solution satisfying given initial conditions $w(0) = w_0$, $z(0) = z_0$.

6. Show that the solution of the complex analytic DE

$$w'(z) = M \left[\frac{1}{2}\left(1 + \frac{w}{K}\right) \right]^{1/n}, \qquad (|z| < K)$$

which satisfies the initial condition $w(0) = 0$, is the function

$$w(z) = K\left[\left(1 + \frac{z}{c}\right)^{n/(n-1)} - 1 \right], \qquad c = \frac{2^{1/n}Kx}{(n-1)M}$$

*7. Using the result of Ex. 6, show that the bound given by Theorem 8 for the domain of existence of a solution is "best possible" for analytic functions of a complex variable.

*12   CONTINUATION OF SOLUTIONS

Even when the function† $\mathbf{X}(\mathbf{x}, t)$ is of class $\mathcal{C}^1$ and is defined for all $\mathbf{x}$ and t, Theorem 8 establishes the existence of solutions only in the neighborhood of a given initial value. In other words, it establishes only the *local* existence of solutions. We shall now study how such local solutions can be joined together to give a *global* solution defined up to the boundary of the domain of definition of the function $\mathbf{X}$.

THEOREM 11.   *Let $\mathbf{X}(\mathbf{x}, t)$ be defined and of class $\mathcal{C}^1$ in an open region $\mathcal{R}$ of $(\mathbf{x},$ $t)$ -space. For any point $(\mathbf{c}, a)$ in the region $\mathcal{R}$, the DE (2) has a unique solution $\mathbf{x}(t)$ satisfying the initial condition $\mathbf{x}(a) = \mathbf{c}$ and defined for an interval $a \leq t < b$ ($b \leq \infty$) such that, if $b < \infty$, either $\mathbf{x}(t)$ approaches the boundary of the region, or $\mathbf{x}(t)$ is unbounded as $t \to b$.*

Proof.   Consider the set S of all local solutions of the system (2) that satisfy the given initial condition $\mathbf{x}(a) = \mathbf{c}$. These are defined on intervals of varying lengths of the form $[a, T)$. Given two solutions $\mathbf{x}$ and $\mathbf{y}$ in this set, defined on intervals I and I' respectively, the function $\mathbf{z}$, defined to be equal to $\mathbf{x}$ or to $\mathbf{y}$ wherever either is defined, and hence also where both are defined, is also a solution defined on their union $I \cup I'$.

We now construct a single solution $\mathbf{x}$, called the *maximal* solution, defined on the union of *all* the intervals in which some local solution is defined, by letting $\mathbf{x}(t)$ be equal to the value of any of the solutions of S defined at the point t. This maximal solution $\mathbf{x}(t)$ is a well-defined function of class $\mathcal{C}^1$, by the Uniqueness Theorem. Furthermore, the interval of definition of this solution is the union of all the intervals of definition and, therefore, is itself an interval of the form $a \leq t < b$.

Consider the limiting behavior of $\mathbf{x}(t)$, as $t \uparrow b$. By the Bolzano–Weierstrass Theorem,‡ any infinite bounded set of points $(\mathbf{x}(t_n), t_n)$ in $\mathbf{x}t$-space must contain a limit point. Hence either $b = +\infty$, or $\lim_{t \uparrow b} |\mathbf{x}(t)| = +\infty$, or at least one finite point $(\mathbf{d}, b)$ is approached by at least one sequence of points $[\mathbf{x}(t_n), t_n]$ on the above solution curve. In the first case, t is unbounded. In the second case, $\mathbf{x}(t)$ is unbounded and the maximal solution may be said to "recede to infinity."

It remains to consider the third case. A typical example is provided by choosing the region $\mathcal{R}$ as the left half-plane $t < 0$ and $x'(t) = t^{-2} \cos t^{-1}$, with general solution $x = C - \sin t^{-1}$.

We shall now prove that, in the third case above, *every* limit point $(\mathbf{d}, b)$ on $t = b$ of the maximal solution curve must lie on the boundary of $\mathcal{R}$. Indeed, suppose that it is in the interior; there would then exist a closed neighborhood

---

† In this section we consider only *real* vectors and functions. The results can, however, be extended to complex-valued and analytic functions, by methods similar to those used in § 1. The continuation so defined is then the *analytic continuation* in the sense of complex function theory, by Theorem 9 (cf. Ch. 9, § 1).

‡ Cf. Courant, Vol. 2, pp. 95 ff., where the Bolzano–Weierstrass Theorem is proved in $\mathbf{R}^n$.

D: $|\mathbf{x} - \mathbf{d}| \le \epsilon$, $|t - b| \le \epsilon$ of $(\mathbf{d}, b)$ also in $\mathcal{R}$. Let $M = \max_D |\mathbf{X}|$. Take $\delta < \min(\epsilon, \epsilon/2M)$, and let $G \subset D$ be the open rectangle $|\mathbf{x} - \mathbf{d}| < \epsilon$, $|t - b| < \delta$. Finally, choose k so that $[\mathbf{x}(t_k), t_k] \in G$. Then, applying Theorem 8 (in G) to the solution through $[\mathbf{x}(t_k), t_k]$, we see that it stays in G until $t = b$. Since this is true for any $\epsilon > 0$, $\lim_{t \to b} \mathbf{x}(t) = \mathbf{d}$. Hence, $\mathbf{x}(t)$ would have to coincide with the unique (by Theorem 1) local solution of (2) through $(\mathbf{d}, b)$. Therefore, $\mathbf{x}(t)$ would not be maximal, a contradiction.

The maximum length $b - a$ of definition of the solution $\mathbf{x}$ is called the *escape time* of the solution for $t > a$. There is a similar notion of the escape time for $t < a$.

A solution with a *finite* escape time is one for which $|\mathbf{x}(t)|$ becomes unbounded or reaches the boundary of $\mathcal{R}$ as $t \to b < \infty$. On the other hand, a solution with an infinite escape time is one that remains within the domain of definition of X for all $t > a$. For example, every solution of the DE $dx/dt = x$ has infinite escape time, whereas every nonzero solution of the DE $dx/dt = x^2$, namely every function $x = 1/(c - t)$, has finite escape time.

*13 THE PERTURBATION EQUATION

It is easy to derive a formula for the dependence on $\mathbf{c}$ of the solution $\mathbf{x} = \mathbf{f}(t, \mathbf{c})$ of the initial-value problem defined by the system $\mathbf{x}'(t) = \mathbf{X}(\mathbf{x}, t)$ and the initial condition $\mathbf{x}(a) = \mathbf{c}$. For simplicity, consider first the case $n = 1$ of a single first-order DE. Assuming that $f(t, c)$ is analytic, that is, that f has a convergent Taylor series expansion, we have

(25)
$$\frac{\partial}{\partial t}\left(\frac{\partial f}{\partial c}\right) = \frac{\partial}{\partial c}\left(\frac{\partial f}{\partial t}\right) = \frac{\partial}{\partial c}[X(f(t, c), t)]$$
$$= \left[\frac{\partial X}{\partial x}(f(t, c), t)\right] \cdot \left[\frac{\partial f}{\partial c}(t, c)\right]$$

When we expand around $c = 0$, this gives formally

(26)
$$f(t, c) = f_0(t) + cf_1(t) + (c^2/2!)f_2(t) + \cdots$$

where, by (25), $f_1(t) = \partial f/\partial c(t, 0)$ satisfies the linear perturbation equation

(27)
$$f'_1(t) = \left[\frac{\partial X}{\partial x}(f_0(t), t)\right]f_1(t), \qquad f_1(0) = 1$$

Hence, if we know $f_0(t)$, we can compute $f_1(t)$ in closed form by quadrature (Ch. 1). Illustrations of this "perturbation method" are given in Exs. F5 and F6 below.

As simple examples show (see Exs. F7–F10 below), the approximate solutions obtained by linear perturbation are accurate near *stable* exact solutions of initial value problems, but they can be misleading near *unstable* solutions.

We now drop the assumption that X is a one-dimensional vector, as well as the assumption that the solution has a convergent Taylor expansion, and we derive analogous results. We show that the solutions of a normal first-order system (1) depend differentiably on the initial values, thus proving (at long last!) that the solution curves of any normal first-order DE or system form a *normal curve family*.

THEOREM 12.   *Let the vector function* $\mathbf{X}$ *be of class* $\mathscr{C}^1$, *and let* $\mathbf{x}(t, \mathbf{c})$ *be the solution of the normal system* (2), *taking the initial value* $\mathbf{c}$ *at* $t = a$. *Then* $\mathbf{x}(t, \mathbf{c})$ *is a continuously differentiable function of each of the components* c_j *of* $\mathbf{c}$.

The proof is subdivided into three steps.

A. Consider the system of DEs for the unknown functions h_i, the components of the vector $\mathbf{h} = (h_1, \ldots, h_n)$:

$$(28) \qquad \frac{dh_i}{dt} = \sum_{k=1}^{n} \frac{\partial X_i(\mathbf{x}(t, \mathbf{c}), t)}{\partial x_k} h_k + H_i(\mathbf{h}, t, \mathbf{c}, \eta_j)$$

where $\mathbf{x}(t, \mathbf{c})$ is the solution of the normal system (2) for which $\mathbf{x}(t, \mathbf{a}) = \mathbf{c}$. We assume that the functions H_i are bounded for $|t - a| \leq T$ and $|\mathbf{h} - \delta^j| \leq K$, where δ^j is the vector whose components are the Kronecker deltas δ_i^j. We also assume that H_i tends to zero as $\eta_j \to 0$, uniformly for $|t - a| \leq T$ and $|\mathbf{h} - \delta^j| \leq K$. We define $\mathbf{h}^j = \mathbf{h}(t, \mathbf{c}, \eta_j)$ as the solution of (28) that satisfied the initial condition $h_i^j(a) = \delta_i^j$. Applying the Corollary of Theorem 3, with $\epsilon = \eta_j$, we find that $\mathbf{h}^j$ tends, as $\eta_j \to 0$, to the solution f^j of the linear system

$$(29) \qquad \frac{df_i}{dt} = \sum_{k=1}^{n} \frac{\partial X_i(x(t,\mathbf{c}), t)}{\partial x_k} \cdot f_i$$

satisfying the same initial conditions, namely, $f_j^i(a) = \delta_i^j$.

In addition, we infer from the same Corollary that the vector functions $\mathbf{h}^j$ remain *bounded* as $\eta_j \to 0$.

B. Set

$$g_i^j(t, \mathbf{c}, \eta_j) = \frac{x_i(t, c_1, c_2, \ldots, c_{j-1}, c_j + \eta_j, c_{j+1}, \ldots, c_n) - x_i(t, \mathbf{c})}{\eta_j}$$

We next find a differential equation satisfied by the vector partial difference $\mathbf{g}_j = (g_1^j, g_2^j, \ldots, g_n^j)$.

By definition,

$$\frac{dg_i^j(t, \mathbf{c}, \eta_j)}{dt} = \eta_j^{-1}[X_i(\mathbf{x}(t, \mathbf{c}) + \eta_j \mathbf{g}^j(t, \mathbf{c}), t) - X_i(t, \mathbf{x}(t, \mathbf{c}))]$$

We now use the assumption that X_i is $\mathcal{C}^1$. By Taylor's theorem for functions of several variables, we infer that the right side equals

$$\sum_{k=1}^{n} \frac{\partial X_i(\mathbf{x}(t, \mathbf{c}), t)}{\partial x_k} \cdot g_k^j(t, \mathbf{c}) + \epsilon_i |\mathbf{g}_k(t, \mathbf{c})|$$

where ϵ_i is a function of t, $\mathbf{c}$, and η_j that tends to zero as $\eta_j \to 0$, uniformly as the variables t and $\mathbf{c}$ range over closed intervals. Setting $H_i(\mathbf{h}, t, \mathbf{c}, \eta_j) = \epsilon_i |\mathbf{h}(t, \mathbf{c})|$, we find that the vector function $\mathbf{g}^j$ is a solution $\mathbf{g}^j = \mathbf{h}$ of a system (28). The function $\mathbf{H}$ satisfies the conditions stated under Step A.

C. The initial conditions satisfied by the g_i^j are, by definition,

$$g_i^j(a, \mathbf{c}, \eta_j) = \frac{x_i(a, c_1, c_2, \ldots, c_{j-1}, c_j + \eta_j, c_{j+1}, \cdots, c_n) - x_i(t, \mathbf{c})}{\eta_j}$$

$$= \begin{cases} \dfrac{c_i - c_i}{\eta_j} = 0 & \text{if } i \neq j \\[2em] \dfrac{c_j + \eta_j - c_j}{\eta_j} = 1 & \text{if } i = j \end{cases}$$

Combining with the results of steps A and B, we conclude that, as $\eta_j \to 0$, the function $\mathbf{g}^j$ tends to the solution $\mathbf{h}^j$ of (25') satisfying the same initial condition. But we know that

$$\lim_{\eta_j \to 0} \mathbf{g}^j(t, \mathbf{c}, \eta_j) = \partial \mathbf{x}(t, \mathbf{c})/\partial c_j$$

We have, therefore, shown that the derivative $\partial \mathbf{x}/\partial c_j$ exists and is indeed a solution of (29), q.e.d.

The linear DE (27) is called the *perturbation equation* or *variational equation* of the normal system (2), because it describes approximately the perturbation of the solution caused by a small perturbation of the initial conditions.

In the course of the preceding argument we have also proved the following result.

COROLLARY. *If $\mathbf{x}(t, \mathbf{c})$ is a solution of the normal system (2) satisfying the initial condition $\mathbf{x}(a) = \mathbf{c}$ for each $\mathbf{c}$, and if each component of the function $\mathbf{X}$ is of class $\mathcal{C}^1$, then for each j the partial derivative $\partial \mathbf{x}(t, \mathbf{c})/\partial c_j$ is a solution of the perturbation equation (28) of the system.*

In the case of *linear* systems $d\mathbf{x}/dt = A(t)\mathbf{x} + \mathbf{b}(t)$, the perturbation equation is the reduced equation $d\mathbf{h}/dt = A(t)\mathbf{h}$ of the given system and is the same for all solutions. But in *nonlinear* systems, the perturbation equations (27) and (28) depend on the particular solution $\mathbf{x}(t, \mathbf{c})$ whose initial value is being varied.

Plane Autonomous Systems. We now apply the preceding results to the trajectories of autonomous systems. The main result is that, near any noncritical point, the trajectories of an autonomous system look like a regular family of parallel straight lines. We give the proof for the case $n = 2$. Recall that a plane autonomous system is one of the form

$$(30) \qquad \frac{dx}{dt} = X(x,y), \qquad \frac{dy}{dt} = Y(x,y)$$

THEOREM 13. *Any plane autonomous system where X and Y are of class $\mathcal{C}^1$ is equivalent under a diffeomorphism, in some neighborhood of any point that is not a critical point, to the system $du/dt = 1$, $dv/dt = 0$.*

Proof. Let the point be (a, b); without loss of generality, we may assume that $X(a, b) \neq 0$. Let the solution of the system for the initial values $x(0) = a$, $y(0) = c$ be $x = \xi(t, c)$, $y = \eta(t, c)$, so that $\partial \xi / \partial t = X$, $\partial \eta / \partial t = Y$. Then by Theorem 12, the transformation $(t, c) \mapsto (\xi(t, c), \eta(t, c))$ is of class $\mathcal{C}^1$. Moreover, since $x(0)$ does not vary with c, the Jacobian

$$\frac{\partial(\xi, \eta)}{\partial(t, c)} = \frac{\partial \xi}{\partial t} \cdot \frac{\partial \eta}{\partial c} - \frac{\partial \xi}{\partial c} \cdot \frac{\partial \eta}{\partial t} = X(a, b) \cdot 1 - 0 \cdot Y(a, b) = X(a, b)$$

is nonvanishing at (a, b). Hence, by the Implicit Function Theorem, the inverse transformation $u = t(x, y)$, $v = c(x, y)$ is of class $\mathcal{C}^1$. In the (u, v)-coordinates, the solutions reduce to $u = t$, $v = c = $ constant; hence, the DE assumes the form stated, q.e.d.

COROLLARY 1. *Any two plane autonomous systems are locally equivalent under a diffeomorphism, except near critical points.*

The system $\dot{u} = 1$, $\dot{v} = 0$ is, therefore, locally a canonical form for plane autonomous systems near noncritical points. In hydrodynamics, the velocity field associated with this system is called a uniform flow.

COROLLARY 2. *If the functions X and Y of the plane autonomous system (30) satisfy local Lipschitz conditions, then its integral curves form a regular curve family in any domain that contains no critical points.*

Proof. By Theorem 1, there is a unique integral curve of (26) passing through each point c, not a critical point. As shown in §12, each such integral curve goes all the way to the boundary. Finally, since Lipschitz conditions imply continuity, the directions of the vectors $(X(x, y), Y(x, y))$ vary continuously with position, except near a critical point, which completes the proof.

EXERCISES F

1. Let $F(x, y)$ be continuous for $|x - a| \leq T$, $|y - c| \leq K$. Show that the set of all solutions of $y' = F(x,y)$, satisfying the same initial condition $f(a) = c$, is equicontinuous.

2. Show that, if $\mathbf{X}(\mathbf{x}, t)$ is continuous and satisfies a Lipschitz condition for $a \leq t \leq b$, every solution of the DE (2) satisfying $\mathbf{x}(a) = \mathbf{c}$ is bounded for $a \leq t \leq b$. Show that the corresponding result is not true for open intervals $a < t < b$.

3. Let $xX(x,y) + y^3Y(x,y) = 0$, where X and Y are of class $\mathcal{C}^1$. Show that the system $x' = X(x,y)$, $y' = Y(x,y)$ has infinite escape time. [HINT: Show that $2x^2 + y^4$ is an integral of the system.]

4. Let the function $\mathbf{X}(\mathbf{x}, t)$ be defined for $0 \leq t < \infty$ and for all $\mathbf{x}$, and let

$$|\mathbf{X}(\mathbf{x}, t) - \mathbf{X}(\mathbf{y}, t)| \leq L(t)|\mathbf{x} - \mathbf{y}|$$

where $\int_0^\infty L(t) \, dt < \infty$. Show that the DE $d\mathbf{x}/dt = \mathbf{X}(\mathbf{x}, t)$ has a solution on $0 \leq t < +\infty$ for every initial condition $\mathbf{x}(a) = \mathbf{c}$. Show that, if one solution is bounded, then all are.

5. Let $\mathbf{X}(\mathbf{x}, t, s)$ be of class $\mathcal{C}^1$ for $|\mathbf{x} - \mathbf{c}| \leq K$, $|t - a| \leq T$, $|s - s_0| \leq S$. Let $\mathbf{x}(t, s)$ be the solution of $\mathbf{x}' = \mathbf{X}(\mathbf{x}, t, s)$ satisfying $\mathbf{x}(a) = \mathbf{c}$. Show that $\mathbf{x}$ is a differentiable function of s.

*6. Under the assumptions of Ex. 5, suppose that $\mathbf{X}(\mathbf{x}, t, s)$ is of class $\mathcal{C}^n$. Show that $\mathbf{x}(t, s)$ has n continuous partial derivatives relative to s.

*7. Show that if there are two distinct solutions f and g of $y' = F(x, y)$ satisfying the same initial condition $c = f(a) = g(a)$ (F continuous in $|x - a| \leq T$, $|y - c| \leq K$), there are infinitely many of them.

*8. Show that there is a *maximal* and a *minimal solution* $f_M(x)$ and $f_m(x)$ of the DE in Ex. 7, such that $f_m(x) \leq f(x) \leq f_M(x)$ for any other solution f such that $f(a) = f_M(a) = f_m(a)$. [HINT: See Ch. 1, Ex. F4.]

*9. Let $F(x, y)$ and $G(x, y)$ be continuous for $a \leq x \leq T$, $|y - c| \leq K$, and $F(x, y) \leq G(x, y)$. Let f be a solution of $y' = F(x, y)$, and let g be the maximal solution of $y' = G(x, y)$. Show that, if $f(a) \leq g(a)$, then $f(x) \leq g(x)$ for $x > a$.

ADDITIONAL EXERCISES

*1. Let $dx/dt = X(x, y, t)$ and $dy/dt = Y(x, y, t)$, where

$$(x - x')[X(x, y, t) - X(x', y', t)] + (y - y')[Y(x, y, t) - Y(x', y', t)]$$

is everywhere negative or zero. Show that, for $t > 0$, the above system has at most one solution satisfying a given initial condition at $t = 0$.

In Exs. 2–4, f'_+ means the right-derivative; prove the implication specified. You may assume the existence of f'_+ and g'_+ freely.

2. If $f'_+(x) \leq g'_+(x)$, then $f(x) - f(y) \leq g(x) - g(y)$ for $x \geq y$.

3. If $|f'_+(x)| \leq K|f(x)|$ then $|f(x)| \leq |f(a)|e^{K|x-a|}$ for $x \geq a$.

4. If $|f'_+(x)| \leq K|f(x)| + \epsilon$, then $|f(x)| \leq |f(a)|e^{K|x-a|} + (\epsilon/K)(e^{K|r-a|} - 1)$.

5. Let $dz_i/dt = Q_i(z_1, \ldots, z_n)$, where the Q_i are quadratic polynomials. Show that, for any initial condition, the nth Picard approximation to the solution is a polynomial in t of degree at most $2^n - 1$.

6. (a) Prove that, if there is a normal kth-order ordinary DE satisfied by two functions u and v and if $n > k$, there is a normal nth-order DE satisfied by both functions. State your differentiability assumptions.

   (b) Prove that, if the given kth-order DE is linear, then the nth-order DE can also be chosen to be linear.

(c) Prove that there is no fourth-order normal DE $u^{iv} = F(u, u', u'', u''', t)$ satisfied by both $u = t^4$ and $v = t^6$ for all real t.

(d) Prove that $u = t^6$ satisfies no normal linear homogeneous DE of order six or less with continuous coefficients.

7. Show that, if $X_1, \ldots, X_n$ satisfy Lipschitz conditions on a compact domain, so does any polynomial function of the X_i.

8. Show that, if $X(t) = \|x_{ij}(t)\|$ is a matrix whose columns are solutions of the homogeneous linear system $X' = A(t)X$, then $\det X(t) = [\det X(a)] \exp \int_a^t \Sigma\, a_{kk}(s)\, ds$.

9. A matrix $X(t)$ is a *fundamental matrix* for $a \le t \le a + T$ of a homogeneous linear system $X' = A(t)X$ if its columns are solutions of the system and $\det (X(t)) \neq 0$. Show that, if the columns of X are solutions of the system and if $\det X(a) \neq 0$, then X is a fundamental matrix.

*10. Show that, if $X(t)$ is a fundamental matrix of the reduced linear system, the function $x(t) = X(t) \int_a^t X^{-1}(s)b(s)\, ds$ is the solution of the inhomogeneous system such that $x(a) = 0$ (X^{-1} is the matrix inverse of X).

CHAPTER 7

APPROXIMATE SOLUTIONS

1 INTRODUCTION

During the past 40 years, the accurate numerical solution of initial value problems for ordinary DEs has become routine, because of the availability of high-speed programmable computers. Even fairly large *systems* of DEs can be treated similarly in many cases, although "stiff" systems involving time scales of different orders of magnitude can be troublesome.

This development has not only made the study of classical numerical methods (e.g., Runge–Kutta methods) more important, as practical substitutes for involved analytical considerations, it has also increased interest in *numerical mathematics* from a theoretical standpoint. In particular, the *power series* methods explained in Chapter 4, together with techniques of numerical linear algebra, have provided the basis for a new field of research.

Because of this changed emphasis, a few simple numerical methods for solving DEs were already described in Chapter 1, §8. In this chapter and the next, we will treat the numerical solution of ordinary DEs and systems of DEs more carefully. This chapter will concentrate on the underlying ideas, while the effective technical implementation of these ideas will be the subject of Chapter 8.

Since these ideas are applicable to *systems* of first-order DEs, we will adopt throughout Chapters 7 and 8 the vector notation introduced in Chapter 5. Thus, we will consider vector DEs of the form

$$(1) \qquad \mathbf{x}'(t) = \mathbf{X}(\mathbf{x}, t), \qquad a \leqq t \leqq a + T$$

However, since writing and "debugging" computer programs for systems of DEs can be very time-consuming, most students will probably find it more satisfactory to interpret all statements and formulas in the conceptually simpler context of $y' = F(x,y)$, the case of a single ordinary first-order DE discussed in Chapter 1.

The basic idea involved, that one can use simple *arithmetic* to compute approximate solutions of DEs, is a very natural one. Indeed, the simple methods to be analyzed in this chapter were mostly known to Euler. However, their rigorous *error analysis* is more recent, having achieved a definitive form only around 1900.

204

Approximate Function Tables. The most effective methods for obtaining approximate solutions of DEs compute in each case [i.e., for each DE (1) and initial value $\mathbf{x}(a) = \mathbf{c}$] an approximate *function table*. Given any *partition*

$$(2) \qquad \pi: \quad a = t_0 < t_1 < t_2 < t_3 < \cdots < t_n = a + T$$

of an interval $[a, a + T]$ of interest by a sequence of *mesh points* t_k, it produces a sequence of *approximate values* $\mathbf{x}_\pi(t_k)$, nearly equal to the "true" values $\mathbf{x}(t_k)$ of the exact solution, whose existence and uniqueness was proved in Chapter 6.

The difference $\mathbf{e}_\pi(t_k) = \mathbf{x}_\pi(t_k) - \mathbf{x}(t_k)$ is the *error* (or "discretization error") of the method, and this chapter will be mainly concerned with the *error analysis* of the methods discussed.

2   ERROR BOUNDS

Cauchy Polygon Method. The simplest way to construct an approximate function table for the solution of the DE (1) satisfying the initial condition $\mathbf{x}(a) = \mathbf{c}$, on a given set of mesh points t_k, is the *Euler method* of Ch. 1, §8. This constructs from $\mathbf{x}_0 = \mathbf{x}(a) = \mathbf{c}$ the sequence of values

$$(3) \qquad \mathbf{x}_0 = \mathbf{c}, \qquad \mathbf{x}_k = \mathbf{x}_{k-1} + \mathbf{X}(\mathbf{x}_{k-1}, t_{k-1})(t_k - t_{k-1}), \qquad k = 1, \ldots, m$$

This formula is *recursive;* each value $\mathbf{x}_k$ can be computed knowing $\mathbf{x}_{k-1}$ alone.

From the approximate function table just defined, one can also construct an *approximate solution* by linear interpolation. This approximate solution is defined by the formula

$$(3') \qquad \mathbf{x}_n(t) = \mathbf{x}_{k-1} + \mathbf{X}(\mathbf{x}_{k-1}, t_{k-1})(t - t_{k-1}) \qquad \text{on} \qquad [t_{k-1}, t_k]$$

Evidently, the graph of the approximate solution (3′) consists of m segments of straight lines; it is a *polygon* in the $(n + 1)$-dimensional $(t, \mathbf{x})$-space. The function defined by (3) and (3′) for each partition π and initial value $\mathbf{c}$ is called the *Cauchy polygon* approximation to the solution, for that partition.†

Example 1. When the DE (1) is of the special form $x' = f(t)$, the preceding method reduces to the Riemann sum formula of Ch. 1, (5′):

$$(4) \qquad \int_a^{t_m} f(t)\, dt \simeq \sum_1^m f(t_{k-1})\, \Delta t_k, \qquad \Delta t_k = t_k - t_{k-1}, \qquad t_0 = a, \qquad t_k < t_{k+1}$$

where the symbol $\simeq$ is to be read "is approximately equal to." The proof of convergence to the exact solution, in this case, is the essence of Riemann's theory of integration.

---

† It was Cauchy who first proved their convergence to exact solutions, though Euler had used "Cauchy polygons" a century earlier.

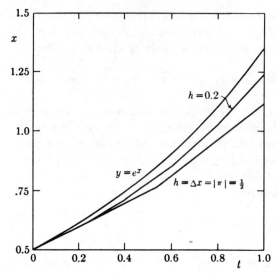

Figure 7.1   Cauchy polygons for $dx/dt = x$, $x(0) = 1/2$.

An error bound for the Euler–Cauchy polygon approximation can be derived in any closed bounded domain D, for any vector DE (1) whose right-hand side $\mathbf{X}(\mathbf{x},t)$ is continuous and satisfies a Lipschitz condition

(5) $$|\mathbf{X}(\mathbf{x},t) - \mathbf{X}(\mathbf{y},t)| \leq L|\mathbf{x} - \mathbf{y}|$$

This bound also depends on the *norm* of the partition π,

(5′) $$|\pi| = \max(\Delta t_1, \ldots, \Delta t_m) = \max_{k=1, \ldots, m} (t_k - t_{k-1})$$

and on the maximum M of $|\mathbf{X}(\mathbf{x},t|$ in D. As the following theorem states, this bound is roughly proportional to L, $|\pi|$, M, and the length $T = t - a$ of the interval of integration.

THEOREM 1.   *Let $\mathbf{X} \in \mathcal{C}^1$ satisfy $|\mathbf{X}| \leq M$, $|\partial\mathbf{X}/\partial t| \leq C$, and (5) in the cylinder $a \leq t \leq a + T$, $|\mathbf{x} - \mathbf{c}| \leq MT$. Then the Cauchy polygon approximation $\mathbf{x}_\pi(t)$ differs from the true solution $\mathbf{x}(t)$ by at most*

(6) $$|\mathbf{x}_\pi(t) - \mathbf{x}(t)| \leq \left[\frac{C}{L} + M\right][e^{L(t-a)} - 1] \cdot |\pi|$$

The proof of Theorem 1 will be presented and its significance explained in §3. Here we emphasize that the inequality (6) only provides an *upper bound* to the error. Because $|\pi|$ is multiplied by a bounded factor, Theorem 1 asserts that the error is $O(|\pi|)$; hence it is $O(h)$ in the case of a uniform mesh with constant step size $\Delta t_k = h$.

However, as examples described in Exercises A show, the magnitude of the true error may be very much smaller than the bound (6). Therefore, in most practical computation, one relies on less general formulas. The basic fact is that, in the important case $t_k = a + kh$ of a *uniform* mesh with mesh length h, the error committed in using the Euler–Cauchy polygon approximation is usually nearly proportional to h; see Figure 7.1.

Example 2.   Consider the DE $y' = y$ on $[0,1]$, for the initial value $y(0) = 1$. The exact solution is e^x, with final value $e = 2.71828182853 \cdots$.

As was stated in Ex. E8 of Ch. 1, the final value of $x_\pi(1)$ computed by Euler's method,

$$(*) \qquad e_n = (1 + h)^n = 1 + nh + n\frac{(n - 1)h^2}{2} + \binom{n}{h}\frac{h^3}{3!} + \cdots$$

is asymptotically $e - (h - \frac{11}{12}h^2 + \cdots) e/2$. This fact can also be deduced from formula (*) (cf. Ex. A2 below).

Note that, in Examples 1 and 2, the error made in each individual step is only $O(h^2)$. Since the number of steps is proportional to $1/h$, the cumulative error is still $O(h)$. More generally, the order of magnitude of the cumulative error made in integrating a first-order DE or system is an infinitesimal of order one less than that of the error per step. It is the same as that of the *relative error* per step, defined as the error divided by the length of the step.

*3   DEVIATION AND ERROR

This section will be devoted to proving Theorem 1, that Euler's method has $O(h)$ accuracy. The proof will be based on a new concept: the *deviation* of a function from a DE. This concept is of theoretical interest in its own right.

DEFINITION.   A vector-valued function $\mathbf{y}(t)$ is an *approximate solution* of the vector DE (1), with *error at most* η, when $|\mathbf{y}(t) - \mathbf{x}(t)| < \eta$ for all $t \in [a, a + T]$. Its *deviation* is at most ϵ when $\mathbf{y}(t)$ is continuous, and satisfies the differential inequality

$$(7) \qquad\qquad |\mathbf{y}'(t) - \mathbf{X}(\mathbf{y}(t), t)| \le \epsilon$$

for all except a finite number of points t of the interval $[a, a + T]$.

Note that the definition requires the function $\mathbf{y}$ to be differentiable, except at a finite, possibly empty, set of points. Such a function is said to be of class $\mathcal{D}^1$.

The following example shows that an approximate solution can have a small deviation without having a small error. It is essentially Example 8 of Ch. 1, §9; note that the DE involved does not satisfy a Lipschitz condition.

Example 3.   The function $y(t) \equiv 10^{-6}$ is an approximate solution of the DE $dx/dt = 3x^{2/3}$ on the interval $[0, \infty)$, with deviation 0.0003. The exact solution to this DE for the initial value $x(0) = 10^{-6}$ is $x = (t + 0.01)^3$. At $t = 1$, it assumes the value $x(1) = 1.030301$ instead of $y(1) = 0.000001$.

Theorem 1 asserts, among other things, that the preceding phenomenon cannot arise if $X(x;t)$ satisfies a Lipschitz condition (5). For such functions, we can always make the *deviation* arbitrarily small by making the norm of the partition sufficiently small. This result is contained in the following theorem.

THEOREM 2.   *Let $X \in \mathscr{C}^I$ satisfy $|X| \leq M$, $|\partial X/\partial t| \leq C$, and (5) in the cylinder D: $|x - c| \leq K$, $a \leq t \leq a + T$. Then any Cauchy polygon in D is an approximate solution of $x'(t) = X(x,t)$ with deviation at most $(C + LM)|\pi|$.*

In proving this theorem, we will use the fact that any Cauchy polygon approximation is continuous in $[a, a + T]$, and is differentiable at all points not mesh points. At these, it still has a left and a right derivative.

Proof.   On each subinterval (t_i, t_{i+1}) of π, it is clear, by (5), that $|X(x(t),t) - X(x_i,t)| \leq L|x(t) - x_i|$;

$$|X(x(t),t) - X(x_i,t)| \leq LM|t - t_i| \leq LM|\pi|$$

since $|x(t) - x_i| = |\int_{t_i}^{t} X(x(s),s)\, ds| \leq M|t - t_i|$. Also

$$|X(x_i,t) - X(x_i,t_i)| = \left| \int_{t_i}^{t} \partial X/\partial t(s)\, ds \right| \leq C|\pi|$$

Adding together the two inequalities just obtained, and using the triangle inequality, we get the desired conclusion:

$$(8) \qquad\qquad |X(x(t),t) - X(x_i,t_i)| \leq (LM + C)|\pi|$$

We now prove a theorem that yields as a corollary an easily computed *a priori* error bound for the Cauchy polygon method in terms of $|\pi|$, $|X|_{max}$, the Lipschitz constant, and the deviation.

THEOREM 3.   *Let $x(t)$ be an exact solution and $y(t)$ an approximate solution with deviation ϵ, of the DE $x'(t) = X(x,t)$. Let X satisfy the Lipschitz condition (5). Then, for $t \geq a$, we have*

$$(9) \qquad |x(t) - y(t)| \leq |x(a) - y(a)|e^{L(t-a)} + \left(\frac{\epsilon}{L}\right)(e^{L(t-a)} - 1)$$

Proof.   Consider $\sigma(t) = |x(t) - y(t)|^2$. Differentiating,

$$\sigma'(t) = 2[X(x(t),t) - X(y(t),t)] \cdot [x(t) - y(t)]$$
$$+ 2[X(y(t),t) - y'(t)] \cdot [x(t) - y(t)]$$

Hence, adding inequalities, we obtain

$$\sigma'(t) \leq 2L\sigma(t) + 2\epsilon\sqrt{\sigma(t)}$$

Now, set $\sigma = v^2$; the foregoing gives $v' \leq Lv + \epsilon$ (for $\sigma > 0$). Applying Theorem 7 of Ch. 1, §12, we get the desired inequality (9), much as in proving the lemma of Ch. 6, §4.

A slight variant of the analysis leading to Theorem 3 yields a closely related bound to the cumulative error of the Cauchy polygon approximation, as follows.

Define the *directional derivative* $\partial X/\partial\xi$ of the vector function $\mathbf{X}(\mathbf{x},t)$ in the direction $\xi = (\xi_0,\xi_1, \ldots , \xi_n)$ in $t,\mathbf{x}$-space, for any vector ξ of unit length, as the sum $\xi_0\,\partial\mathbf{X}/\partial t + \sum_{k=1}^{n} \xi_k\,\partial\mathbf{X}/\partial x_k$. It follows as in the proof of the lemma of Ch. 6, §2, that

$$|\mathbf{X}(t,\mathbf{x}) - \mathbf{X}(u,\mathbf{y})| \leq |\partial\mathbf{X}/\partial\xi| \cdot |(t,\mathbf{x}) - (u,\mathbf{y})|$$

where ξ is the unit vector in $(t,\mathbf{x})$-space pointing in the direction $(t - u, \mathbf{x} - \mathbf{y})$. This inequality gives a bound on the change in $\mathbf{X}(\mathbf{x},t)$ along any side of a Cauchy polygon, which we now use to complete the proof of Theorem 1.

Proof of Theorem 1. The inequality (6) of Theorem 1 is an immediate corollary of Theorems 2 and 3. Under the hypotheses of Theorem 1, the deviation of $\mathbf{x}_\pi(t)$ is by Theorem 2 at most $\epsilon = (C+LM)|\pi|$. Since $\mathbf{x}_\pi(a) = \mathbf{x}(a)$, the first term of the inequality (9) vanishes if we let $\mathbf{y}(t)$ be the Cauchy polygon approximation for the initial value $\mathbf{x}(a)$ in Theorem 3, and so (9) simplifies to

(*) $$|\mathbf{x}(t) - \mathbf{x}_\pi(t)| \leqq \left[(C + LM)|\pi| \right] [e^{L(t-a)} - 1]$$

This yields (6) by elementary algebra.

In particular, by setting

$$N = \left[\frac{C}{L} + M \right] [\exp LT - 1]$$

we obtain the following simple corollary of (6).

COROLLARY. *Under the hypotheses of Theorems 1 and 2, let the interval $[a, a + T]$ be divided into n equal parts of length $h = T/n$. Then the error of the Cauchy polygon approximation is bounded by Nh, where N is a constant independent of h.*

EXERCISES A

1. (a) What is the deviation of the approximate solution $x = t^2/2 - t^4/24$ of the initial value problem defined by $dx/dt = \sin t$, $x(0) = 0$ on the interval $0 \leq t \leq 1$?
   (b) Compare the difference $1 - (\cos 1) - \frac{11}{24}$ with the bound given by formula (*), for the deviation computed in (a).
   (c) For the initial value $x(0) = 1$, bound the difference between the solutions of $dx/dt = \sin t$ and $dx/dt = t - (t^3/6)$.

In Exs. 2–5, for the initial value problem specified: (a) use the Cauchy polygon method to compute an approximate function table for $t_k = 0.1, 0.2, \ldots, 1.0$; (b) find the deviation of the approximate solution obtained from this table by linear interpolation; (c) find the exact solution; (d) find the error.

2. $\dot{x} = x, \quad x(0) = 1$ 3. $\dot{x} = 1 - 2x, \quad x(0) = 0$

4. $\dot{x} = y, \quad \dot{y} = -x, \quad x(0) = 0, \quad y(0) = 1$ 5. $\dot{x} = y, \quad \dot{y} = x, \quad x(0) = 1, \quad y(0) = 0$

6. (a) Find the deviation of the approximate solution $y \equiv 10^{-10}$ of the DE $x'(t) = 5x^{4/5}$.

    (b) What is the exact solution of this DE on $[0, \infty)$ for the initial value $x(0) = 10^{-10} = y(0)$?

    (c) Prove in detail the uniqueness of this solution.

7. On the interval $[0, 1]$, for any $\epsilon > 0$, construct an approximate solution with deviation ϵ to a suitable first-order DE, for which the exact solution with the same initial value is unbounded.

*8. For the DE $x'(t) = f(t) - x$, show that

$$|x_n(t) - x(t)| \le C|\pi|, \qquad \text{where} \qquad C = \sup |f'(t)|$$

*9. (a) Sharpen (9) and (*) in the stable case $[X(x, t) - Y(x, t)] \cdot [x - y] \le 0$. Compare with the limiting case $L = 0$ of these formulas.

    (b) When $X(x, t)$ satisfies the one-sided Lipschitz condition $[X(x, t) - X(y, t)] \cdot [x - y] \le L |x - y|^2$, how can (9) and (*) be sharpened? [HINT: See Ex. 8.]

4   MESH-HALVING; RICHARDSON EXTRAPOLATION

In practical computation, one can often reduce the error by a large factor by accepting as a *working hypothesis*, the theoretical result that, in a wide variety of situations, the truncation error† under repeated mesh-halvings is of the form

(10) $$z(x_k, h) - y(x_k) = Ch^\nu + O(h^{\nu+1})$$

As has just been emphasized, the *order of accuracy* $\nu = 1$ for the Euler–Cauchy polygon method. For the modified and improved Euler's methods to be discussed later in this chapter, $\nu = 2$. For other methods to be discussed in Chapter 8, $\nu = 4$.

If one knows ν *a priori*, as one does for the methods of Euler just mentioned, one can determine the unknown constant C in (10) with fair accuracy, by comparing the computed value Y_0 for a given partition π_0, with the corresponding value Y_1 for the partition π_1 obtained from it by mesh-halving.

This is because formula (10) implies that

(11) $$Y_1 - y(x_k) = \frac{1}{2^\nu} [(Y_0 - y(x_k)] + O(h^{\nu+1})$$

---

† The truncation (or "discretization") error is the error that would occur if computer floating point arithmetic were exact. See the discussion of roundoff error at the end of this section.

Comparing with (10), we obtain,

$$(12) \qquad y(x_k) = \frac{[2^\nu Y_1 - Y_0]}{(2^\nu - 1)} + O(h^{\nu+1})$$

The approximate value of $y_k = y(x_k)$ obtained by suppressing the $O(h^{\nu+1})$ term in (12) is said to be obtained by *Richardson extrapolation*. This name is given to honor the inventor of the method, L. F. Richardson, who called it "deferred approach to the limit."

Note that formula (12) *corrects* each computed value Y_h by adding $(Y_h - Y_{2h})/(2^\nu - 1)$ to it. Thus, suppose we compute e^x as the solution of $y' = y$ for the initial value $y(0) = 1$ by the Euler-Cauchy polygon method for $h = 2^{-m}$ ($m = 0,1,2,3,4$). Then $\nu = 1$, so that the *corrected* value is

$$(13a) \qquad Y = 2Y_h - Y_{2h}.$$

The resulting approximate values of $e^1 = (1 + h)^{2m}$ are tabulated in Table 7.1, together with their errors and the better approximations obtained using (13a).

For the improved Euler method (Heun's method), the approximate value $Y_h = (1 + h + h^2/2)^{2m}$ has $0(h^2)$ accuracy. Since $\nu = 2$, the corrected value is

$$(13b) \qquad Y = Y_h + \frac{1}{3}(Y_h - Y_{2h}) = \frac{1}{3}(4Y_h - 3Y_{2h})$$

The improvement made by applying Richardson extrapolation to this method is shown in Table 7.2.

The final error is reduced by a factor of nearly $8 = 2^3$ each time that the mesh-length is halved.

Checking ν. A good practical check on the reliability of Richardson approximation consists in *verifying* that $Y_{4h} - Y_{2h}$ is indeed about 2^ν times $Y_{2h} - Y_h$. When ν is unknown, one can also estimate it by *assuming* this same formula, for all h. Summing the geometric series $\sum_{k=}^{\infty} 2^{-\nu} = 1/(2^\nu - 1)$, we obtain after some algebraic manipulation the following extrapolated approximation Y to the limiting value y of the series $Y_h, Y_h/2, Y_h/4, \ldots$

$$(14) \qquad Y = Y_h - \frac{(Y_h - Y_{2h})^2}{Y_{4h} - 2Y_{2h} + Y_h}$$

Table 7.1. Richardson Extrapolation of Euler's Method

$h =$	$\frac{1}{2}$	$\frac{1}{4}$	$\frac{1}{8}$	$\frac{1}{16}$
Y_h	2.25	2.44141	2.56574	2.63793
Error	.46828	.27687	.15254	.08035
$2Y_h - Y_{2h}$	2.5	2.63282	2.69007	2.71072
Error	.21828	.08645	.02821	.00816

Table 7.2.   Richardson Extrapolation of Heun's Method

$h =$	$\frac{1}{2}$	$\frac{1}{4}$	$\frac{1}{8}$	$\frac{1}{16}$
Y_h	2.640625	2.694856	2.711841	2.716593
Error	.077657	.023426	.006441	.001689
$(4Y_h - 3Y_{2h})/3$	2.6875	2.712933	2.717503	2.718177
Error	.030782	.005349	.000799	.000105

Caution.   Although valid for sufficiently small h, formula (14) with $h = \frac{1}{4}$ overcorrects the computed value 2.694856 in Table 7.2, and overcorrects 2.44141 very badly in Table 7.1.

Roundoff Errors.   The preceding discussion has set no limits to the fineness of the mesh used in solving DEs numerically, and it has been tacitly assumed that all arithmetric operations and function evaluations are *exact*. Actually, however, the floating-point arithmetic on many computers has an accuracy of only around 10^{-7}. On such computers, the dominant source of error when $h = 1/1024$ (say) may well be due to so-called "roundoff errors" in floating-point arithmetic. This is especially likely if values of the x_k that are not exact "binary decimals" are used—e.g. if $h = .001$ is used instead of $h = 1/1024$.

Roundoff errors will be discussed again in Chapter 8, §6.

5   MIDPOINT QUADRATURE

As we have observed (Example 2 above), the relative error made in computing e by solving $y' = y$ on $[0, 1]$ for the initial condition $y(0) = 1$ by the improved Euler method of Ch. 1, §8, is $-h^2/3 + h^4/4 + O(h^4)$. In this section, we shall derive some much more accurate error formulas for evaluating definite integrals by the midpoint and trapezoidal formulas (i.e., for solving the DE $y' = F(x)$). This can be viewed as lending further credence to the Richardson extrapolation method of §4.

The simplest formula for numerical quadrature having a higher order of accuracy than the Cauchy polygon formula (4) is the *midpoint quadrature* formula

$$(15) \qquad \int F(x)\, dx \simeq M_\pi[F] = \sum_{i=1}^{n} F(m_i)\, \Delta x_i, \qquad m_i = \frac{(x_{i-1} + x_i)}{2}$$

Given the partition π of the interval of integration $[a, b]$ by points of subdivision $a = x_0 < x_1 < \cdots < x_n = b$, the midpoint approximation $M_\pi[F]$ is easily computed; it takes its name from the fact that m_i is the midpoint of the ith interval of subdivision. We now derive an *error bound* for the midpoint quadrature formula (15).

THEOREM 4.   *If $F \in \mathcal{C}^2$, then*

(16)
$$\left| \int_a^b F(x)\, dx - \sum_{i=1}^n F(m_i)\, \Delta x_i \right| \leq \frac{|F''|_{\max} |\pi|^2 (b-a)}{24}$$

Proof.   On each interval $[x_{i-1}, x_i] = [m_i - \Delta x_i/2,\ m_i + \Delta x_i/2]$, Taylor's formula implies that

$$F(m_i + t) - F(m_i) - t F'(m_i) = \frac{t^2 F''(m_i + r)}{2}$$

where r is between 0 and t. But $F''(m_i + r)$ is bounded below by the minimum $F''_{\min}$ of $F''(x)$ on $[a, b]$ and above by its maximum value $F''_{\max}$. Therefore, we have

$$\frac{F''_{\min}\, t^2}{2} \leq F(m_i + t) - F(m_i) - t F'(m_i) \leq \frac{F''_{\max}\, t^2}{2}$$

Integration of this inequality over $-\Delta x_i/2 \leq t \leq \Delta x_i/2$ gives

$$\frac{F''_{\min}\, \Delta x_i^3}{24} \leq \int_{x_{i-1}}^{x_i} F(x)\, dx - F(m_i)\, \Delta x_i \leq \frac{F''_{\max}\, \Delta x_i^3}{24}$$

Summing over i and noting that $0 \leq \Delta x_i^2 \leq |\pi|^2$, we get (16).

   Theorem 4 shows that the midpoint quadrature formula (15) has order of accuracy $O(h^2)$, one order higher than the Cauchy polygon method.

Error Estimate.   In the case of subdivisions into intervals of constant length h, we can obtain a much more accurate estimate of the error in the midpoint quadrature formula by considering the higher-order terms in Taylor's formula.

THEOREM 5.   *Any function for $F \in \mathcal{C}^6$ on a uniform mesh with constant mesh length $\Delta x_i = 2k = h$,*

(17)
$$M_\pi[F] = \int_a^b F(x)\, dx - \frac{h^2}{24} [F'(b) - F'(a)]$$
$$+ \frac{7h^4}{5760} [F''(b) - F''(a)] + O(h^6)$$

Proof.   By Taylor's formula with remainder, since $F \in \mathcal{C}^6$, we have

$$F(m_i + t) = \sum_{r=0}^5 \frac{F^{(r)}(m_i) t^r}{(r!)} + \frac{F^{(6)}(\xi) t^6}{720}$$

where ξ is some number between m_i and $m_i + t$. On each ith interval (x_{i-1}, x_i), the final term ("remainder") is bounded in magnitude by $Mk^6/720$, where $M = \max |F^{(6)}(\xi)|$, the maximum being taken on the entire interval $a \leq \xi \leq b$. Integrating over $-k \leq t \leq k$, we get

$$\int_{x_{i-1}}^{x_1} F(x)\, dx = 2kF(m_i) + \frac{k^3 F''(m_i)}{3} + \frac{k^5 F^{iv}(m_i)}{60} + O(k^6)\, \Delta x_i$$

where the factor $O(k^6)$ is bounded in magnitude by $Mk^6/720$. When we sum over i, there results the estimate

$$(18) \quad \sum_{i=1}^{n} F(m_i)\, \Delta x_i = \int_a^b F(x)\, dx - \left(\frac{k^2}{6}\right) \sum_{i=1}^{n} F''(m_i)\, \Delta x_i$$

$$- \left(\frac{k^4}{120}\right) \sum_{i=1}^{n} F^{iv}(m_i)\, \Delta x_i + O(k^6).$$

An application of (18) to the function $F''(x) \in \mathcal{C}^4$ gives similarly (one term being dropped because of the loss in differentiability),

$$(18') \quad \sum_{i=1}^{n} F''(m_i)\, \Delta x_i = \int_a^b F''(x)\, dx - \left(\frac{k^2}{6}\right) \sum_{i=1}^{n} F^{iv}(m_i)\, \Delta x_i + O(k^4)$$

Applied to $F^{iv}(x) \in \mathcal{C}^2$, this gives

$$(18'') \qquad \sum_{i=1}^{n} F^{iv}(m_i)\, \Delta x_i = \int_a^b F^{iv}(x)\, dx + O(k^2)$$

Substituting from (18') and (18'') back into (18), and combining terms, we get

$$\sum_{i=1}^{n} F(m_i)\, \Delta x_i = \int_a^b F(x)\, dx - \frac{k^2}{6} \int_a^b F''(x)\, dx + \frac{7k^4}{360} \int_a^b F^{iv}(x)\, dx + O(k^6)$$

When we set $k = h/2$, formula (17) follows immediately.

Note that the error estimate (17) implies the very accurate *corrected* midpoint formula

$$(19) \quad \int_a^b F(x)\, dx = \sum_{i=1}^{n} F(m_i)\, \Delta x_i + \frac{h^2[F'(b) - F'(a)]}{24}$$

$$- \frac{7h^4[F''(b) - F''(a)]}{5760} + O(h^6)$$

EXERCISES B

In each of Exs. 1–4, a numerical quadrature formula is specified for approximately evaluating $\int_{-h}^{h} f(x)\, dx$. In each case: (a) compute the truncation error for $f(x) = x^n$, $n = 0$, 1, 2, 3, . . . , and (b) find the order of accuracy of the formula, using Taylor's formula with remainder, assuming $f(x)$ to be analytic.

1. Simpson's rule: $S[f] = \dfrac{h}{3}\,[f(-h) + 4f(0) + f(h)]$.

2. Cotes' rule: $C[f] = \dfrac{h}{4}\left[f(-h) = 3f\left(\dfrac{-h}{3}\right) + 3f\left[\dfrac{h}{3}\right] + f(h) \right]$.

3. Weddle's rule:

$$W[f] = \dfrac{h}{10}\left[f(-h) + 5f\left(-\dfrac{2h}{3}\right) + f\left(-\dfrac{h}{3}\right) + 6f(0) + f\left(\dfrac{h}{3}\right) + 5f\left(\dfrac{2h}{3}\right) + f(h) \right]$$

4. Hermite rule: $H[f] = h[f(h) + f(-h)] - \dfrac{h^2}{3}\,[f'(h) - f'(-h)]$.

*5. In Ex. 3, find weighting coefficients w_k such that the approximation

$$w_0 f(-h) + w_1 f\left(-\dfrac{2h}{3}\right) + w_2 f\left(-\dfrac{h}{3}\right) + w_3 f(0) + w_4 f\left(\dfrac{h}{3}\right) + w_5 f\left(\dfrac{2h}{3}\right) + w_6 f(h)$$

to $\int_{-h}^{h} f(x)\, dx$ has a maximum order of accuracy. Compare with Weddle's rule.

6. (a) Show that if $F(x) = 1/x$, then $(F'(2) - F'(1))/24 = \frac{1}{32}$ and $[F'''(2) - F'''(1)]/5760 = 1/1024$.
   (b) Infer that $\ln 2 = M_\pi(f) + h^2/32 - 7h^4/1024 + O(h^6)$.
   (c) Knowing that $\ln 2 = .69317408$, compare with numerical experiments.

7. (a) Show that all odd-ordered derivatives $F^{(2n+1)}(0)$ of $F(x) = 1/(1 + x^2)$ vanish when $x = 0$.
   (b) Show that $F'(1) = -\frac{1}{2}$ and $F'''(1) = \frac{3}{4}$.
   (c) Knowing that $\pi/4 = \arctan 1 = \int_0^1 dx/(1 + x^2)$, derive the formula

$$\pi/4 = M_\pi[F] + \dfrac{h^2}{48} - \dfrac{7h^4}{7680} + O(h^6)$$

*8. Derive formulas similar to those of Exs. 6–7 for
   (a) $\int_0^1 \sqrt{1 + x^4}\, dx$,   and   (b) $\int_0^1 \sin(x^2)\, dx$.

6   TRAPEZOIDAL QUADRATURE

The formula for *trapezoidal quadrature* is

(20) $$\int_a^b F(x)\, dx \simeq T_\pi[F] = \sum_{i=1}^{n} [F(x_{i-1}) + F(x_i)]\, \Delta x_i/2.$$

We will now use the concept of the Green's function for a two-endpoint problem, as defined in Ch. 2, §11, to obtain an exact expression for the error in trapezoidal quadrature over a single interval. Consider the linear function

$$(21) \qquad L(x) = F(a) + (x - a)[F(b) - F(a)]/h, \qquad h = b - a,$$

defined by linear interpolation between the values $F(a)$ and $F(b)$, and let $R(x) = F(x) - L(x)$. Then $R''(x) \equiv F''(x)$, and $R(a) = R(b) = 0$.

Now consider a single interval of length $h_i = x_i - x_{i-1} = 2k$, and translate coordinates so that (x_{i-1}, x_i) becomes the interval $(-k, k)$. As in Ch. 2, §11, we have

$$(22) \qquad R(x) = \int_{-k}^{k} G(x, \xi)R''(\xi)\, d\xi = \int_{-k}^{k} G(x, \xi)F''(\xi)\, d\xi,$$

in which $R(x)$, defined as above to be the difference between the function $F(x)$ and its trapezoidal approximation $L(x)$, vanishes at the endpoints and satisfies $R'' = F''$. The Green's function $G(x, \xi)$ for $F'' = r(x)$ is given by

$$(23) \qquad G(x, \xi) = \begin{cases} (\xi x/k + \xi - x - k)/2, & x \le \xi, \\ (\xi x/k - \xi + x - k)/2, & \xi \le x. \end{cases}$$

The error in trapezoidal quadrature over $(-k, k)$ is

$$T_\pi[F] - \int_{-k}^{k} F(x)\, dx = \int_{-k}^{k} [L(x) - F(x)]\, dx = \int_{-k}^{k} R(x)\, dx.$$

Substituting for $R(x)$ the integral expression displayed above and interchanging the order of integration in the resulting double integral, we get

$$T_\pi[F] - \int_{-k}^{k} F(x)\, dx = - \int_{-k}^{k} \left\{ \int_{-k}^{k} G(x, \xi)\, dx \right\} F''(\xi)\, d\xi.$$

But by direct calculation, $\int_{-k}^{k} G(x, \xi)\, dx = -(k^2 - \xi^2)/2$. Hence

THEOREM 6.   *The error in trapezoidal quadrature over a single interval* $(-k, k)$ *is exactly* $\int_{-k}^{k} (k^2 - \xi^2)F''(\xi)\, d\xi/2.$

Furthermore, since $G(x, \xi) \le 0$ for all $x, \xi \in (-k, k)$, we can use the Second Mean Value Theorem of the Calculus to obtain as a corollary that the error is

$$F''(\xi) \int_{-k}^{k} (k^2 - t^2)\, dt/2 \qquad \text{for some} \qquad \xi \in (-k, k)$$

The integral is easily evaluated as $2k^3/3 = h_i^3/12$.

COROLLARY.   *The error in trapezoidal quadrature over a single interval of length h_i is $h_i^3 F''(\xi_i)/12$, for some ξ_i in the interval.*

Since $h_i^3 \leq |\pi|^2 h_i$, summation over i now gives our final result.

THEOREM 7.   *The error bound for trapezoidal quadrature is given by the inequality*

$$(24) \qquad \left| T_\pi[F] - \int_a^b F(x) \, dx \right| \leq \frac{1}{12} \, |F''|_{\max}(b - a)|\pi|^2$$

We shall next obtain an analog of Theorem 5 for trapezoidal quadrature in the case that all the intervals of subdivision have the same length, $\Delta x_i = 2k = h$. For any $F \in \mathcal{C}^6$, Taylor's formula with remainder gives, much as in the proof of Theorem 5,

$$F(m_i + k) + F(m_i - k) = 2F(m_i) + k^2 F''(m_i) + \frac{k^4}{12} F^{iv}(m_i) + O(k^6)$$

Multiplication by $\Delta x_i/2$, followed by summation over i, now gives the further estimate

$$(25) \qquad T_\pi[F] = M_\pi[F] + \frac{k^2}{2} M_\pi[F''] + \frac{k^4}{24} M_\pi[F^{iv}] + O(k^6)$$

The right side of (25) can be evaluated by repeated use of the midpoint quadrature formula error estimate (17). The conclusion is the truncated Euler–Maclaurin formula.

THEOREM 8.   *For $F \in \mathcal{C}^6$, let all intervals of subdivision have the same length $\Delta x_i = 2k = h$. Then*

$$(26) \qquad T_\pi[F] = \int_a^b F(x) \, dx + \frac{h^2}{12} [F'(b) - F'(a)]$$

$$- \frac{h^4}{720} [F'''(b) - F'''(a)] + O(h^6)$$

Proof.   Replacing h by $2k$ in (17) and then substituting from (17) into (25), we obtain, as the contribution from the first term on the right-hand side of (25),

$$\int_a^b F(x) \, dx - \frac{k^2}{6} [F'(b) - F'(a)] + \frac{7k^4}{360} [F'''(b) - F'''(a)] + O(k^6)$$

From the second term we obtain

$$\left(\frac{k^2}{2}\right)\left\{\int_a^b F''(x)\,dx - \left(\frac{k^2}{6}\right)[F'''(b) - F'''(a)] + O(k^4)\right\}$$

while the third term gives $(k^4/24)[F'''(b) - F'''(a)] + O(k^6)$. Adding these three contributions together, simplifying, and writing $k = h/2$, we get (26).

Simpson's Rule.   Comparing the error estimates (17) and (26) for midpoint and trapezoidal quadrature, we are led to an error estimate for Simpson's rule. For a given partition π, this is defined as

$$(28) \quad S_\pi[F] = \frac{2}{3} M_\pi[F] + \frac{1}{3} T_\pi[F] = \frac{1}{6}\sum_{i=1}^{n} [F(x_{i-1}) + 4F(m_i) + F(x_i)]\,\Delta x_i$$

Forming the linear combination indicated for subdivision into double steps of constant length $2k = h$, we obtain

$$(29) \qquad S_\pi[F] = \int_a^b F(x)\,dx + \frac{k^4}{180}[F'''(b) - F'''(a)] + O(k^6)$$

Simpson's rule will be studied further in Ch. 8, §9.

EXERCISES C

1. Use (26) to estimate the difference

$$\ln 2 - \left[\frac{1}{20} + \frac{1}{40} + \sum_{k=1}^{9}\frac{1}{10+k} + \frac{1}{400}\right]$$

In Exs. 2–5, use (26) with $h = 0.2$ to evaluate the following numbers approximately:

2. $\ln 2 = \int_1^2 dx/x$
3. $\arctan 1 = \int_0^1 dx/(1 + x^2)$
4. $\int_0^1 \sqrt{1 + x^c}\,dx$
5. $\int_0^1 \sin(x^2)\,dx$

In Exs. 6–9, use Simpson's rule (28) with double step $2k = h = 0.2$ to evaluate approximately the numbers defined in Exs. 2–5, respectively.

10. For a subdivision into $2n$ intervals of length $h = (b - a)/2n$, Simpson's approximation to $\int_a^b f(x)\,dx$ is $\sum_{i=1}^{n} (h/3)[f(x_{2i-1}) + 4f(x_{2i-1}) + f(x_{2i})]$. Show that the truncation error is $(h^5/90)\sum_{i=1}^{n} f^{iv}(x_{2i-1}) + O(h^6)$.

11. Show that $\int_{-h}^{h} F(x)\,dx = 2hF(0) + \frac{1}{6}[\int_{-h}^{h} (|h| - |x|)^3 F''(x)\,dx]$. [HINT: Construct the Green's function for the initial value problem defined by $u'' = F''(x)$ and $F(0) = F'(0) = 0$, and study the proof of Theorem 6.]

*7   TRAPEZOIDAL INTEGRATION

The rest of this chapter will be devoted to the theoretical analysis of three classical methods for integrating first-order ordinary DEs (and systems of DEs). Like (uncorrected) midpoint and trapezoidal quadrature, these methods have only $O(h^2)$ accuracy. Since Runge–Kutta and other algorithms having at least $O(h^4)$ accuracy are readily available and easy to use, readers who are primarily interested in *applications* of numerical methods may wish to proceed directly to Chapter 8, which will take up such more efficient methods.

For various reasons, the *errors* arising from the use of these more efficient methods cannot in practice be predicted purely theoretically. Therefore, the discussion to follow will have no parallel in Chapter 8.

Our theoretical analysis will first take up *trapezoidal integration*. For any system $dx/dt = \mathbf{X}(\mathbf{x}, t)$ of first-order DEs, this is defined *implicitly*† by the recursion formula (difference equation)

$$(30) \qquad \mathbf{y}_k = \mathbf{y}_{k-1} + [\mathbf{X}(\mathbf{y}_{k-1}, t_{k-1}) + \mathbf{X}(\mathbf{y}_k, t_k)] \, \Delta t_k/2$$

where $\Delta t_k = t_k - t_{k-1}$. From a given initial value $\mathbf{y}_0 = \mathbf{c}$ and partition π, formula (30) defines a sequence of values $\mathbf{y}_k = \mathbf{y}_\pi(t_k)$, that is, a *function table* describing approximately the solution of the DE $\dot{x} = \mathbf{X}(\mathbf{x}, t)$ satisfying the initial value $\mathbf{x}(a) = \mathbf{c}$.

Note that when $\mathbf{X} = \mathbf{X}(t)$, formula (30) is equivalent to the trapezoidal quadrature formula (20). Also note that, as in Ch. 6, §5, formula (30) can be extended to DEs and systems of arbitrary order.‡ Last and most important, note that if $\mathbf{X}(\mathbf{x}, t) = A(t)\mathbf{x} + \mathbf{b}(t)$ is *linear*, then (30) is equivalent to

$$(30') \qquad (I - \theta_k A_k)\mathbf{y}_k = (I + \theta_k A_{k-1})\mathbf{y}_{k-1} + \theta_k(\mathbf{b}_{k-1} + \mathbf{b}_k)$$

where $\theta_k = \Delta t_k/2$, $A_j = A(t_j)$, and $\mathbf{b}_j = \mathbf{b}(t_j)$. Hence, for Δt_k small enough, the system (30') can be solved for $\mathbf{y}_k$, given $\mathbf{y}_{k-1}$, by Gaussian elimination.

Example 4. Consider the solution of the *linear* DE

$$(31) \qquad \frac{dx}{dt} + 2tx = 1$$

taking the initial value $x(0) = 0$. By the formula of Ch. 1, §6, the solution is the function $x = e^{-t^2} \int_0^t e^{s^2} \, ds$. Looking up values of the definite integral in a table,†† we get the first row of entries in the following display.

t	0.1	0.2	0.3	0.4	0.5	0.6	0.7	0.8	0.9
x	0.0993	0.1948	0.2826	0.3599	0.4244	0.4748	0.5105	0.5321	0.5407
y	0.0990	0.1941	0.2818	0.3590	0.4235	0.4739	0.5097	0.5315	0.5404
z	0.099	0.194	0.282	0.3586	0.424	0.4739	0.509	0.531	0.5383

---

† For large Δt_k, Eq. (30) may have more than one solution. But usually, $\mathbf{y}_k$ can be computed by iterating (30) two or three times.

‡ This does not mean that reduction to a first-order system is recommended in numerical integration.

†† E. Jahnke and F. Emde, *Tables of Functions*, Dover, 1943, p. 32; W. L. Miller and A. R. Gordon, *J. Phys. Chem.* **35** (1931), p. 2878.

The second row of entries gives values y_k of the approximate function table constructed using the trapezoidal integration formula (30), with $x(0) = 0$ and constant mesh length $h = 0.1$. With this value of h, the formula in Example 4 reduces to

$$y_k = \frac{1 + (10 - t_{k-1})y_{k-1}}{10 + t_k}$$

The entries tabulated in the second row were calculated from the preceding formula, rounding off all numbers to six decimal digits and then rounding off the final values to four decimals. The last row of the table gives values z_k computed by the "improved Euler method" of §8 below.

Above, we used a table of $\int_0^t e^{s^2} ds$; we next ask: How should one construct this table? For large t, numerical quadrature formulas tend to be inefficient, because the integrand e^{t^2} varies so rapidly (by more than 10% between 5 and 5.01, for example). For this reason, rather than computing $x(t)$ as we did, it is more efficient to solve the DE (31), for the initial value $x(0) = 0$, by an accurate numerical method (see Ch. 8), and then to compute $\int_0^t e^{s^2} ds = e^{t^2}x(t)$ as a product, than to compute $x(t)$ as we did.

The preceding discussion illustrates an important principle. Reductions to quadratures and substitutions in special formulas do not necessarily help one to obtain accurate numerical values for solutions of DEs.

Asymptotic Expansion. On the other hand, asymptotic expansions and other analytical devices may be very helpful for analyzing the singularities of solutions in DEs, their behavior for very large values of the independent variables, and their dependence on parameters. Thus, to evaluate the function of Example 4 for very large t (say, $t \geq 25$), it is best to set $s = t - r$ in Example 4 and to expand e^{r^2} in a Taylor's series. We then obtain

$$e^{-t^2} \int_0^t e^{s^2} ds = \int_0^t e^{-2tr}\left(1 + r^2 + \frac{r^4}{2!} + \frac{r^6}{3!} + \cdots\right) dr$$

For large t, the kth term gives the integral

$$\frac{t^{-2k-1}}{(k!)2^{2k+1}} \int_0^{2t^2} e^{-\rho} \rho^{2k} d\rho \sim \frac{(2k)!t^{-2k-1}}{(k!)2^{2k+1}}, \qquad \rho = 2tr$$

Moreover, by truncating the series at the kth term, an asymptotic error estimate can be obtained.

Knowing (or having guessed) the form of the asymptotic series $x(t) \sim \sum_{k=0}^{\infty} a_k/t^{2k+1}$, we also can derive from (31) purely formally, by the method of undetermined coefficients, that $a_0 = 1/2$ and $a_{k+1} = (2k + 1)a_k/2$, from which (asymptotically) we obtain

$$x \sim \frac{1}{2t} + \frac{1}{4t^3} + \frac{3}{8t^5} + \frac{15}{16t^7} + \cdots \qquad \text{as} \qquad t \to \infty$$

Though the series is (ultimately) divergent, the first few terms give an extremely accurate approximation to $x(t)$ for $t \geq 25$, which is much more accurate than could be obtained by general numerical methods.

Error Bound. Finally, we derive a bound on the cumulative error of the function table constructed by the trapezoidal integration formula (30), in terms of the properties of $X(x, t)$. To do this, we first construct an *approximate solution* of the DE $x'(t) = X(x, t)$ from the approximate function table of points (y_k, t_k) defined by (30). Since $\Delta y_k/\Delta t$ in (30) is the arithmetic mean of the slopes $y_j' = X_j = X(y_j, t_j)$ for $j = k - 1, k$, the quadratic function

$$y(t) = y_{k-1} + y_{k-1}'\tau + \frac{(y_k' - y_{k-1}')\tau^2}{2\,\Delta t}, \qquad \tau = t - t_{k-1}$$

interpolates not only to y_{k-1} and y_k, but also to y_{k-1}' and y_k' at t_{k-1} and t_k, respectively.† This gives us a *piecewise quadratic* approximate solution of class $\mathcal{C}^1$, which satisfies the given DE exactly at all points t_k.

We next bound the deviation (§1) $|y'(t) - X(t, y(t))|$ of this approximate solution. Since the deviation is zero at t_{k-1} and t_k, while $y'(t)$ is linear in the interval $[t_{k-1}, t_k]$, it follows (by Theorem 1 of Ch. 8) that the deviation there is at most $\tau(h - \tau)|\ddot{X}|_{\max}$, where $h = t_k - t_{k-1}$, and $|\ddot{X}|_{\max}$ signifies the maximum absolute value of the second time derivative of $X(t)$. On the other hand, we have

$$(32) \qquad \ddot{X} \sim \frac{d^3x}{dt^3} = \frac{\partial^2 X}{\partial t^2} + 2X\frac{\partial^2 X}{\partial x\,\partial t} + X^2\frac{\partial^2 X}{\partial x^2} + \frac{\partial X}{\partial x}\frac{\partial X}{\partial t} + X\left(\frac{\partial X}{\partial x}\right)^2$$

In Example 4, this gives $\ddot{X} = -4 + 12xt + 4t^2 - 8xt^3$.

As in the proof of Theorem 3, we now set $\sigma(t) = |x(t) - y(t)|^2$ and differentiate, to get in any interval of length h,

$$(32') \qquad \sigma'(t) \leq 2L\sigma(t) + 2\epsilon(t)\sqrt{\sigma(t)}$$

where, in any subinterval $[t_{k-1}, t_k]$ of length h or less $|\epsilon(\tau)| \leq \frac{1}{2}\tau(h - \tau)|X|_{\max}$, $\tau = t - t_{k-1}$. A more careful repetition of the proof of Theorem 3 shows that, since $\int_0^h t(h - t)\,dt = h^3/6$ and $|x(a) - y(a)| = 0$, the *cumulative* (truncation) error must satisfy

$$(33) \qquad |x(b) - y(b)| \leq \frac{h^2}{6L}|\ddot{X}|_{\max}\{e^{L(b - a)} - 1\}$$

As a corollary, the order of accuracy (§4) of trapezoidal integration is $0(h^2)$.

---

† It is actually the "cubic" Hermite interpolant to the y_j and y_j' (to be discussed in Ch. 8), but this happens to be quadratic in the present case.

EXERCISES D

In Exs. 1–4, compute the trapezoidal approximations to the solutions of the initial value problems specified, over the range $0 \le t \le 1$ with $h = 0.1$.

1. $\dot{x} = -tx$, $x(0) = 1$ 2. $\dot{x} = 1 + x^2$, $x(0) = 0$

3. $\dot{x} = y$, $\dot{y} = 0$, $x(0) = 0$, $y(0) = 1$

4. $\dot{x} = y$, $\dot{y} = x + y$, $x(0) = 0$, $y(0) = 1$

5. For the DE $y' = F(x, y)$, show that the truncation error of the Cauchy polygon method, in one step $0 \le x \le h$, is

$$\frac{h^2}{2} \left[\frac{\partial F}{\partial x} (0, y_0) + F(0, y_0) \frac{\partial F}{\partial y} (0, y_0) \right] + O(h^3), \qquad y_0 = y(0)$$

In Exs. 6–9, compute the truncation error for trapezoidal integration over one interval of length h, in terms of the Taylor series expansions of the functions involved:

6. $\dot{x} = p(t)x$, p analytic   7. $\dot{x} = 1 + x^2$

8. $\dot{x} = y$, $\dot{y} = 0$ 9. $\dot{x} = y$, $\dot{y} = x + y$

10. For the DE $y'' = F(x, y)$, what is the order of accuracy of the formula

$$y_{n+1} = y_{n-1} + 2hF(x_n, y_n)$$

if all $\Delta x_k = h$?

8   THE IMPROVED EULER METHOD

The trapezoidal method is very convenient for getting approximate solutions to *linear* DEs, because one can solve algebraically for $\mathbf{y}_k$. Thus in Example 4, formula (30) is equivalent to the recursive formula $y_k = [1 + (10 - t_{k-1})y_{k-1}]/[10 + t_k]$.

But the trapezoidal method is awkward when it comes to *nonlinear* DEs because of the difficulty of solving (30) for $\mathbf{y}_k$. In general, formula (30) does not define the vectors $\mathbf{y}_k$ recursively but only implicitly. To determine each $\mathbf{y}_k$, we have to solve an equation† (30) where $\mathbf{y}_k$ is the unknown, $\mathbf{y}_{k-1}$ having been previously determined.

For small Δt_k, we can do this by *iteration*: start with a trial value of $\mathbf{y}_k$ [say, $\mathbf{y}_{k-1} + \mathbf{X}(\mathbf{y}_{k-1}, t_{k-1}) \, \Delta t_k$], substitute this trial value $\mathbf{y}_k^0$ into the right-hand side of (30) to get a better approximation $\mathbf{y}_k^1$, and then repeat the process

(33) $$\mathbf{y}_k^{r+1} = \mathbf{y}_{k-1} + [\mathbf{X}(\mathbf{y}_{k-1}, t_{k-1}) + \mathbf{X}(\mathbf{y}_k^r, t_k)] \, \Delta t_k/2$$

until (30) is satisfied up to the error tolerated.

---

† The vector equation (30) is, of course, equivalent to a system of n simultaneous equations in the components. For linear systems, these equations can be solved by Gauss elimination, instead of iteration.

Instead of solving the implicit equation (30) precisely by using many iterations, one usually gets greater accuracy for the same amount of work by using a finer mesh and stopping after one or two iterations. If one stops after a single interation, one has the *improved Euler method,* which is adequate for many nonlinear engineering problems requiring moderate accuracy (say, two significant decimal digits).

Predictor-corrector Methods. The improved Euler method typifies an important class of so-called predictor-corrector methods, whose underlying philosophy is as follows.

First, by extrapolation or otherwise, one tries to make a reasonably good first guess $\mathbf{y}_k^{(0)}$ as to what $\mathbf{y}_k$ should be; in the present case, this guess is provided by the Cauchy polygon construction applied to $\mathbf{y}_{k-1}$:

$$(34) \qquad \mathbf{y}_k^{(0)} = \mathbf{y}_{k-1} + \mathbf{X}(\mathbf{y}_{k-1}, t_{k-1})\,\Delta t_k$$

This guess is called the *predictor.*

One then considers the implicit equation to be solved, for example, Eq. (30), as a *corrector:*

$$(34') \qquad \mathbf{y}_k^{(r+1)} = \mathbf{y}_{k-1} + [\mathbf{X}(\mathbf{y}_{k-1}, t_{k-1}) + \mathbf{X}(\mathbf{y}_k^{(r)}, t_k)]\,\Delta t_k/2$$

to be solved iteratively if necessary. In most cases, the full order of accuracy of the implicit equation [$O(h^2)$ in the present instance] is achieved after one iteration!

The improved Euler method consists in computing the sequence of z_k by performing these two substitutions in alternation. In the case $X(x, t) = F(t)$ of quadrature, it is equivalent to the trapezoidal method.

Applied to the initial value problem of Example 4, the improved Euler method gives the approximate solution tabulated in the last row of the table of §7; in this example, the work was carried to three decimal places to reduce the cumulative error to 0.01.

To apply the improved Euler method to first-order systems, simply substitute vectors for scalars in formulas (34), (34'). We illustrate the procedure by an example.

Example 5. Consider the initial value problem defined by the nonlinear system

$$(35) \qquad \frac{dx}{dt} = x^2 + y^2, \qquad \frac{dy}{dt} = 1 + x^2 - y^2, \qquad x(0) = y(0) = 0$$

There is little hope that formal methods of integration will help in the computation, but a straightforward application of the improved Euler method enables one to calculate an approximate solution.

The calculations to be performed give, for this example, the double sequence of numbers x_k, y_k defined by $x_0 = y_0 = 0$ and by the formulas

(36)
$$p_k = x_{k-1} + (x_{k-1}^2 + y_{k-1}^2)\, \Delta t_k$$
$$q_k = y_{k-1} + (1 + x_{k-1}^2 - y_{k-1}^2)\, \Delta t_k$$

and

(36')
$$x_k = x_{k-1} + (x_{k-1}^2 + y_{k-1}^2 + p_k^2 + q_k^2)\, \Delta t_k/2$$
$$y_k = y_{k-1} + (2 + x_{k-1}^2 - y_{k-1}^2 + p_k^2 - q_k^2)\, \Delta t_k/2$$

Each step requires squaring four numbers and performing 14 additions and subtractions and four multiplications.

For a subdivision into intervals of constant length h, the relative error committed in using the improved Euler method is $O(h^2)$, provided that the function $X(x, t)$ is of class $\mathcal{C}^2$. For, expanding the exact solution $x(t)$ of $dx/dt = X(x, t)$ by Taylor's Theorem with remainder, we have

(37) $$x_k = x(t_k) = x_{k-1} + hX_{k-1} + \frac{h^2}{2}\left[X\frac{\partial X}{\partial x} + \frac{\partial X}{\partial t} \right]_{k-1} + O(h^3)$$

where X_{k-1} denotes $X(x_{k-1}, t_{k-1}) = Xx(t_{k-1}), t_{k-1})$ and the subscript $k - 1$ on the term in square brackets has a similar meaning. The improved Euler method of (34)–(34') gives

(38) $$y_k = y_{k-1} + hY_{k-1} + \frac{h^2}{2}\left[Y\frac{\partial Y}{\partial x} + \frac{\partial Y}{\partial t} \right]_{k-1} + O(h^3)$$

where Y_{k-1} denotes $X(y_{k-1}, t_{k-1})$, and so on. The *relative* error committed in substituting (38) for (37) is thus $O(h^2)$.

A more explicit error bound is deduced in §10.

*9   THE MODIFIED EULER METHOD

The improved Euler method has the advantage over the trapezoidal method of being explicit. Various other explicit methods about as accurate as the improved Euler and trapezoidal methods can also be constructed. For instance, one can use the following adaptation of the midpoint quadrature formula:

(39) $$w_k = w_{k-1} + hX\left(w_{k-1} + \frac{hX_{k-1}}{2}, t_{k-1} + \frac{h}{2} \right), \qquad h = \Delta t_k$$

This *midpoint* or *modified* Euler method is about twice as accurate as the trapezoidal and improved Euler methods in the special case $dx/dt = F(t)$ of quadrature, as a comparison of formulas (17) and (26) shows. (In this case, the improved Euler method *is* the trapezoidal method.)

But for first-order DEs generally, no such simple error comparison holds. To see this, let $X = \Sigma b_{jk} t^j x^k$ be expanded into a double power series, and let $x(t)$ satisfy $\dot{x} = X(x, t)$. Then, just as in Ch. 4, §2, we have $\ddot{x} = X_t + XX_x$, and

$$\frac{d^3 x}{dt^3} = X_{tt} + 2XX_{tx} + X^2 X_{xx} + X_t X_x + XX_x^2$$

We now introduce the abbreviations

$$C = b_{10} + b_{00} b_{01}, \qquad B = b_{20} + 2b_{11} b_{00} + b_{02} b_{00}^2 \qquad B^* = b_{10} b_{01} + b_{00} b_{01}^2$$

Expanding out to infinitesimals of the fourth order, we find that the *exact* solution of the DE $dx/dt = X(x, t)$ for the initial condition $x(0) = 0$ has the expansion

$$(40) \qquad x(h) = hb_{00} + \frac{h^2 C}{2} + \frac{h^3 (B + B^*)}{6} + O(h^4)$$

With the trapezoidal approximation (30), we obtain

$$(41) \qquad y(h) = hb_{00} + \frac{h^2 C}{2} + \frac{h^3 (2B + B^*)}{4} + O(h^4)$$

giving a truncation error $h^3(B/3 + B^*/12) + O(h^4)$. With the improved Euler approximation (34)–(34′), we get

$$(42) \qquad z(h) = hb_{00} + \frac{h^2 C}{2} + \frac{h^3 B}{2} + O(h^4)$$

with error $h_3(2B - B^*)/6 + O(h^4)$. With the midpoint approximation (39), we finally obtain

$$(43) \qquad w(h) = hb_{00} + \frac{h^2 C}{2} + \frac{h^3 B}{4} + O(h^4)$$

so that the error is $h^3(2B^* - B)/12 + O(h^4)$

Corrected Trapezoidal Method. Theorem 8, when combined with Theorem 5 of Ch. 6, shows that the exact solution $x(t)$ of the first-order DE $dx/dt = X(x, t)$ satisfies

$$(44) \qquad x_k = x_{k-1} + \frac{[X_{k-1} + X_k]\, \Delta t_k}{2} + \frac{[\dot{X}_{k-1} - \dot{X}_k]\, \Delta t_k^2}{12} + O(\Delta t_k^4)$$

where X_k denotes $X(x_k, t_k)$ and $\dot{X}$ denotes $\partial X/\partial t + X\,\partial X/\partial x$. Dropping the last term, we get a *corrected* trapezoidal integration formula, which may be expected to have a cumulative error of only $O(|\pi|^3)$.

For instance, when applied to the inhomogeneous linear DE $\dot{x} = 1 - 2tx$ of Example 4, in which $\dot{X} = -2x - 2t + 4xt^2$, this formula gives the approximate recursion formula

$$x_k[1 + t_k\,\Delta t_k + (-1 + 2t_k^2)\,\Delta t_k^2/6]$$

$$= x_{k-1}[1 - t_{k-1}\,\Delta t_k + (-1 + 2t_{k-1}{}^2)\,\Delta t_k{}^2/6] + \Delta t_k + \frac{\Delta t_k{}^3}{6}$$

with absolute error $O(\Delta t_k{}^4)$ and relative error $O(\Delta t_k{}^3)$.

EXERCISES E

In Exs. 1–4, compute approximate function tables on $[0, 1]$, with $\Delta t_k = 0.1$, by the improved Euler method for the following initial value problems:

1. $\dot{x} = -tx$, $x(0) = 1$ 2. $\dot{x} = (1 + x^2)$, $x(0) = 0$

3. $\dot{x} = y, \dot{y} = 0, x(0) = 0, y(0) = 1$ 4. $\dot{x} = y, \dot{y} = x + y, x(0) = 0, y(0) = 1$

In Exs. 5–8, compute approximate function tables for the data of Exs. 1–4, using the *midpoint* (or modified Euler) method, instead of the improved Euler method.

9. Obtain an expression through terms in h^5 for the error committed in applying (a) the improved Euler method and (b) the midpoint method to the DE $\dot{x} = p(t)x$, $p(t)$ analytic.

10. For the analytic DE $y' = F(x, y)$ and one interval $0 \le x \le h$, let the exact solution be given by $y(h) = a_0 + a_1 h + a_2 h^2 + a_3 h^3 + O(h^4)$ and let

$$y = c_0 + c_1 h + c_2 h^2 + c_3 h^3 + O(h^4)$$

be the approximate value given by trapezoidal formula. Show that $c_0 = a_0, c_1 = a_1$, $c_2 = a_2$, and $c_3 = 3a_3/2$.

11. Show that through the terms computed in Ex. 10, the improved Euler method gives the same result as trapezoidal integration for c_0, c_1, c_2.

*10 CUMULATIVE ERROR BOUND

All the methods for constructing approximate function tables that have been described in this chapter have had one feature in common. Namely, the kth entry in the table has been constructed from the immediately preceding entry alone, the $(k - 1)$st entry, without reference to the earlier entries. Such methods are called *one-step* (or "two level") methods.

Given a one-step method for numerically integrating the DE $dx/dt = X(x, t)$, that is, for constructing an approximate function table with entries $y_k = y(t_k)$, we can express the preceding property by writing

(45)
$$y_k = \phi(y_{k-1}, t_{k-1}, t_k; X)$$
.

Here ϕ is the function expressing y_k in terms of y_{k-1} and the data of the problem. Bounds for the errors associated with one-step methods can be obtained by using the following general theorem, which applies equally to the Cauchy polygon method, the trapezoidal method, and the improved Euler and midpoint methods.

THEOREM 9. *In any one-step method for numerical integration of $dx/dt = X(x, t)$, where X satisfies the Lipschitz condition*

$$|X(x, t) - X(y, t)| \leq L|x - y|$$

let the relative error at each step be at most ϵ. Then, over an interval of length T, the cumulative error is at most $(\epsilon/L)(e^{LT} - 1)$.

Proof. For any partition π, let ϵ_k denote the error introduced at the kth step. That is, if $\tilde{x}_k(t)$ is that exact solution of the given DE satisfying the initial condition $\tilde{x}_k(t_k) = y_k$, where y_k is the value of the computed approximate solution at t_k, let

$$\epsilon_k = |y_k - \tilde{x}_{k-1}(t_k)| = |\tilde{x}_k(t_k) - \tilde{x}_{k-1}(t_k)|$$

By the definition of "relative error," we have $\epsilon_k \leq \epsilon \, \Delta t_k$. The magnitude of the cumulative error is, by definition,

$$|y_m - x(t_m)| = |\tilde{x}_m(t_m) - \tilde{x}_0(t_m)| = \left| \sum_{k=1}^{m} [\tilde{x}_k(t_m) - \tilde{x}_{k-1}(t_m)] \right|$$

$$\leq \sum_{k=1}^{m} |\tilde{x}_k(t_m) - \tilde{x}_{k-1}(t_m)|$$

But $|\tilde{x}_k(t_m) - \tilde{x}_{k-1}(t_m)|$ is the magnitude of the difference, at $t = t_m$, of two solutions of the given DE that differ by ϵ_k at $t = t_k$. By Theorem 2 of Ch. 6, this is at most

$$\epsilon_k e^{L(t_m - t_k)} \leq \epsilon \, e^{L(t_m - t_k)} \, \Delta t_k$$

since $\epsilon_k \leq \epsilon \, \Delta t_k$. Here ϵ is an upper bound to the relative error. Summing over k, we get the following upper bound to the cumulative error:

$$|y_\pi(t_m) - x(t_m)| \leq \epsilon \sum_{k=1}^{m} [e^{L(t_m - t_k)} \, \Delta t_k] \leq \epsilon \, e^{Lt_m} \sum_{k=1}^{m} [e^{-Lt_k} \, \Delta t_k]$$

But the final sum is the Riemann lower sum approximation to the definite integral $\int_{t_0}^{t_m} \exp(-Lt)\, dt = [\exp(-Lt_0) - \exp(-Lt_m)]/L$. Hence

$$|y_\pi(t_m) - x(t_m)| \leq \frac{\epsilon[e^{LT} - 1]}{L}$$

and Theorem 9 follows.

Trapezoidal Integration. For trapezoidal integration, the discussion of §7 shows that the truncation error ϵ_k at the kth step is bounded by

$$\frac{(|\pi|^2|\dddot{X}|_{\max})\,\Delta t_k}{12[1 - (L\,\Delta t_k/2)]}$$

Therefore, in this case, an error bound is given by the following corollary.

COROLLARY. *The error in the approximate function table constructed by the trapezoidal formula (30) is at most*

(46)
$$\frac{|\pi|^2|\dddot{X}|_{\max}(e^{Lt} - 1)}{12L[1 - L|\pi|/2]} \quad \text{if} \quad |\pi| \leq \frac{1}{L}$$

Similar error bounds can be found for the midpoint and improved Euler approximate integration methods.

Approximate Solutions. The error bound (46) refers to the approximate function table constructed by the trapezoidal integration formula (30). Using linear interpolation between successive values, we can obtain from this function table a continuous approximate solution to the DE $dx/dt = X(x, t)$. Since the error in linear interpolation is bounded by $|\ddot{x}|_{\max}|\pi|^2/2$, where $\ddot{x} = \dot{X} = \partial X/\partial t + X\,\partial X/\partial x$, we see that the order of accuracy of this approximate solution is also $O(|\pi|^2)$.

EXERCISES F

1. Let $x(t)$ and $y(t)$ be approximate solutions of the system (1) with deviations ϵ_1 and ϵ_2, defined for $a \leq t \leq b$. Show that (7) implies that

$$|x(t) - y(t)| \leq |x(a) - y(a)|e^{L|t-a|} + \frac{\epsilon_1 + \epsilon_2}{L}[(e^{L|t-a|} - 1)]$$

2. Let $F(x, y)$ and $G(x, y)$ be everywhere continuous; let F satisfy a Lipschitz condition with Lipschitz constant L, and let $|F(x, y) - G(x, y)| \leq K$. Show that if $f(x)$ and $g(x)$ are approximate solutions of the DEs $y' = F(x, y)$ and $z' = G(x, z)$ with deviations ϵ and η, then

$$|f(x) - g(x)| \leq |f(a) - g(a)|e^{L|x-a|} + \frac{K + \epsilon + \eta}{L}[(e^{L|x-a|} - 1)]$$

*3. Assume that, for equally spaced subdivisions of mesh length h, the truncation error of a given approximate method $I_h[f]$ is $Mh^n + O(h^{n+1})$, where M is independent of h. Prove that the *extrapolated* estimate

$$I_h^*[f] = (2^n I_{h/2}[f] - I_h[f])/(2^n - 1)$$

has a truncation error $O(h^{n+1})$.

*4. (a) Show that the extrapolation of Ex. 3 gives the trapezoidal approximation from the Cauchy polygon approximation, and Simpson's rule from the trapezoidal approximation.
   (b) Show that Simpson's rule satisfies the hypotheses of Ex. 3 with $n = 4$. Derive an extrapolation estimate for Simpson's rule.

*5. For the DE $dx/dt + tx = 0$ and the mesh length $h_n = 1/(10|t_n| + 1)$, show that the truncation error of the trapezoidal method tends to zero as $t \to \infty$, regardless of the initial value $x(0)$. What is the limiting truncation error as $t \to -\infty$?

*6. Let $f(t)$ be an analytic function, and $f(t + 1) = f(t)$. Show that trapezoidal integration of $\int_0^1 f(t)\, dt$ for $h = 1/n$ has an infinite order of accuracy as $n \to \infty$. [HINT: Expand $f(t)$ in Fourier series.]

7. Show that the error is $O(h^7)$ in the extended Simpson's rule

$$\int_{x-h}^{x+h} F(x)\, dx \simeq \frac{h}{90}\{114F(x) + 34[F(x + h) + F(x - h)] - [F(x + 2h) + F(x - 2h)]\}$$

8. (a) Let $z_k = z_0 + i^{k-1}h$, $i = \sqrt{-1}$. Show that, if $F(z)$ is any complex polynomial of degree five or less, then we have

(*) $$\int_{z_0}^{z_1} F(z)\, dz = \frac{h}{15}[24F_0 + 4(F_1 + F_3) - (F_2 + F_4)]$$

   (b) Infer that, if $F(z)$ is a complex analytic function, (*) holds with an error that is $O(h^7)$.

*9. Let $F(x, y)$ be bounded and continuous on the strip $0 \le x \le 1$, $-\infty < y < +\infty$; let $\{\pi_n\}$ be any sequence of partitions of $[0, 1]$ with $|\pi_n| \to 0$; and let $f_n(x)$ be the Cauchy polygon approximate solution defined by π_n for the initial value $y(0) = 0$.
   (a) Show that, if the $f_n(x)$ converge to a limit function $f(x)$, then $f(x)$ is a solution of $y' = F(x, y)$.
   (b) Show that, in any case, a uniformly convergent subsequence $\{f_{n(i)}(x)\}$ can be found, $n(i) < n(i + 1)$. [HINT: See Ch. 6, §13.]

*10. In Ex. 9, show that, if the DE $y' = F(x, y)$ admits only one solution for the initial value $y(0) = 0$, any sequence of Cauchy polygon approximations defined for $y(0) = 0$ by partitions whose norms tend to zero must converge to the exact solution.

CHAPTER 8
EFFICIENT NUMERICAL INTEGRATION

1 DIFFERENCE OPERATORS

In Ch. 7, we analyzed theoretically a number of simple methods for computing approximate solutions for normal first-order systems

$$
(1) \qquad\qquad \frac{dx}{dt} = X(x, t)
$$

of ordinary DEs. The simplicity of the methods considered, all due to Euler, facilitated a rigorous theoretical analysis of their errors. In general, they had $O(h^2)$ accuracy for $X \in \mathcal{C}^2$.

In this chapter, we will describe some more efficient methods having higher order accuracy, usually $O(h^4)$ for $X \in \mathcal{C}^4$. We will explain the guiding ideas that motivated the construction of the algorithms used but will not push the analysis to the point of getting rigorous error bounds. This is partly to avoid lengthy discussions of complicated formulas, but mostly because errors are usually estimated in practice by studying the numerical output.

Such higher order methods are almost always used in practice when more than two or three significant digits are wanted. If their errors are accurately and reliably known, the errors should be subtracted to obtain improved results, as in Richardson extrapolation (Ch. 7, §5).

Like the schemes already analyzed in Ch. 7, many of the schemes for numerical integration to be studied later will refer to an assumed partition π of the interval $[a, b]$ of integration by a finite number of points (the *mesh*),

$$
(2) \qquad \pi\colon a = t_0 < t_1 < t_2 < \cdots < t_n = b = a + T
$$

Typically, the partition is made into *steps* $\Delta t_k = t_k - t_{k-1}$ of constant length h, so that $t_r = a + rh$; we then speak of a *uniform mesh*.

On the mesh (2), the DE (1) is approximated by a suitable *difference equation*. This difference equation is then solved step by step in hand computations, using ordinary arithmetic supplemented by readings from available function tables. In machine computations, however, function tables are usually replaced by simple subroutines that give accurate approximations by rational functions.

Using Taylor's formula with remainder, it is easy to derive higher order approximations to derivatives by difference quotients (see §2). Thus for $f \in \mathcal{C}^5$, we have

$$f(a + s) = c_0 + c_1 s + c_2 s^2 + c_3 s^3 + c_4 s^4 + O(s^5)$$

For $t_{n+k} = a + kh$ and $x_j = f(t_j)$, this gives

(3) $$[-x_{n+2} + 8x_{n+1} - 8x_{n-1} + x_{n-2}]/12h = f'(t_n) + O(h^4)$$

This suggests trying to solve $x'(t) = X(x, t)$ by simply substituting the difference quotient of (3) for the left side of the DE.

Unfortunately, as will be shown in §4, this procedure is highly unstable. Stability is only one of several ideas and techniques, few of which were known in Euler's time, that must be learned before one can understand, even superficially, the efficient schemes of numerical integration that are most commonly used today. The object of §§2–7 to follow will be to explain some of these ideas and techniques; the remainder of the chapter will be devoted to deriving some truly efficient schemes of numerical integration.

Much of our preliminary discussion will be concerned with difference operators and difference equations (or ΔEs, as we will write for short). Basic to these are the *forward* difference operator Δ, the *backward* difference operator ∇, and the *central* difference operator δ, defined by the formulas

(4a) $$\Delta f(x) = f(x + h) - f(x)$$

(4b) $$\nabla f(x) = f(x) - f(x - h)$$

(4c) $$\delta f(x) = f(x + \tfrac{1}{2}h) - f(x - \tfrac{1}{2}h)$$

In the preceding formulas, the symbols Δ, ∇, δ stand for *linear operators* that transform functions into functions. Unlike the linear differential operators of Ch. 2, §5, they apply to *all* functions.

These operations are useful in obtaining approximate solutions because they yield approximations to the derivative $f'(x)$. If $f \in \mathcal{C}^1$, the derivative $f'(x)$ is the limit of the difference quotients:

$$f'(x) = \lim_{h \to 0} \frac{\Delta f}{h} = \lim_{h \to 0} \frac{\nabla f}{h} = \lim_{h \to 0} \frac{\delta f}{h}$$

For obvious reasons, these are called the forward, backward, and central divided difference approximations to $f'(x)$.

The difference operators (4a)–(4c) can be applied to any *function table* defined on a uniform mesh with step h, consisting of the equally spaced points

(5) $$x_r = x_0 + rh, \qquad r = 0, \pm 1, \pm 2, \ldots; \qquad h > 0$$

Using the standard abbreviations $y_r = f_r = f(a + rh)$, we obtain the identity

$$f_1 - f_0 = f(a + h) - f(a) = \Delta f_0 = \nabla f_1 = \delta f_{1/2}$$

This shows that the usual difference notation is highly redundant.†

This redundancy is also apparent when we *iterate* the difference operators (4a)–(4c) to define the *second differences*

(6a) $\Delta^2 f(x) = \Delta(\Delta(fx)) = \Delta(f(x + h) - f(x)) = f(x + 2h) - 2f(x + h) + f(x)$

(6b) $\nabla^2 f(x) = \nabla(\nabla f(x)) = \nabla(f(x) - f(x - h)) = f(x) - 2f(x - h) + f(x - 2h)$

(6c) $\delta^2 f(x) = f(x + h) - 2f(x) + f(x - h)$

We easily verify the identities $\delta^2 f(x) = \Delta(\nabla f(x)) = \nabla(\Delta f(x))$.

In formulas (6a)–(6c), the exponent 2 describes the effect of applying the operators of formulas (4a)–(4c) twice, or "squaring" them. More generally, we can form *polynomials* (with constant coefficients) of difference operators, like $\Delta^2 - 3\Delta + 2$. Such linear difference operators with constant coefficients commute (are permutable); those with variable coefficients do not commute (cf. Ex. A4 below).

2. POLYNOMIAL INTERPOLATION

The difference notation of §1 permits us to write down simple formulas for the polynomials of least degree interpolated through given values on any uniform mesh. These interpolation formulas make it possible to construct accurate approximating functions from accurate function tables.

Simplest is the *linear* interpolation formula (for fixed h and variable k)

(7) $$p(x_0 + k) = y_0 + \frac{k}{h}(y_1 - y_0), \qquad 0 < k < h$$

Next simplest is the formula for quadratic or *parabolic* interpolation. Using the second central difference notation of formula (6c),

$$\delta^2 y_i = y_{i+1} - 2y_i + y_{i-1}$$

we obtain the quadratic interpolation formula

(8) $$q(x_1 + k) = y_1 + \frac{k}{2h}\left(y_2 - y_0 + \frac{k}{h}\delta^2 y_1\right)$$

---

† It is also inconsistent with the usual notation $\Delta t_i = t_i - t_{i-1}$ employed in writing Riemann sums, used in Ch. 7.

If second differences are tabulated, as they are in many tables, to use this formula requires only two multiplications and three additions.†

In a similar way, we can derive the *quartic* (fourth order) *interpolation formula*

$$(9) \quad f(x_2 + k) = y_2 + \frac{k}{2h}\left(y_3 - y_1 + \frac{k}{n}\delta^2 y_2\right)$$

$$+ k(k^2 - h^2)\left[\frac{(\delta^2 y_3 - \delta^2 y_1)}{12h^3} + \frac{k\delta^4 y_2}{24h^4}\right]$$

Here $\delta^4 y_2 = \delta^2(\delta^2 y_2) = y_4 - 4y_3 + 6y_2 - 4y_1 + y_0$.

The preceding formulas are based on central differences. For polynomial interpolation between $n + 1$ successive values on a uniform mesh, one often uses the *Gregory–Newton interpolation formula*,

$$(10) \qquad p(x) = f(x_0) + \sum_{k=1}^{n} \frac{\Delta^k f(x_0)}{h^k(k!)}\left[\prod_{j=0}^{k-1}(x - x_0 - jh)\right]$$

where $n = (x_n - x_0)/h$, and where Δ^k is the iterated forward difference operator. This formula gives an approximation to $f(x)$ in terms of the differences of n equally spaced values of f. This formula is a difference analog of Taylor's formula, without a remainder term.

Lagrange Interpolation Formula.   The Gregory–Newton formula, in turn, can be regarded as a special case of a very general interpolation formula due to Lagrange. Given the numbers $x_0 < x_1 < \cdots < x_n$ and $y_0, y_1, \ldots, y_n$, one can show‡ that there exists a unique polynomial $p(x)$ of degree n or less which satisfies $p(x_k) = y_k$ for $k = 0, 1, \ldots, n$, that is, which assumes the $n + 1$ given values at the points specified. Let

$$Q(x) = (x - x_0)(x - x_1) \cdots (x - x_n)$$

$$p_k(x) = \frac{Q(x)}{(x - x_k)} = \prod_{j \neq k}(x - x_j)$$

Then the polynomial

$$(11) \qquad p(x) = \sum_{k=0}^{n} \frac{p_k(x)}{p_k(x_k)}y_k = Q(x)\sum_{k=0}^{n} \frac{y_k}{Q'(x_k)(x - x_k)}$$

---

† We do not count the division required to calculate k/h, since this requires only a decimal-point shift in most tabulations.

‡ Birkhoff and MacLane, p. 60.

takes exactly the values $p(x_k) = y_k$, $0 \leq k \leq n$. Indeed, $p_k(x_j) = 0$ for $j \neq k$, so that substituting $x = x_j$ into (11), we have

$$p(x_j) = \sum_{k=0}^{n} \frac{p_k(x_j)}{p_k(x_k)} y_k = \frac{p_j(x_j)}{p_j(x_j)} y_j = y_j$$

Formula (11) is the Lagrange interpolation formula. Since the polynomial $p(x)$ is (11) is unique, (11) is equivalent to (10) if $x_j = a + jh, j = 0, 1, \ldots, n$. Hence as interpolants to a function tabulated at equal intervals, (10) and (11) have the same error.

Hermite Interpolation. The limiting case of Lagrange interpolation of order $2m$, obtained by letting $x_0, \ldots, x_{m-1}$ approach a and $x_m, \ldots, x_{2m-1}$ approach b, is called *osculatory* interpolation of order m on $[a, b]$. The limit exists if $f \in C^{m-1}[a, b]$, and is that polynomial $p(x)$ of degree $2m - 1$ that satisfies $p^{(j)}(a) = f^{(j)}(a)$ and $p^{(j)}(b) = p^{(j)}(b)$ for $j = 0, \ldots, m - 1$. *Hermite* interpolation of order m means, for any partition π of $[a, b]$, just osculatory interpolation of order m on each subinterval.

The case $m = 2$ of *cubic* Hermite interpolation is especially useful. On $[0, 1]$, [writing $f(0) = y_0, f'(0) = y_0', f(1) = y_1, f'(1) = y_1'$], this gives

$$p(x) = y_0 + y_0'x + [3y_1 - 3y_0 - 2y_0' - y_1']x^2 + [y_1' + y_0' - 2y_1 + 2y_0]x^3$$

The preceding formula is easily applied to the solutions of first-order DEs $y' = F(x, y)$ (and of first-order systems), because the $y_k' = F(x_k, y_k)$ are then known and usually already computed.

Spline Interpolation. In problems whose formulation uses empirical data, or whose solution will only be found on a coarse mesh, Lagrange and Hermite interpolation can often be advantageously replaced by cubic *spline* interpolation. This interpolates a *piecewise* cubic polynomial with continuous first and *second* derivatives through any set of mesh points (x_k, y_k); see Exs. A9–A11.

The concepts defined above are fundamental to the understanding of two-step and *multistep* methods for solving ordinary DEs and systems (see §9).

Moreover, the study of their properties is very attractive from a theoretical standpoint. However, the one-step Runge–Kutta methods to be explained in §8 are based on power series considerations, which are very different. Therefore, some readers may wish to postpone the study of these properties until they have become familiar with Runge–Kutta methods.

EXERCISES A

1. (a) Show that, if the function $f(x) = \Sigma a_k x^k$ is a polynomial of degree m, then $\Delta^m f = \nabla^m f = \delta^m f = h^m f^{(m)}$, where $f^{(m)}$ denotes the mth derivative of f.
   (b) Show that, if $y \in C^r$, then $\delta' y = 0(h')$ and $\Delta' y = 0(h')$.

2. Define the *divided differences* $[u_0, u_1]$ and $[u_0, u_1, u_2]$ as $(u_1 - u_0)/(x_1 - x_0)$ and $([u_1, u_2] - [u_0, u_1])/(x_2 - x_0)$, respectively. Show that $\lim_{x_2 \downarrow x_1, x_0 \uparrow x_1}[u_0, u_1, u_2] = u''(x_1)$.

3. Show that

$$\delta^r y_j = \sum_{k=0}^{r} (-1)^k \frac{r!}{k!(r-k)!} y_{j-k+r/2}$$

4. Show that, if $f(x) = x^2$, then $x(\Delta f) = 2hx^2 + h^2x$, yet $\Delta(xf) = 3x^2h + 3xh^2 + h^3$.

5. Solve the ΔE $u_{n+1} = 2u_n - u_{n-1}$, for the initial conditions $u_0 = 1$, $u_1 = -1$.

6. The nth Fibonacci number F_n is the value at n of the solution of the ΔE

$$F_{n+1} = F_n + F_{n-1}$$

for the initial conditions $F_0 = 0$, $F_1 = 1$.

(a) Show that $F_n = (\rho^n - \sigma^n)\sqrt{5}$, where $\rho = (\sqrt{5} + 1)/2$, $\sigma = (1 - \sqrt{5})/2$.

(b) What is the solution G_n of the Fibonacci ΔE for the initial conditions $G_0 = 2$, $G_1 = 1$?

7. Estimate the largest h such that parabolic interpolation in a six-place table of $\log_{10} x$ on $2 \leq x \leq 3$, with mesh length h, will yield five-place accuracy.

8. Show that, with parabolic interpolation between $y(-h)$, $y(0)$, and $y(h)$, the maximum error is normally near $x = \pm h/\sqrt{3}$ and is about $h^3 |f''(x)|/9\sqrt{3}$ there.

9. Show that the cubic Hermite interpolants to given values of $f(x)$ and $f'(x)$ at the endpoints of the intervals $(a-h,a)$ and $(a,a+h)$ define a "cubic spline" function $f \in \mathcal{C}^2$ $[a-h,a+h]$ if and only if

(*)　　　　　　(*) $h[f'(a-h) + 4f'(a) + f(a+h)] = 3\delta^2 f(a)$

*10. Let $h = (b - a)/n$ and $x_i = a + ih$. Prove that, given $y_i[i = 0, \ldots, n]$ and y_0', y_n', there is one and only one cubic spline function $s(x) \in \mathcal{C}^2(a,b)$ which satisfies: (i) the interpolation conditions $s(x_i) = y_i$ for $i = 0, \ldots, n$; (ii) $s(a) = y_0'$, $s(b) = y_n'$; and (iii) is a cubic polynomial in $[x_{i-1}, x_i]$ for $i = 1, \ldots, n$.
(Hint: Use Ex. 9.]

*11. (a) Show that if $s(x)$ is the (piecewise) spline interpolant to given y_i and y_0', y_n' specified in Ex. 10, and $f(x) = s(x) + v(x) \in \mathcal{C}^2(a,b)$ is any other interpolant with the same properties, then

$$\int_a^b [f''(x)]^2 dx = \int_a^b [s''(x)]^2 dx + \int_a^b [v''(x)]^2 dx$$

(b) Infer that the cubic spline interpolant minimizes the mean square value of the second derivative on (a,b), among all interpolants $f \in \mathcal{C}^2(a,b)$.

(Hint: Show that $\int_a^b s''(x)v''(x)dx = 0$ for all continuous piecewise *linear* functions $v''(x)$.)

*3 INTERPOLATION ERRORS

Among the many interesting properties of interpolation schemes, their errors are clearly most basic. We therefore take them up first. The *order of magnitude* of such errors can often be determined algebraically, by simply finding the polynomial of least degree for which the formula ceases to be exact.

In this section, we shall do much better, by giving explicit expressions for the *magnitude* of the error, by formulas involving an appropriate derivative of the function. More precisely, let a function $f(x)$ be tabulated at $n + 1$ points $x_0 < x_1 < \cdots < x_n$, and let $p(x)$ be the unique interpolation polynomial, of degree n or less, which satisfies $p(x_k) = f(x_k)$, $k = 0, 1 \ldots, n$. How big is the error of $p(x)$, considered as an approximation to $f(x)$? An answer to this question, when $f(x)$ is sufficiently smooth, is provided by the following result.

THEOREM 1.   *Let* $p(x) = a_0 + a_1x + \cdots + a_nx^n$ *be the polynomial satisfying* $p(x_1) = f(x_1)$, *for* $x_0 < x_1 < \cdots < x_n$. *If* $f \in \mathscr{C}^{n+1}[x_0, x_n]$, *then for every* x *in* $[x_0, x_n] = I$ *there exists* ξ *in* I, *such that*

$$(12) \qquad f(x) - p(x) = \frac{(x - x_0) \cdots (x - x_n)}{(n + 1)!} f^{(n+1)}(\xi)$$

Proof.   Let $e(x) = f(x) - p(x)$ denote the error function. Since $p(x)$ is a polynomial of degree n, $p^{(n+1)}(x) \equiv 0$. Therefore

$$e^{(n+1)}(x) = f^{(n+1)}(x)$$

for all x, and $e(x_0) = e(x_1) = \cdots = e(x_n) = 0$. Consider now the function

$$(13) \qquad \phi(t) = Q(x)e(t) - Q(t)e(x)$$

where Q is the polynomial $Q(x) = (x - x_0)(x - x_1) \cdots (x - x_n)$. We consider ϕ as a function of t on I, for x fixed. Clearly $\phi \in \mathscr{C}^{n+1}$; moreover, $\phi(x_k) = 0$, $k = 0, 1, \ldots, n$, and in addition $\phi(x) = 0$. By Rolle's Theorem, between any two points where ϕ vanishes there is at least one point where ϕ' vanishes. Since the function ϕ vanishes at $n + 2$ points (if $x \neq x_k$ for all k), the function ϕ' vanishes for at least $n + 1$ points. Repeating the same argument for higher derivatives, we eventually conclude that $\phi^{(n+1)}(t)$ vanishes for at least one point ξ in the interval I. Differentiating (13) relative to t, $n + 1$ times, we obtain

$$0 = \phi^{(n+1)}(\xi) = Q(x)e^{(n+1)}(\xi) - (n + 1)!e(x)$$

Since $e^{(n+1)}(\xi) = f^{(n+1)}(\xi)$ and $e(x) = f(x) - p(x)$, this gives

$$f(x) - p(x) = e(x) = \frac{Q(x)}{(n + 1)!} f^{(n+1)}(\xi), \qquad \text{q.e.d.}$$

COROLLARY.   *If* p *is the Lagrange interpolation polynomial of a function* $f(x)$ *of class* $\mathscr{C}^{n+1}$ *in the interval* $[x_0, x_n]$ *and* $x_0 < x_1 < \cdots < x_n$, *the error at any point* $x \in [x_0, x_n]$ *is at most*

$$|f(x) - p(x)| \leq \frac{1}{(n + 1)!} N_n M_n$$

where $M_n = \max_{x_0 \leq \xi \leq x_n} |f^{(n+1)}(\xi)|$ *and* $N_n = \max_{x_0 \leq x \leq x_n} |Q(x)|$.

When the mesh points are equally spaced, we can compute N_n explicitly. Thus, we have $N_1 = h^2/4$; if $x_1 < x < x_2$, $N_3 = 9h^4/16$, and so on. Moreover, similar arguments show that for $k \leq n$, the error in the kth derivative of the Lagrangian interpolant is $O(h^{n-k+1})$. It follows (though we shall not prove it) that one can develop multipoint ΔEs that approximate DEs to an arbitrarily high order of accuracy.

Applications. For example, if $x = x_j + k$, $0 < k < h$, the magnitude of the error in linear interpolation is $|k(h - k)f''(\xi)|/2 \leq h^2|f''(\xi)|/8$, for some $\xi \in [x_j, x_j + h]$. Likewise, parabolic interpolation through $f(x_j - h)$, $f(x_j)$, and $f(x_j + h)$ gives an approximate value differing from $f(x_j + k)$ by $|k(h^2 - k^2)f'''(\xi)|/6$. For $|k| \leq h/2$, the error is therefore bounded by $h^3|f'''|_{max}/16$. Since we would naturally choose j to minimize $|x - x_j|$, this bounds the error in parabolic interpolation. The maximum error in the interval $(x_j - h, x_j + h)$ is slightly larger; see Ex. 8.

Ordinarily, parabolic interpolation is sufficiently accurate. For example, with $\sin x$, $|f'''|_{max} = 1$; hence the error is bounded by $h^3/16$ in radian units. Therefore, parabolic interpolation give four-place accuracy in a table at 6° intervals! More generally, unless $|f'''|_{max} > 10$, six-place tables can be extended by parabolic interpolation to all x if $h = 0.01$, without an appreciable loss of accuracy. The same is true of nine-place tables if $h = 0.001$ (and of three-place tables if $h = 0.1$).

For these reasons, higher order interpolation is unnecessary for most tables in common use.

Caution. The *approximations* to a given function $f(x)$ on a fixed interval, defined by polynomial *interpolation* over that interval, are not necessarily *good* approximations to $f(x)$, even if $f(x)$ is analytic. Thus, the approximations to the analytic function $f(x) = 1/(1 + x^2)$, obtained by the Gregory-Newton interpolation formula (10), *do not converge*† to $f(x)$ on the interval $-5 \leq x \leq 5$, but oscillate more and more wildly as the step length h tends to zero.

This shows that Newtonian interpolation cannot be used to define the approximating polynomials referred to in the Weierstrass Approximation Theorem. To get the best such *uniform* polynomial approximations, one must use a very different method due to Chebyshev (see Ch. 11, §7) and Remez.

4 STABILITY

The accurate numerical solution of initial value problems for ordinary DEs (and systems) involves much more than interpolation error bounds and estimates. For one thing, the relative error involves *stability* considerations. These are most easily explained in the special case of linear DEs with constant coefficients, previously discussed in Ch. 3.

---

† See J. F. Steffensen, *Interpolation*, Williams & Wilkins, Baltimore, 1927, pp. 35–38 and the references given there.

A linear ΔE with constant coefficients is one of the form

(14) $$y_{n+m} = a_0 y_n + a_1 y_{n+1} + \cdots + a_{m-1} y_{n+m-1}$$

where a_k are given constants. The solutions of such a ΔE can be obtained by a substitution similar to the exponential substitution of Ch. 3, §1. Try the sequence $y_r = \rho^r$, where ρ is a number to be determined. This gives from (14) the characteristic equation

(15) $$\rho^m - a_0 - a_1 \rho - \cdots - a_{m-1} \rho^{m-1} = 0$$

For each root ρ_k of this characteristic equation, the sequence $y_r = \rho_k^r$ is the solution of the linear ΔE (14), which satisfies the initial conditions $y_0 = 1$, $y_1 = \rho_k, \ldots, y_{m-1} = \rho_k^{m-1}$.

Midpoint Method.   The midpoint method for solving $y' = F(x, y)$ consists in first computing y_1 (perhaps by Taylor series) and then using the formula

(16) $$y_{n+1} = y_{n-1} + 2hF(x_n, y_n)$$

For example, consider again the DE $y' = y$ for the initial condition $y(0) = 1$, whose exact solution is $y = e^x$. In this case (16) reduces to

(16') $$y_{n+1} = y_{n-1} + 2hy_n$$

For $h = 0.1$ and $y_0 = y(0) = 1$, the exponential series truncated after five terms gives $y_1 = 1.1052$, rounded off to four decimal places. Substituting into (16') we can compute the approximate function table for $y_r = \exp(r/10)$:

x	0.1	0.2	0.3	0.4	0.5	0.6	0.7
y	1.1052	1.2210	1.3494	1.4909	1.6476	1.8204	2.0117

and so on. After 10 steps, this gives the approximate value $e = 2.714$, whose error is about -0.0043.

In (16), the characteristic equation is $\rho^2 = 1 + 2h\rho$, with distinct roots

$$\rho_i = h \pm \sqrt{1 + h^2} = \pm 1 + h \pm \frac{h^2}{2} \mp \frac{h^4}{8} \pm \cdots$$

For $h = 0.1$, this gives $\rho_1 = 1.10499$, $\rho_2 = -0.90499$ when rounded off to five decimal places. The first root differs from the growth factor $e^{0.1} = 1.105171$ of the exact solution by about 0.00018. This is about 0.27% of 0.105, which explains why the $O(h^2)$ approximation (16) gives only two-digit accuracy for $h = 0.1$.

Stability. By analogy with Ch. 3, §4, we will say that the homogeneous linear nth order ΔE with constant coefficients (14) is *stable* when all solutions y_r are *bounded* sequences and are *strictly stable* when all solutions are sequences tending to zero, as $r \to \infty$.

Since we can obtain a basis of solutions of (14) of the form $r^j \rho_k^r$, the ΔE (14) is strictly stable if and only if all roots of the characteristic equation (15) are less than one is absolute value. This condition is obviously necessary; it is sufficient because $\lim_{r \to z} r^j \rho^r = 0$ whenever $|\rho| < 1$.

The concept of stability brings out a significant aspect of the effectiveness of the central difference approximation (16) for integrating numerically the DE $y' = y$. The general solution of (16) is $A\rho_1^r + B\rho_2^r$, where $\rho_i = h \pm (1 + h^2)^{1/2}$ as above, and A, B are arbitrary constants. The positive root ρ_1, which is approximately equal to $e^{0.1}$, is *dominant* in the sense that $|\rho_1| > |\rho_2|$. Therefore, the term $B\rho_2^r$ can be neglected in comparison with $A\rho_1^r$ for large r, provided that $A \neq 0$.

Example 1.   For the DE $y' = -y$, the situation is reversed: the (central) difference approximation $y_{k+1} - y_{k-1} = -2hy_r$ to the *stable* DE $y' = -y$ is *unstable*.

Thus for $h = 0.1$, we have $\rho_1 = 0.904988$, $\rho_2 = -1.10499$. Although ρ_1 approximates the exact growth factor $\rho = 0.90584$ reasonably well, it is dominated by the "extraneous root" ρ_2 which is introduced in approximating a first-order DE by a *second*-order ΔE. As a result, the "approximate solution" will ultimately grow like e^x, whereas the true solution decays like e^{-x}.

Example 2.   Consider the difference approximation (16) to the initial value problem defined by the DE $y' = 2x$ and initial $y(0) = 0$. For the true solution $y = x^2$, (16) is *exact*, but it still gives very bad results because of *roundoff errors*.

This will be explained in §5; here we simply consider the characteristic polynomial of the ΔE specified:

$$(17) \qquad y_{k+2} = 8y_{k+1} - 8y_{k-1} + y_{k-2} - 12hy_k', \qquad y_k' = 2x_k$$

The characteristic polynomial of this ΔE is

$$(18) \qquad \rho^4 - 8\rho^3 + 8\rho - 1 = (\rho^2 - 1)(\rho^2 - 8\rho + 1)$$

whose roots are ± 1 and $4 \pm \sqrt{15}$. One of these is near 8, so that errors tend to grow by a factor 8 per time step. The ΔE is thus very unstable!

EXERCISES B

In Exs. 1–4, verify the formulas indicated for $f \in \mathcal{C}^4$, a uniform mesh with step h, $0 < \theta < 1$, and $\bar{\theta} = (1 - \theta)$.

1. $f(x_0 + \theta h) = y_0 + \theta \Delta y_0 - (\theta\bar{\theta}/2) \Delta^2 y_0 + O(h^2)$ (Newton)

2. $f(x_0 + \theta h) = y_0 + \theta\delta y_{1/2} - (\theta\bar{\theta}/4)(\delta^2 y_0 + \delta^2 y_1) + O(h^3)$ (Bessel)

3. $f(x_0 + \theta h) = (\bar{\theta} y_0 + \theta y_1) + [(\bar{\theta}^3 - \bar{\theta})\delta^2 y_0 + (\theta^3 - \theta)\delta^2 y_1]/6 + O(h^4)$ (Everett)

4. $f(x_0 + \theta h) = \frac{1}{2}(y_0 + y_1) + (\theta - \frac{1}{2})\delta y_{1/2} - (\theta\bar{\theta}/4)(\delta^2 y_0 + \delta^2 y_0)$
   $- [\theta\bar{\theta}(2\theta - 1)/12]\,\delta^3 y_{1/2} + O(h^4)$ (Bessel)

5. Find the truncation errors of the formulas of Exs. 1–3, for quartic polynomials $q(x) = a + bx + cx^2 + dx^3 + ex^4$.

6. (a) Find the cubic polynomial $c(x)$ that satisfies $c(0) = y_0$, $c'(0) = y_0'$, $c(h) = y_1$, $c'(h) = y_1'$.
   (b) Derive your formula as a limiting case of the four-point Lagrange interpolation formula.

7. Test the following ΔEs for stability or instability, by calculating the roots of their characteristic equations:
   (a) $u_{n+1} = 2u_n - u_{n+1}$ (b) $u_{n+1} = u_n + u_{n-1}$
   (c) $u_{n+1} - 5u_n + 6u_{n+1} = 0$ (d) $u_{n+1} = -u_{n-1}$

*8. Show in detail that $|a + ab| + b^2 < 1$ is a necessary and sufficient condition for the strict stability of the ΔE $y_{n+2} = ay_{n+1} + by_n$.

9. Derive necessary and sufficient conditions for the strict stability of a general linear third-order ΔE with constant coefficients.

*5   NUMERICAL DIFFERENTIATION; ROUNDOFF

Using Taylor series with remainder, one can in principle obtain approximations to the derivatives of tabulated functions having arbitrarily high orders of accuracy, from suitably designed difference quotient and divided difference formulas.

For example, whereas the usual forward difference quotient formula $\Delta f/h$ has only $O(h)$ accuracy, and even the central difference quotient formula $\delta f/h$ has only $O(h^2)$ accuracy, formula (3) of §1 has $O(h^4)$ accuracy. Likewise, we have the truncation error estimates

(19a)
$$\frac{\Delta f}{h} - f'(x) = \frac{f''(x)h}{2} + \frac{f'''(x)h^2}{6} + \frac{f^{iv}(\xi)h^3}{24}$$

(19b)
$$\frac{\delta f}{h} - f'(x) = \frac{f'''(x)h^2}{24} + \frac{f^{v}(\xi)h^4}{1920}$$

where ξ is in the interval over which the difference is being taken. This illustrates the general principle that central difference quotients give more accurate approximations to derivatives than forward or backward difference quotients of the same order. We can obtain *truncation error bounds* similarly:

(19c)
$$\left| \frac{[f(x + h) - f(x - h)]}{2h} - f'(x) \right| \le \frac{|f'''|_{max}h^2}{6}$$

Note that the interval in (19c) is twice as long as in (19b); hence, the truncation error is multiplied by about four.

Similar approximations can be made to $f''(x)$, using second difference quotients. For $f \in \mathcal{C}^2$, we have

$$\delta^2 f(x) = f(x + h) - 2f(x) + f(x - h)$$

$$= [f(x) + hf'(x) + \frac{h^2}{2}f''(\xi_1)] - 2f(x) + \frac{1}{2}[f(x) - hf'(x) + h^2f''(\xi_2)]$$

$$= \frac{h^2}{2}[f''(\xi_1) + f''(\xi_2)]$$

for some numbers ξ_1 and ξ_2 in the intervals $[x, x + h]$ and $[x - h, x]$, respectively. Hence $\delta^2 f/h^2$ lies between the minimum and maximum values of $f''(\xi)$ for ξ in the interval $x - h \leq \xi \leq x + h$. Since a continuous function assumes all values between its minimum and maximum, and since $f''(x)$ is continuous in the interval, we conclude that $\delta^2 f = h^2 f''(\xi)$ for some ξ in $[x - h, x + h]$. This shows that the difference quotient $\delta^2 f/h^2$ is a good approximation to the derivative f'' for small h. Since $\delta^2 f/h^2$ can be computed by consulting a numerical table of $f(x)$, the preceding formula may again be regarded as one for *approximate numerical differentiation*. This is written in the form $f''(x) \simeq \delta^2 f(x)/h^2$, where the symbol $\simeq$ means "is approximately equal to," as in Ch. 7, §2.

For $f \in \mathcal{C}^4$, the preceding analysis can be refined to give an estimate of the truncation error. Taylor's formula with remainder gives

$$f(x \pm h) = f(x) \pm hf'(x) + \frac{h^2 f''(x)}{2} \pm \frac{h^3 f'''(x)}{6} + \frac{h^4 f^{iv}(\xi)}{24}$$

where $x - h \leq \xi \leq x + h$. Since $f^{iv}(\xi)$ assumes all values between its minimum and maximum values on this interval, we can write

(20) $$\delta^2 f - h^2 f''(x) = h^4 f^{iv}(\xi)/12, \qquad x - h \leq \xi \leq x + h$$

This formula gives the truncation error estimate

$$f''(x) - \frac{\delta^2 f}{h^2} = -\frac{h^2}{12}f^{iv}(\xi)$$

in the formula $f''(x) \simeq \delta^2 f(x)/h^2$ for numerical differentiation. This formula shows that the truncation error is of the order of h^2, and tends to zero fairly rapidly when the step h is taken smaller and smaller.

Higher-order Derivatives. The preceding truncation error estimates and bounds are special cases of a general result, namely:

THEOREM 2.   *If $f \in C^{(n+2)}$ on $I = [x - nh/2, \, x + nh/2]$, then*

(21)
$$f^{(n)}(x) = \frac{\delta^n f}{h^n} - \frac{nh^2}{24} f^{(n+2)}(\xi), \qquad \xi \in I$$

Proof.   The cases $n = 1, 2$ have been treated above. Proceeding by induction, we get

$$\delta f(x) = \int_{-h/2}^{h/2} f'(x + t) \, dt, \qquad \delta^2 f(x) = \int_{-h/2}^{h/2} dt \int_{-h/2}^{h/2} f''(x + t + u) \, du, \; \cdots$$

$$\delta^n f(x) = \int_{-h/2}^{h/2} dt_1 \int_{-h/2}^{h/2} dt_2 \cdots \int_{-h/2}^{h/2} f^{(n)}(x + t_1 + \cdots + t_n) \, dt_n$$

This is a multiple integral over an n-dimensional domain D with center $\mathbf{t} = \mathbf{0}$ and volume h^n, symmetric under the reflection $t_1 \rightarrow -t_j$ $(j = 1, \ldots, n)$. The arithmetic mean of the integrands at symmetrically placed points $x - T$ and $x + T$ is, by Taylor's formula,

$$\frac{1}{2} \left[f^{(n)}(x + T) + f^{(n)}(x - T) \right] = f^{(n)}(x) + \left(\frac{T^2}{2} \right) f^{(n+2)}(x + \theta T)$$

for some θ, $-1 \leq \theta \leq 1$. Hence, setting $T = t_1 + \cdots + t_n$ and integrating over D, we have

$$\frac{m}{2} \int T^2 \, dt_1 \cdots dt_n \leq \delta^n f - h^n f^{(n)}(x) \leq \frac{M}{2} \int T^2 \, dt_1 \cdots dt_n$$

where m and M are the least and greatest values of $f^{(n+2)}(\xi)$ for ξ in the given interval. Since

$$\int_{-h/2}^{h/2} t_k^2 \, dt_k = \frac{h^3}{12} \qquad \text{and} \qquad \int_{-h/2}^{h/2} t_\ell t_k \, dt_\ell = 0, \qquad k \neq \ell$$

and since $f^{(n+2)}(\xi)$, being continuous, assumes all values between its extreme values m and M, formula (21) follows.

COROLLARY.   *Under the hypothesis of Theorem 2, we have the truncation error bound*

(22)
$$|f^{(n)}(x) - \delta^n f / h^n| \leq \frac{nMh^2}{24}$$

where M is the maximum of $|f^{(n+2)}(\xi)|$ as ξ ranges over the given interval.

Roundoff Errors.   In the preceding discussion, as in the error estimates and error bounds derived in Ch. 7, it has been tacitly assumed that all arithmetic operations and readings from function tables were *exact,* to an unlimited number of decimal places. In reality, however, only a finite number of significant digits are used. This leads to a source of error called the *roundoff error,* which has been ignored previously in this book. The errors discussed previously are referred to technically as *truncation errors,* or *discretization errors.*

In polynomial interpolation of low order, the roundoff error is not a serious problem. For $|k| < h$, formulas (12)–(14) show that the roundoff error affects only the last decimal place tabulated. For instance, writing $k/h = r$, parabolic interpolation gives $f(x_1 + rh) = w_0 y_0 + w_1 y_1 + w_2 y_2$, where $w_0 = (r^2 - r)/2$, $w_1 = 1 - r^2$, $w_2 = (r + r^2)/2$, and $|r| < \frac{1}{2}$, if the three nearest tabulated values are used. Since

$$(23) \qquad |w_0| + |w_1| + |w_2| \leq 1 + |r| + 2r^2 \leq 2$$

and tabulated values are correct to $\frac{1}{2}$ in the last decimal place, the roundoff error is at most one in the last decimal place.

In numerical quadrature formulas, which also have the form $\Sigma w_k y_k$ per step with Σw_k equal to the mesh length, the *maximum* roundoff error is similarly bounded by the length of the interval multiplied by the maximum tabulation error (ordinarily $\frac{1}{2}$ in the last decimal place).

However, the effect of roundoff can be dramatic in other cases, as Example 2 of §4 demonstrates. The truncation error is zero in this example; hence if $h = \frac{1}{8}$ or some other binary fraction, the computer printout will also be exact. But if $h = 0.1$ (which is not binary), the small initial roundoff error is amplified by a factor 8 at each step, and dwarfs the true solution after 10 or 20 steps.

Empirically, roundoff errors are nearly independent, and randomly distributed in the first untabulated decimal place with a mean nearly zero. Hence,† the cumulative roundoff error has a roughly normal distribution on a Gaussian curve, and the *probable* cumulative roundoff error with n equal subdivisions is only $O(1/\sqrt{n})$ times the maximum cumulative roundoff error.

Similar results hold for the numerical integration formulas to be considered. The roundoff errors may be thought of as "noise," superimposed on the systematic truncation error. Both are amplified in the course of the calculation by a factor of at most $e^{L(b-a)}$, where $L = \sup \partial F/\partial x$ is the one-sided Lipschitz constant, and $(b - a)$ is the interval of integration.‡ The reason for this is that, for L as defined above,

$$(24) \qquad [y(x) - z(x)]' = F(x, y) - F(x, z) \leq L(y - z) \qquad \text{if} \qquad y > z$$

see also Ch. 1, Theorem 5.

---

† By the central limit theorem of probability theory.

‡ For a careful analysis of the cumulative roundoff error, see Henrici.

It is interesting to compare the truncation error bound (19c) with the corresponding *roundoff error* bound, which is $10^{-m}/2h$ if an m-place table is used. When $h < (3/10^m|f'''|_{max})^{1/3}$, therefore, the roundoff error exceeds the truncation error. To minimize the sum $(|f'''|_{max}h^2/6) + (10^{-m}/2h)$, which is the maximum total error if both terms have the same sign, set $2h^3 = 3/10^m|f'''|_{max}$. This shows that the maximum total error with parabolic interpolation into an m-place table cannot be reduced below

$$\tfrac{1}{4}\sqrt[3]{18|f'''|_{max}/10^{2m}}$$

and that one loses accuracy if $|f''_{max}| < 2$, by choosing h smaller than 0.04, if a four-place table is used to approximate $f'(x)$ by a central difference quotient.

For example, for the function $f(x) = \sin x$, where $f^{iv}(x) = \sin x$, since $f^{iv}(x)$ ranges between -1 and 1, the maximum truncation error is approximately $h^2/12$, and the maximum roundoff error is $2 \times 10^{-5}/h^2$, using five-place tables. To minimize the greater of the truncation error (which tends to zero with h) and the maximum roundoff error, we must make $h^2/24 \simeq 10^{-5}/h$. Hence, we minimize the maximum total error of $f'' \simeq \delta^2 f/h^2$ near $h^4 = 2.4 \times 10^{-4}$ or $h \simeq 0.13$ radian $\simeq 8°$, roughly, a surprisingly large interval!

Roundoff errors are not considered further in this chapter. This is partly because, with high-speed computing machines, truncation errors are usually bigger unless h is very small (most modern machines carry at least ten decimal digits), and partly because the analysis of roundoff errors involves difficult statistical considerations.

EXERCISES C

1. Show that the effect of roundoff errors on tenth differences is bounded by about 500 $\times 10^{-n}$ in n-place tables.

2. Show that $hf'(x_0 + h/2) = \delta y_{1/2} - (1/24)\delta^3 y_{1/2} + O(h^5)$.

3. Show that $hf'(x_0 + h/2) = \delta y_{1/2} - \delta^3 y_{1/2}/24 + \delta^5 y_{1/2}/1920 + O(h^7)$.

4. Given a six-place table of $\sin x$ (x in radians), show that the approximate formula $f''_0 \simeq \delta^2 f_0/h^2$ has a combined truncation and roundoff error bounded by $2/10^6 h^2 + h^2/12$, and that this expression has a minimum of about 0.0008, assumed for h about 0.07.

5. (a) Show that $h^2 f'' = \delta^2 f - \delta^4 f/12 + O(h^6)$.
   *(b) Show that $h^2 f'' = \delta^2 f - \delta^4 f/12 + \delta^6 f/90 + O(h^8)$.

6. Show that for $y \in \mathcal{C}^8$, we have $\delta^2 y_0 = h^2[y''_0 + \delta^2 y''_0/12 - \delta^4 y''_0/240] + O(h^8)$.

7. Show that, for small h, the ΔE (10) provides a strictly stable approximation.

*6 HIGHER ORDER QUADRATURE

As we have emphasized repeatedly, it is especially easy to derive accurate numerical formulas for solving DEs of the form $u' = u(x)$—that is, for numerical *quadrature*. In this section, we will derive a rigorous error bound for Simpson's Rule (Ch. 1, §8), by a method which is applicable to a wide variety of numerical quadrature formulas. In the next section, we will discuss two other remarkable quadrature formulas, which seem to have no analogs for other DEs.

Let $0 \leq r_0 < r_1 < r_n \leq 1$ be given, and let $x_i = a + r_i h$, $h > 0$, so that

$$(25) \qquad a \leq x_0 < x_1 < \cdots < x_n \leq a + h = b$$

For any function $f \in \mathcal{C}^{n+2}$, let $p(x)$ be the Lagrange polynomial interpolant of degree n to the $y_i = f(x_i)$ at the x_i. Then the approximation of $f(x) \simeq p(x)$ is associated with a formula for numerical quadrature, namely

$$(25') \qquad \int_a^b f(x)\, dx \simeq \int_a^b p(x)\, dx = \sum_{i=0}^n w_i y_i = \sum_{i=0}^n w_i f(x_i)$$

The coefficients w_i are the integrals of the polynomials $p_i(x)/p_i(x_i)$ in (5);

$$(26) \qquad w_i = \frac{1}{p_r(x_i)} \int_a^b p_i(x)\, dx, \qquad p_i(x) = \prod_{j \neq i} (x - x_j)$$

Formula (25') is exact for all polynomials for degree $\leq n$, since then $f(x) = p(x)$. For other functions, the error in (25') is $- \int_a^b e(x)\, dx$, where $-e(x) = p(x) - f(x)$. Hence, by Theorem 1, the error in (26) for any given n is of the order of h^{n+2} at most. We have proved the following theorem.

THEOREM 3. *For any choice of real numbers r_i with $0 \leq r_0 < r_1 < \cdots < r_n \leq 1$ and weights w_i defined by formula (26), we have, for all $f \in \mathcal{C}^{n+2}\ [a, b]$:*

$$\int_a^b f(x)\, dx - \sum_{i=1}^n w_i f(a + r_i h) = O(h^{n+2})$$

Setting $n = 1$ and $r_0 = 0$, $r_1 = 1$ in the preceding formulas, we obtain $p_0(x) = x - x_1$, $p_1(x) = x - x_0$. Therefore

$$w_1 = \int_{x_0}^{x_1} \frac{(x - x_0)\, dx}{x_1 - x_0} = \frac{x_1 - x_0}{2} = \frac{h}{2}$$

A similar calculation gives $w_0 = h/2$, so that the formula for *trapezoidal quadrature* (Ch. 7, §6),

$$(27) \qquad \int_a^{a+h} f(x)\, dx \simeq \frac{h}{2}\, [f(a) + f(a + h)] = \frac{h}{2}\, (y_0 + y_1)$$

is obtained as a special case of (25') and (26). In this special case, the error $e(x)$ satisfies by Theorem 1, applied to the linear interpolant $L(x)$ to $f(x)$

$$e(x) = L(x) - f(x) = \tfrac{1}{2}(x - x_0)(x_1 - x)f''(\xi)$$

for some $\xi = \xi(x)$, in the interval $x_0 \leq \xi \leq x_0 + h = x_1$. Since

$$\int_{x_0}^{x_1} L(x) \, dx = \frac{h(y_0 + y_1)}{2}$$

the error $\int_{x_0}^{x_1} e(x) \, dx = \int_0^h e(x_0 + t) \, dt$ in trapezoidal quadrature satisfies

$$\frac{1}{2} f''_{\min} \int_0^h t(h - t) \, dt \leq \int_{x_0}^{x_1} e(x) \, dx \leq \frac{1}{2} f''_{\max} \int_0^h t(h - t) \, dt$$

Since $\int_0^h t(h - t) \, dt = h^3/6$, we obtain formula (24) of Ch. 7:

$$(28) \qquad \frac{h}{2}(y_0 + y_1) = \int_{x_0}^{x_1} f(x) \, dx = \frac{h^3}{12} f''(\xi), \qquad x_0 < \xi < x_1$$

Applying (28) to each component of any *vector*-valued function $x(t) \in C^2$, we have

$$\int_{t_0}^{t_1} x_k(t) \, dt - \frac{h}{2} [x_k(t_0) + x_k(t_1)] = -\frac{h^3}{12} x_k''(\tau)$$

for some τ in the interval $[t_0, t_0 + h]$, and for each component x_k. By choosing one axis parallel to the error vector $\int_{t_0}^{t_1} x(t) \, dt - h(x_0 + x_1)/2$, we obtain, as a special case of the preceding result, the inequality

$$\left| \int_{t_0}^{t_1} x(t) \, dt - \frac{h}{2}(x_0 + x_1) \right| \leq \frac{h^3}{12} |\sup x''(\tau)|, \qquad t_0 < \tau < t_1$$

The *relative* truncation error is thus $O(h^2)$, as is to be expected from a formula that neglects quadratic terms.

Note that the vector analogs of the Mean Value Theorem and of (28) are false. For example, let $x(t) = (t^3, t^4)$, $t_0 = 0$, and $t_1 = 1$. Then

$$\int_0^1 x(t) \, dt - \frac{1}{2}(x_0 + x_1) = -\left(\frac{1}{4}, \frac{3}{10} \right)$$

is not equal to $-x''(\tau)/12 = -(6\tau, 12\tau^2)/12 = -(\tau, 2\tau^2)/2$ for any τ in $[0, 1]$.

Simpson's Rule.   Formula (8) for parabolic interpolation leads similarly to Simpson's rule:

$$(29) \qquad \int_{x_0}^{x_1} f(x)\, dx \simeq \frac{h}{3}\, [y_0 + 4y_1 + y_2] = \frac{h}{3}\, [f(x_0) + 4f(x_1) + f(x_2)]$$

$h = x_2 - x_1 = x_1 - x_0$. This formula is exact for quadratic polynomials. Since $\int_{-h}^{h} x^3\, dx = 0$, Simpson's rule also is exact for cubic polynomials; this coincidence makes Simpson's rule especially practical.

The error estimate for Simpson's rule (29) will now be derived for any $f \in \mathcal{C}^4$. Consider the cubic polynomial

$$p(x) = a_0 + a_1 x + a_2 x^2 + a_3 x^3$$

satisfying $p(x_0) = y_0$, $p(x_1) = y_1$, $p'(x_1) = y_1' = f'(x_1)$, $p(x_2) = y_2$, with $x_0 = x_1 - h$, $x_2 = x_1 + h$. These conditions amount to $a_0 = y_1$, $a_1 = y_1'$, $a_2 = \delta^2 y_1 / 2h^2$, and $a_3 = -a_1/h^2 + (y_2 - y_0)/2h^3$. Hence they can be satisfied for any y_0, y_1, y_2.

To estimate the error $e(x) = p(x) - f(x)$, translate coordinates so that $x_0 = -h$, $x_1 = 0$, $x_2 = h$. Consider the function of t for *fixed* x

$$(30) \qquad \phi(t) = x^2(x^2 - h^2)e(t) - t^2(t^2 - h^2)e(x)$$

analogous to the function (13) used in proving Theorem 1. We have $\phi(0) = \phi(\pm h) = \phi(x) = 0$, and further, $\phi'(0) = 0$, since $e'(0) = 0$. By Rolle's Theorem, the function $\phi'(t)$ vanishes at three places besides $t = 0$ in the interval $-h < t < h$. Hence $\phi''(t)$ vanishes at least three times in $-h < t < h$, $\phi'''(t)$ vanishes twice, and $\phi^{iv}(t)$ vanishes once, at some point $t = \xi$. We thus have, much as in (27),

$$(31) \qquad 0 = \phi^{iv}(\xi) = x^2(x^2 - h^2)e^{iv}(\xi) - 24e(x)$$

Since $p(x)$ is a cubic polynomial, $p^{iv}(x) \equiv 0$; therefore $-e^{iv}(\xi) = f^{iv}(\xi)$. Substituting in (31) and solving for $e(x)$, we get the error estimate

$$(32) \qquad e(x) = \frac{x^2(h^2 - x^2)f^{iv}(\xi)}{24}, \qquad -h \le \xi \le h$$

Integrating (32) with respect to x, since $x^2(h^2 - x^2) \ge 0$, it follows that the truncation error in using Simpson's rule for quadrature over $-h \le x \le h$ lies between $m = \min f^{iv}(\xi)$ and $M = \max f^{iv}(\xi)$ times the definite integral

$$\int_{-h}^{h} \frac{x^2(h^2 - x^2)\, dx}{24} = \frac{h^5}{90}$$

Since $f \in \mathcal{C}^4$, $f^{iv}(\xi)$ assumes every value between m and M, which gives the following theorem.

THEOREM 4.   *If $f(x) \in \mathcal{C}^{iv}$, the truncation error for Simpson's rule on the interval $-h \leq x \leq h$ is equal to $h^5 f^{iv}(\xi)/90$ for some ξ in the interval $[-h, h]$.*

The *relative* truncation error is therefore $h^4 f^{iv}(\xi)/180$. For example, to achieve five decimal places of accuracy in computing $\ln 2 = \int_1^2 dx/x$, about 10^4 points must be taken if Riemann sums are used, about 100 with trapezoidal quadrature, whereas 10 are sufficient using Simpson's rule!

*7   GAUSSIAN QUADRATURE

In formulas (25) and (26) for numerical quadrature, with the x_i as free parameters, we have $2n + 2$ adjustable constants in all. This suggests the hope that by properly locating the x_i, we can get a formula which is exact for all polynomials of degree $2n + 1$ or less, since these form a $(2n + 1)$-parameter family.

Such a formula was obtained by Gauss; in deriving it, it is convenient to renormalize to the interval $(-1, 1)$ and to label the points $\xi_1, \ldots, \xi_m$, so that $m = n + 1$. The formula uses two properties of the Legendre polynomials $P_m(x)$ defined in Ch. 4, §2, which will be proved in Ch. 11, §6. These properties are:

(i) $P_m(x)$ has m distinct zeros $x = \xi_1 < \xi_2 < \cdots < \xi_m$ in the interval $(-1, 1)$, whence $p_m(x) = c_m(x - \xi_1)(x - \xi_2) \cdots (x - \xi_m)$ for some constant c_m.

(ii) $P_m(x)$ is orthogonal to any polynomial of lower degree:

$$\int_{-1}^1 x^n P_m(x) \, dx = 0, \qquad \text{if} \qquad n < m$$

We shall assume these results, and also the following definition.

DEFINITION.   The Gaussian quadrature formula of order m is the special case of formulas (25′) and (26), in which $r_{i-1} = (1 + \xi_i)/2$, $i = 1, \ldots, m$, and ξ_i is the ith zero of the Legendre polynomial $P_m(x)$.

THEOREM 5.   *Gaussian quadrature of order m is exact if $f(x)$ is a polynomial of degree $2m - 1$ or less.*

Proof.   By centering the origin and change of scale, we can assume the interval of integration to be $[-1, 1]$ without loss of generality. Let $f(x)$ be any polynomial of degree $2m - 1$ or less; let $p(x)$ be the Lagrange interpolation polynomial of degree at most $m - 1$ satisfying $p(\xi_i) = f(\xi_i)$, $i = 1, 2, \ldots, m$. Then $e(x) = f(x) - p(x)$ vanishes at $\xi_i, \ldots, \xi_m$. Hence,† we have $e(x) = (x - \xi_1) \cdots (x - \xi_m)b(x)$, where $b(x)$ is a polynomial of degree at most $m - 1$.

---

† This follows by the Remainder Theorem; see Birkhoff-MacLane, p. 75.

Therefore, by (i) above, $e(x) = c_m^{-1}P_m(x)b(x) = s(x)P_m(x)$, where $s(x)$ is a polynomial of degree at most $m - 1$. But, by (ii), $P_m(x)$ is orthogonal to any polynomial of degree less than m. Hence

$$\int_{-1}^{1} e(x) \, dx = \int_{-1}^{1} s(x)P_m(x) \, dx = 0$$

so that (25′) and (26) are exact for $e(\xi)$ by the choice of the ξ_i. Moreover, Gaussian quadrature is exact for $p(x)$ by the choice of the ξ_i; hence, it is also exact for the given $f(x)$, completing the proof.

Let $f(x) \in \mathcal{C}^{2m}[a, a + h]$, and let $q(x)$ be the polynomial of degree $2m - 1$ (or less) satisfying $q(x_j) = f(x_j)$, for $x_j = a + [jh/(2m + 1)], j = 1, \ldots, 2m$. Then, by Theorem 1, we have $e(x) = q(x) - f(x) = O(h^{2m})$ and so

$$\int_{a}^{a+h} q(x) \, dx - \int_{a}^{a+h} f(x) \, dx = \int_{a}^{a+h} e(x) \, dx = O(h^{2m+1})$$

But, by Theorem 5, we have

$$\int_{a}^{a+h} q(x) \, dx = \sum_{j=1}^{m} w_j q(a + r_j h) = \sum_{j=1}^{m} w_j f(a + r_j h)$$

Substituting back into the preceding equation, we get

$$(33) \qquad \int_{a}^{a+h} f(x) \, dx - \sum_{j=1}^{m} w_j f(a + r_j h) = O(h^{2m+1}) \qquad \text{if} \qquad r_j = \frac{(1 - \xi_j)}{2}$$

This proves the following result.

COROLLARY. *For* $f(x) \in \mathcal{C}^{2m}$, *Gaussian quadrature of order* m *has an absolute error* $O(h^{2m+1})$ *and a relative error* $O(h^{2m})$.

Romberg Quadrature.† For more than a century, Gaussian quadrature was the ultimate in ingenious quadrature methods. Then, around 1960, an *extrapolation* method based on the idea of successive mesh-halving (as in Richardson extrapolation) to *trapezoidal* quadrature turned out to be even more accurate in many cases.

Let $T_0^{(k)}$ be the trapezoidal sum

$$(33') \qquad T_0^{(k)} = \frac{h}{2} \left\{ F(a) + F(b) + 2 \sum_{j=1}^{\ell - 1} F(a+jh) \right\}, \quad h = (b-a)/\ell, \ \ell = 2^k.$$

and let $T_m^{(k)} = [4^m T_{m-1}^{(k+1)} - T_{m-1}^{(k)}]/(4^m - 1)$. Then the $T_k^{(k)}$ converge extremely rapidly to $\int_a^b F(x) \, dx$.

---

† See F. L. Bauer, H. Rutishauser, and E. Stiefel, *Proc. XV Symposium on Applied Math.*, Am. Math. Soc. 1963, 199–218.

EXERCISES D

1. Using a five-place table of sin x, x in radians [but not tables of Si(x)], evaluate

$$\int_0^x t^{-1} \sin t \, dt \qquad \text{for} \qquad x = 0.1, 0.2, \ldots, 1.0$$

by Simpson's rule, with $h = 0.1$.

2. Show that $\int_1^2 dx/x = \ln 2$ is given by various numerical quadrature formulas, with $h = 1/10$, as follows: (a) initial point 0.73654401, (b) trapezoidal 0.69377139, (c) midpoint 0.6928354, (d) Simpson 0.6931474.

3. Use Weddle's rule (Ex. B3, Ch. 7) with 12 subdivisions to compute the approximation 0.69314935 to $\ln 2 = 0.69314718056$.

4. Use Cotes' rule (Ex. B2, Ch. 7) with 12 subdivisions, to compute the approximation $\ln 2 \simeq 0.69319535$.

*5. Show that

$$\int_{x_0}^{x_1} f(x) \, dx = \frac{h}{2} \left[f_0 + f_1 - \frac{1}{12} (\delta^2 f_0 + \delta^2 f_1) + \frac{11}{720} (\delta^4 f_0 + \delta^4 f_1) \right] + O(h^7)$$

*6. Show that, for $n = 3$, the Gauss quadrature formula on $(-h, h)$ is

$$\frac{h}{9} \{ 8f(0) + 5 \, [f(-h \sqrt{\tfrac{3}{5}}) + f(h \sqrt{\tfrac{3}{5}})] \}$$

with truncation error $h^6 f^{vi}(\xi)/15{,}750$, where $-h < \xi < h$.

*7. Show that the error in Hermite's tangent cubic quadrature formula,

$$\int_0^h y \, dx \simeq \frac{h}{2} (y_0 + y_1) - \frac{h^2}{12} (y_0' - y_1')$$

is $h^6 f^{iv}(\xi)/720$, where $0 < \xi < h$, if $y = f(x) \in \mathcal{C}^4$.

8. (Simpson's Five-Eight Rule). Show that, if $f \in \mathcal{C}^3$, then

$$\int_0^h f(x) \, dx = \frac{h}{12} [5f(h) + 8f(0) - f(-h)] + O(h^4)$$

*9. Show that, if $f(\theta)$ is an analytic periodic function, then trapezoidal quadrature over a complete period has $O(h^n)$ accuracy for all n.

8 FOURTH-ORDER RUNGE–KUTTA

In principle, it is easy to derive formulas of numerical integration for $y' = F(x, y)$ having an arbitrarily high order of accuracy, if F is sufficiently smooth.

Simply evaluate the successive derivatives of y as in Ch. 4, §8, getting $y' = F$, $y'' = F' = F_x + FF_y$

$$y''' = F_{xx} + 2FF_{xy} + F^2 F_{yy} + y'' F_y$$
$$y^{iv} = F_{xxx} + 3FF_{xxy} + 3F^2 F_{xyy} + F^3 F_{yyy} + y''(3F_{xy} + 3FF_{yy}) + y''' F_y$$

and so on. If F is linear or a polynomial of low degree, the preceding formulas may even be practical for computation.

Then evaluate Taylor's formula, valid for $F \in \mathcal{C}^n$,

$$(34) \qquad y(x_k + h) = y_k + hy'_k + \frac{h^2 y''_k}{2} + \cdots + \frac{h^n y_k^{(n)}}{n!} + O(h^{n+1})$$

ignoring the remainder $O(h^{n+1})$. The relative error in the recursion relation so obtained

$$(35) \qquad y_{k+1} = y_k + hy'_k + \frac{h^2 y''_k}{2} + \cdots + \frac{h^n y_k^{(n)}}{(n!)}$$

is obviously of order n. Hence, by Theorem 9 of Ch. 7, so is the cumulative error. Moreover, this *one-step explicit* method has the advantages of permitting a variable mesh length and of being stable for strictly stable DEs.

The calculation of (35), however, is rather cumbersome since it involves many terms. Furthermore, in some cases the derivatives of F can be computed only approximately by numerical differentiation, which amplifies roundoff errors (§5). A one-step method of integration that gives a high order of accuracy and avoids these defects will be described next.

This approach is based on the idea of obtaining as high an order of accuracy as possible, using an *explicit, one-step* method. It consists in extending the approximations of the improved Euler method (Ch. 7, §8) further, so as to obtain a one-step formula having a higher order of accuracy. One-step methods have the advantage of permitting a change of mesh length at any step, because no starting process is required.

The most commonly used one-step method with high order of accuracy is the Runge–Kutta† method. We now describe the ΔE used in this method, for the first-order system

$$(36) \qquad \frac{d\mathbf{x}}{dt} = \mathbf{X}(\mathbf{x}, t), \qquad a \le t \le b$$

with mesh points $a = t_0 < t_1 < t_2 < \cdots$. Let $\mathbf{y}_0 = \mathbf{x}(a)$ be the initial value.

---

† *Zeits, Math. Phys.* 46 (1901), 435–453; C. Runge and H. König, *Numerische Rechnung*, 1924, Ch. 10.

The approximate function table of values $\mathbf{y}_i$ corresponding to the points t_i is defined by the ΔE

$$\mathbf{y}_{i+1} = \mathbf{y}_i + \frac{h}{6} (\mathbf{k}_1 + 2\mathbf{k}_2 + 2\mathbf{k}_3 + \mathbf{k}_4)$$

(37) $$\mathbf{k}_1 = \mathbf{X}(\mathbf{y}_i, t), \qquad \mathbf{k}_2 = \mathbf{X}\left(\mathbf{y}_i + \frac{h\mathbf{k}_1}{2}, t_i + \frac{h}{2}\right)$$

$$\mathbf{k}_3 = \mathbf{X}\left(\mathbf{y}_i + \frac{h\mathbf{k}_2}{2}, t_i + \frac{h}{2}\right), \qquad \mathbf{k}_4 = \mathbf{X}(\mathbf{y}_i + h\mathbf{k}_3, t_i + h)$$

where the mesh length $h = h_i$ may vary with i.

We now show that the preceding Runge–Kutta method has an error of only $O(h^5)$ per step.‡ For simplicity, we restrict attention to the first-order DE

(38) $$\frac{dx}{dt} = X(x, t), \qquad X \in \mathcal{C}^4, \qquad a \leq t \leq b$$

and to the initial condition $x(0) = 0$. Formula (37) reduces in this case to

(39) $$y_{i+1} = y_i + \left(\frac{h}{6}\right)(k_1 + 2k_2 + 2k_3 + k_4)$$

where

(39′) $$k_1 = X(y_i, t_i), \qquad\qquad k_2 = X\left(y_i + \frac{hk_1}{2}, t_i + \frac{h}{2}\right)$$

$$k_3 = X\left(y_i + \frac{hk_2}{2}, t_i + \frac{h}{2}\right), \qquad k_4 = X(y_i + hk_3, t_i + h)$$

Let $x(t)$ be the exact solution of the DE satisfying $x(0) = y_0 = 0$. Set $h/2 = \theta$. Then k_2 can be written as

$$k_2 = X(x_{1/2}, \theta) + X_x(x_{1/2}, \theta)[x_0 - x_{1/2} + \theta k_1]$$

$$+ \frac{1}{2} X_{xx}(x_{1/2}, \theta)[x_0 - x_{1/2} + \theta k_1]^2 + \cdots,$$

where $x_0 = x(0)$, $x_{1/2} = x(\theta)$, and subscripts x stand for partial derivatives. Using primes to indicate total derivatives with respect to t, so that $X' = \partial X/\partial t +$

---

‡ The proof that follows was constructed by Robert E. Lynch.

$X\,\partial X/\partial x$, we get

$$x_{1/2} = x_0 + X(0,0)\theta + \frac{X'(0,0)\theta^2}{2} + \frac{X''(0,0)\theta^3}{6} + \frac{X'''(0,0)\theta^4}{24} + O(h^5)$$

Since $k_1 = X(0,0)$, it follows that

$$x_0 - x_{1/2} + \frac{\theta k_1}{2} = -\frac{X'(0,0)\theta^2}{2} - \frac{X''(0,0)\theta^3}{6} + O(h^4)$$

so that

$$k_2 = X(x_{1/2}, \theta) - X_x(x_{1/2}, \theta)\left[\frac{X'(0,0)\theta^2}{2} + \frac{X''(0,0)\theta^3}{6}\right] + O(h^4).$$

For k_3, we have, similarly,

$$k_3 = X(x_{1/2}, \theta) + X_x(x_{1/2}, \theta)[x_0 - x_{1/2} + \theta k_2] + \cdots$$

From this formula, since

$$x_0 = x_{1/2} - X(x_{1/2}, \theta)\theta + \tfrac{1}{2}X'(x_{1/2}, h/2)\theta^2 - \tfrac{1}{6}X''(x_{1/2}, \theta)\theta^3$$
$$+ \tfrac{1}{24}X'''(x_{1/2}, \theta)\theta^4 + O(h^5)$$

we obtain

$$k_3 = X(x_{1/2}, \theta) + \tfrac{1}{2}X_x(x_{1/2}, \theta)X'(x_{1/2}, \theta)\theta^2 - \tfrac{1}{2}X_x(x_{1/2}, \theta)X_x(x_{1/2}, \theta)X'(0,0)\theta^3$$
$$- \tfrac{1}{6}X_x(x_{1/2}, \theta)X''(x_{1/2}, \theta)\theta^3 + O(h^4)$$

Similarly, we have

$$k_4 = X(x_1, h) + X_x(x_1, h)[x_0 - x_1 + hk_3] + \cdots$$

and

$$x_0 - x_1 = -hX(x_{1/2}, \theta) - \tfrac{1}{3}X''(x_{1/2}, \theta)\theta^3 + O(h^4)$$

so that

$$k_4 = X(x_1, h) - \tfrac{1}{3}X_x(x_1, h)X''(x_{1/2}, \theta)\theta^3 + X_x(x_1, h)X_x(x_{1/2}, \theta)X'(x_{1/2}, \theta)\theta^3 + O(h^4)$$

Finally, we have the relations

$$X(x_{1/2}, \theta) = X(0, 0) + X'(0, 0)\theta + \tfrac{1}{2}X''(0, 0)\theta^2 + \tfrac{1}{6}X'''(0, 0)\theta^3 + O(h^4)$$

$$X(x_1, h) = X(0, 0) + X'(0, 0)h + \tfrac{1}{2}X''(0, 0)h^2 + \tfrac{1}{6}X'''(0, 0)h^3 + O(h^4)$$

$$X_x(x_1, h) = X_x(x_{1/2}, \theta) + X_x'(x_{1/2}, \theta)\theta + O(h^2)$$

Combining these results, we find that

$$y_1 = y_0 + \left(\frac{h}{6}\right)[k_1 + 2k_2 + 2k_3 + k_4]$$

$$= y_0 + X(0, 0)h + X'(0, 0)\frac{h^2}{2} + X''(0, 0)\frac{h^3}{6} + X'''(0, 0)\frac{h^4}{24} + O(h^5)$$

and $y_0 = x_0$. Since $x(h)$ is given by

$$x(h) = x_1 = x_0 + X(0, 0)h + X'(0, 0)h^2/2$$
$$+ X''(0, 0)h^3/6 + X'''(0, 0)h^4/24 + O(h^5)$$

we see that

$$|y_1 - x_1| = O(h^5)$$

Hence, the relative error is of order four. Therefore, by Theorem 9 of Ch. 7, so is the cumulative error. The method of proof consists in comparing various Taylor series.

The main defect of the Runge–Kutta method is the need for evaluating $k_j = X(x_j, t_j)$ for four values of (x_j, t_j) per time step. If X is a complicated function, this may be quite time-consuming. To avoid this repetitious evaluation, many computer programs use Adams-type methods instead; see Exercise F8.

EXERCISES E

1. (a) Derive a power series expansion for $f(a + h)$ through terms in h^3 for the solution of $y' = 1 + y^2$ satisfying $f(a) = c$.
   (b) Truncating the preceding series after terms in h^3, evaluate approximately in three steps the solution of $y' = 1 + y^2$ satisfying $f(0) = 0$, setting $x_1 = 0.5$, $x_2 = 0.8$, $x_3 = 1$. What is the truncation error? [HINT: Consider tan x.]

2. Same question for $y' = x^2 + y^2$. (The exact solution, rounded off to five decimal places, is 0.35023.)

3. (a) Apply the Picard process to the DE $y' = x^2 + y^2$ for the initial value $y_0 = 0$ and initial trial function $y^{(0)} \equiv 0$. Calculate the first four iterates.
   (b) Using the power series method of the text, calculate the Taylor series of the solution through terms in x^{17}, and check against the answer to (a).
   (c) Evaluate $y(1)$ numerically at $x = 1$, using the preceding truncated power series, and compare with the answer of Ex. 2.

4. (a) Apply the Runge–Kutta method to the DE $y' = 1 + y^2$ for the initial value $y(0) = 0$, setting $x_0 = 0$, $x_1 = 0.5$, $x_2 = 0.8$, $x_3 = 1$.

   (b) Same question for the DE $y' = x^2 + y^2$, with $y(0) = 0$, and the same mesh.

5. For the first-order linear system $dx/dt = A(t)\mathbf{x}$, show that the Runge–Kutta method is equivalent to

$$\mathbf{x}_{i+1} = \left[I + \frac{h}{6} (A_0 + 4A_1 + A_2) + \frac{h^2}{6} (A_1 A_0 + A_1^2 + A_2 A_1) \right.$$
$$\left. + \frac{h^3}{12} (A_1^2 A_0 + A_2 A_1^2) + \frac{h^4}{24} (A_2 A_1^2 A_0) \right] \mathbf{x}_i.$$

where $A_0 = A(t_i)$, $A_1 = A(t_{i+1/2})$, $A_2 = A(t_{i+1})$.

6. (a) Show that the system $u' = v$, $v' = -u$ is neutrally stable, and indeed that $|(u(t), v(t))| = $ const. for any solution.

   (b) Show that the Runge–Kutta method is strictly stable, and satisfies

$$|(u(t + h), v(t + h))| = \left(1 - \frac{h^6}{72} + \frac{h^8}{576} \right) |(u(t), v(t))|$$

In Exs. 7–11, let $y' = F(x, y) = \sum_{j,k=0}^{\infty} b_{jk} x^j y^k$ and $y(0) = 0$.

7. Show that

$$y'' = F_x + FF_y \quad \text{and} \quad y''' = F_{xx} + 2FF_{xy} + F^2 F_{yy} + F_x F_y + FF_y^2$$

In Exs. 8–11, let $B = b_{20} + b_{11} b_{00} + b_{02} b_{00}^2$ and $B^* = b_{10} b_{01} + b_{00} b_{01}^2$.

8. Show that, if $y(0) = 0$, then we have

$$y(h) = h b_{00} + h^2 (b_{10} + b_{00} b_{01})/2 + h^3 (B/3 + B^*/6) + O(h^4)$$

9. Show that, with midpoint integration, $y(h)$ is given by the approximate formula

$$y_M(h) = h b_{00} = \frac{h^2}{2} (b_{10} + b_{00} b_{01}) + \frac{h^3 B}{4} + O(h^4)$$

and that the truncation error is $-h^3(B/12 + B^*/6) + O(h^4)$.

10. Show that the improved Euler method gives

$$y_E(h) = h b_{00} + \frac{h^2}{2} (b_{10} + b_{00} b_{01}) + \frac{h^3 B}{2} + O(h^4)$$

with truncation error $h^3(B - B^*)/6 + O(h^4)$.

11. Show that the trapezoidal approximation to $y(h)$ is

$$y_T(h) = h b_{00} + \frac{h^2}{2} (b_{10} + b_{00} b_{01}) + h^3 \left(\frac{B}{2} + \frac{B^*}{4} \right) + O(h^4)$$

with truncation error $h^3(B/6 + B^*/12) + O(h^4)$.

12. Check the formulas of Ex. E8 against those of Exs. 9–10 above in the special case $t_{i-1} = 0$, $t_i = h$, and $A(t) = p(t) = p_0 + p_1 t + p_2 t^2 + \cdots$ of a first-order linear DE.

*13. For the linear DE $dx/dt = p(t)x$, $p(t) = \sum_{k=0}^{\infty} p_k(t-a)^k$, evaluate $x(a+h)$ through terms in h^6 by the Runge–Kutta method. Compare this with the Taylor series for the exact solution.

*9 MILNE'S METHOD

A very different method for solving initial value problems with fourth-order accuracy is due to W. E. Milne. Whereas Runge–Kutta methods are based directly on power series expansions, and the Euler methods of Ch. 7 (and Ch. 1, §8) basically approximate derivatives by difference quotients, Milne's method replaces $\mathbf{x}'(t) = \mathbf{X}(\mathbf{x}; t)$ by the equivalent (vector) integral equation:

$$(40) \qquad \mathbf{x}(t) = \mathbf{x}(a) + \int_a^t \mathbf{X}(\mathbf{x}, s)\, ds$$

as in Ch. 6, (11). If the integral in (40) is evaluated by Simpson's rule, we get a very simple implicit, two-step ΔE

$$(41) \qquad \mathbf{x}_{k+2} = \mathbf{x}_k + \frac{h}{3}\{\mathbf{X}(\mathbf{x}_k, t_k) + 4\mathbf{X}(\mathbf{x}_{k+1}, t_{k+1}) + \mathbf{X}(\mathbf{x}_{k+2}, t_{k+2})\}$$

due to W. E. Milne.† It is perhaps the simplest scheme for achieving $O(h^4)$ accuracy. Moreover, in the case of linear systems $\mathbf{x}'(t) = A(t)\mathbf{x} + \mathbf{b}(t)$, one can solve for $\mathbf{x}_{k+2}$ in (41) algebraically.

Example 3. Consider the linear DE $y' = 1 - 2xy$ of Ch. 7, §7, with initial value $y(0) = 0$ and mesh-length $h = 0.1$. In this case (41) reduces to

$$y_{k+2} = [15 + x_{k+2}]^{-1}[3 + (15 - x_k)y_k - 4x_{k+1}y_{k+1}]$$

Evaluating $y(0.1) = 0.09934$ by power series, Milne's formula gives $y(0.2) = 0.19475$. This approximate value agrees to five places with the value of $y(\frac{1}{5})$ obtained by power series expansion. The comparison suggests that the mesh length $h = 0.1$ is adequate for four-place accuracy. Repeated use of (41) then gives the following *approximate function table*.

x	0.1	0.2	0.3	0.4	0.5
y	0.09934	0.19475	0.28264	0.36000	0.42444

---

† W. H. Milne, *Numerical Solution of Differential Equations,* John Wiley & Sons, 1953.

The truncation error is about 10^{-5}, as can be verified by use of the power series solution

$$y = x - \frac{2}{3}x^3 + \frac{4}{15}x^5 + \cdots = \sum_{k=0}^{\infty} a_{2k+1}x^{2k+1}, \qquad \frac{a_{2k+1}}{a_{2k-1}} = \frac{-2}{2k+1}$$

which gives $y(1) = 0.538079$.

Having been successfully used in a wide variety of DEs, Milne's method provides an excellent illustration of *implicit, two-step* methods.

Starting Process.   Given the initial value $y_0 = c = f(a)$, one must compute y_1 by a *one-step* method before one can begin to apply a *two-step* method. For analytic F, it is usually best to calculate $y_1 = f(a + h)$ by expanding $f(x)$ in a Taylor series as in Ch. 3, §7. When F is not analytic, but fairly smooth (say, if $F \in \mathcal{C}^4$), good approximations to $f(a + h)$ are often obtained by repeated *mesh-halving* of the interval $[a, a + h]$, using a one-step method with a lower order of accuracy. For instance, we might first compute $f(a + h/8)$ by midpoint integration, and then use Milne's formula to get $f(a + h/4) = y_{1/4}$ from y_0 and $y_{1/8}$. Next we compute $y_{1/2}$ from y_0 and $y_{1/4}$ by a second application of Milne's formula with mesh length $h/4$, finally getting y_1 from y_0 and $y_{1/2}$ by a third application of the same process.

Iterative Solution.   Although (41) can be solved algebraically for linear DEs and systems, for *nonlinear* DEs one must resort to iterative methods to compute $\mathbf{x}_{k+2}$ from $\mathbf{x}_k$ and $\mathbf{x}_{k+1}$. One can do this by a method analogous to Picard's method of successive approximation (Ch. 6, §7), as follows. First rewrite Milne's equation (40) in the following form

$$(42) \qquad \mathbf{x}_{k+2} = U(\mathbf{x}_{k+2}) \equiv \mathbf{x}_k + \frac{h}{3}\{\mathbf{X}(\mathbf{x}_k, t_k) + 4\mathbf{X}(\mathbf{x}_{k+1}, t_{k+1}) + \mathbf{X}(\mathbf{x}_{k+2}, t_{k+2})\}$$

where all quantities are known except $\mathbf{x}_{k+2}$.

Regarded as an equation in the unknown vector $\mathbf{x}_{k+2}$, (42) has the form $\mathbf{x}_{k+2} = U(\mathbf{x}_{k+2})$, where the function U is computable. For any initial *trial value* $\mathbf{x}_{k+2}^{(0)}$, one can hope that the sequence

$$(42') \qquad \mathbf{x}_{k+2}^{(1)} = U(\mathbf{x}_{k+2}^{(0)}), \qquad \mathbf{x}_{k+2}^{(2)} = U(\mathbf{x}_{k+2}^{(1)}), \qquad \mathbf{x}_{k+2}^{(3)} = U(\mathbf{x}_{k+2}^{(2)}), \ldots$$

will converge fairly rapidly to the true solution.

In the case of a single DE $x'(t) = X(x, t)$, we now show that this will be the case if X satisfies a Lipschitz condition, where L is the Lipschitz constant and h is small. More precisely, iteration *converges* if $h < 3/L$, and it converges *rapidly* if $h \ll 3/L$. This results from (42'), in which

$$|\mathbf{x}_{k+2}^{(r+1)} - \mathbf{x}_{k+2}^{(r)}| = \frac{h}{2}|\mathbf{X}(\mathbf{x}_{k+2}^{(r)}, t_{k+2}) - \mathbf{X}(\mathbf{x}_{k+2}^{(r-1)}, t_{k+2})| \leq \frac{hL}{3}|\mathbf{x}_{k+2}^{(r)} - \mathbf{x}_{k+2}^{(r-1)}|$$

Hence, if $\theta = hL/3$, we have by induction on r

$$|\mathbf{x}_{k+2}^{(r+1)} - \mathbf{x}_{k+2}^{(r)}| \leq \theta^r |\mathbf{x}_{k+2}^{(1)} - \mathbf{x}_{k+2}^{(0)}|$$

For $h < 3/L$, $\theta < 1$ and so the sequence of $\mathbf{x}_{k+2}^{(r)}$ is a Cauchy sequence; let $\mathbf{x}_{k+2}$ be its limit. Moreover U is a *contraction* which shrinks all distances by a factor θ or less, and so is continuous. Hence, passing to the limit on both sides of the equation $\mathbf{x}_{k+2}^{(r+1)} = U(\mathbf{x}_{k+2}^{(r)})$, we get (42).

*10   MULTISTEP METHODS

Milne's method (41) is evidently a *two-step* method in the sense that each new value of an approximate solution is computed using the two preceding values. In this section, we will study Milne's method more critically, and describe other multistep "Adams-type" methods.

Multistep methods are usually best executed as predictor-corrector methods, in the sense of Ch. 7, §8. An explicit "predictor" formula based on extrapolation is made to yield higher order accuracy by one or two iterations of an implicit "corrector" formula.

Milne's Predictor.   Thus with Milne's method, we can use the predictor

$$(43) \qquad \mathbf{y}_{k+2} = \mathbf{x}_k + 2h\mathbf{X}(\mathbf{x}_{k+1}, t_{k+1})$$

which has $O(h^3)$ absolute, and $O(h^2)$ relative accuracy. To get $O(h^4)$ relative accuracy from this, we must iterate *twice* with the corrector (41). Alternatively, for $k \geq 2$, one can use the four-step (five level) predictor

$$(43') \qquad \mathbf{z}_{k+2} = \mathbf{z}_{k-2} + \left(\frac{4h}{3}\right)\{2\mathbf{X}_{k-1} - \mathbf{X}_k + 2\mathbf{X}_{k+1}\}$$

which has $O(h^4)$ accuracy, and apply the corrector (41) once.

Stability.   Unfortunately, like the two-step approximation $y_{k+1} = y_{k-1} + 2hy_k'$ discussed in §3, Milne's method can give an *unstable* difference approximation to a stable DE—and in fact it does this in the case of $y' + y = 0$. This can be verified by solving the relevant characteristic equation, which is

$$(44) \qquad \rho^2 + \frac{4h\rho}{3 + h} + \frac{3 - h}{3 + h} = 0$$

Setting $\rho = 1 - h + h^2/2! - h^3/3! + h^4/4!$, it is easily verified that Eq. (44) holds through terms in h^4, confirming that one characteristic root $\rho_1 = e^{-h} + O(h^5)$. This corresponds to *relative* $O(h^4)$ accuracy. Unfortunately, the other root $\rho_2 = -1 - h/3 + O(h^2)$; hence, the magnitude of the error will grow exponentially like $e^{t/3}$ with alternating sign. Therefore, computed values will ultimately oscillate with increasing amplitude, whereas the exact values tend smoothly to zero.

Adams-type Methods.   The preceding instability is avoided by more sophisticated multistep methods. One of the most successful of these uses the *predictor* formula of Adams and Bashforth (1883)

$$(45) \qquad \tilde{\mathbf{y}}_{k+1} = \mathbf{y}_k + h \sum_{m=0}^{3} \beta_m \nabla^m \mathbf{X}_k$$

where $\beta_0 = 1$, $\beta_1 = \frac{1}{2}$, $\beta_2 = \frac{5}{12}$, $\beta_3 = \frac{3}{8}$. This is followed by the explicit *corrector* formula of Moulton (1926).

$$(46) \qquad \mathbf{y}_{k+1} = \mathbf{y}_k + h \sum_{m=0}^{3} \gamma_m \nabla^m \tilde{\mathbf{X}}_k$$

where $\gamma_0 = 1$, $\gamma_1 = -\frac{1}{2}$, $\gamma_2 = -\frac{1}{12}$, $\gamma_3 = -\frac{1}{24}$, and

$$\tilde{\mathbf{X}}_{k+1} = \mathbf{X}(\tilde{\mathbf{y}}_{k+1}, t_{k+1})$$

By combining the results of §4 and §10, it is possible to derive an *a priori* error bound for Milne's method. Specifically, we can prove that the cumulative truncation error, over any fixed interval $[a, a + T]$, is bounded by Mh^4 for some finite constant M, that is, independent of the mesh length h. This is true, provided that both y_0 and y_1 are accurate to $O(h^5)$.

Indeed, let $y(x)$ be any exact solution of the DE $y' = F(x, y)$, and let $\phi(x) = F(x, y(x))$. Then by Theorem 4, we have

$$y_{k+2} = y_k + \frac{h}{3} \left[\phi(x_k) + 4\phi(x_{k+1}) + \phi(x_{k+2}) \right] + e_{k+2}$$

where $|e_{k+2}| \leqq |\phi^{iv}|_{\max} h^5/90$. From this, a discussion like that of Ch. 7, §10, yields the bound

$$(47) \qquad M = |\phi^{iv}|_{\max}(e^{LT} - 1)/90L \left(1 - \frac{hL}{3} \right)$$

Finally, we can express $\phi^{iv}(x)$ in terms of F and its derivatives, just as in Ch. 4, §8:

$$\phi' = F_x + FF_y, \qquad \phi'' = F_{xx} + 2FF_{xy} + F^2 F_{yy} + F_x F_y + FF_y^2$$

and so on. Combining these results, we can compute M *a priori* in terms of the values of F and its derivatives, thus getting an explicit error bound.

A priori error bounds, like the preceding, are seldom useful for methods having $O(h^4)$ accuracy. One reason is that they are so complicated. In practice, reliance is usually placed on *a posteriori* error estimates, which utilize computed values. Another reason is that they neglect *roundoff* errors.

EXERCISES F

1. (a) Show that, for the DE $y' = y$ and $h = 0.1$, Milne's method amounts to using the $\Delta E\ y_{k+2} = (31y_k + 4y_{k+1})/29$.
   (b) Integrate the DE $y' = y$ from $x = 0$ to $x = 1$ by Milne's method with $h = 0.1$, for the starting values $y_0 = 1$ and $y_1 = 1.1052$.

2. Same question for the DE $y' = 1 + y^2$, using the starting values $y_0 = 0$ and $y_1 = 0.1003$.

3. Same question for the DE $y' = x^2 + y^2$, with the starting values $y_0 = 0$ and $y_1 = 0.00033$.

Exercises 4–6 concern the *two-level midpoint method,* defined by (16).

4. (a) Show that, if $y' = F(x, y)$, where $F \in \mathcal{C}^2$, then $y = f(x)$ satisfies the two-level midpoint formula (16) with discrepancy $O(h^3)$.
   (b) Show that, if $F \in \mathcal{C}^4$ and y_{n-1}, y_n are exact, the truncation error is $Kh^3 + O(h^4)$.

5. (a) Integrate $y' = y$ approximately by the two-level midpoint method with $h = 0.1$, taking $y_0 = 1, y_1 = 1.1052$ as starting values and integrating to $y_{10} \simeq 2.7145$.
   (b) Estimate the discrepancy and the cumulative truncation error in (a), comparing them with the roundoff error.

6. (a) Do the same as in Ex. 5a but for the system $y' = z - 2y$, $z' = y - 2z$ and the starting values $y_0 = 1, z_0 = 0, y_1 = 0.8228, z_1 = 0.0820$, computed by the Taylor series method.
   (b) Show that the system in question is stable but that the approximating ΔE is not. Explain why the computed table is approximately correct, although the method is unstable.

7. (a) For general $h > 0$, set

$$y_0 = 1, \qquad y_1 = 1 + h + h^2\left(\frac{1}{2} + \frac{h}{6} + \frac{h^2}{24}\right) \qquad \text{and} \qquad y_{n+1} = y_{n-1} + 2hy_n$$

   Show that $|y_n - e^{nh}| = O(h^2)$, as $h \to 0$ with nh constant.
   (b) If $h = 10^{-p}$ and the roundoff error is 10^{-p}, infer that the cumulative *total* error is $O(10^{p-h})$.

8. *Adams three-level methods* for integrating $y' = F(x, y)$ are

$$(A_3) \qquad y_{n+1} = y_n + \frac{h}{12}[23F_n - 16F_{n-1} + 5F_{n-2}] \qquad \text{(explicit)}$$

$$(A_3') \qquad y_{n+1} = y_n + \frac{h}{24}[9F_{n+1} + 19F_n - 5F_{n-1} + F_{n-2}] \qquad \text{(implicit)}$$

   Show that the truncation error of (A_3) is $O(h^4)$ per step, while that of the implicit method (A_3') is $O(h^5)$.

CHAPTER 9

REGULAR SINGULAR POINTS

1 INTRODUCTION

We briefly discussed the complex exponential function of a complex variable, $w = e^{\lambda z}$, in Chapter 3, §2. There we used its properties to explain the behavior of $\ln z = \int dt/t$, the natural logarithm of z, and the power function $z^\lambda = e^{\lambda \ln z}$ for an arbitrary real or complex exponent $\lambda = \mu + i\nu$, as $z = x + iy$ ranges over the complex plane. In the rest of that chapter, however, we assumed the independent variable to be *real*.

In the present chapter, we shall consistently be considering complex-valued functions $w = f(z)$ of a complex independent variable z, and the behavior of such functions as z varies in the complex domain. Specifically, we shall usually be studying functions that satisfy some *second-order, linear* homogeneous ordinary DE of the form

$$p_0(z)w'' + p_1(z)w' + p_2(z)w = 0$$

We shall be particularly interested in the way in which their behavior depends on the coefficient-functions $p_j(z)$ $(j = 0, 1, 2)$, much as (in Ch. 2) we considered the analogous questions for *real* t and $p_j(t)$, extending the results to higher-order DEs in Chapter 3.

Throughout, we shall be exclusively concerned with *analytic* functions. Here an *analytic,* or *holomorphic,* function $w = f(z)$ of a *complex* variable $z = x + iy$ is one having a complex derivative $f'(z) = dw/dz$ at every point.† This is equivalent to the definition given in Ch. 4, §5, as is proved in books on complex analysis: any complex analytic function can be expanded in a convergent power series. An analytic DE is one in which the functions involved are all analytic. Its solutions are then necessarily also analytic (Ch. 6, §11).

The solutions of analytic DEs are best studied as functions of a complex variable, because their isolated singular points are surrounded by connected domains in the complex z-plane. This permits one to continue solutions beyond and around isolated singular points, whereas on the real line, solutions terminate abruptly at singular points.

---

† Ahlfors, p. 24; Hille, p. 72. Some knowledge of complex function theory is assumed in this chapter.

For instance, consider the DE $du/dx = u^2$ for real x and u. The formula $u = -1/x$ defines *two* real solutions of this DE; one defined for $x > 0$; the other for $x < 0$. As $x \to 0$, one solution tends to $+\infty$, and the other to $-\infty$; the behavior of these two solutions near $x = 0$ seems to be unrelated if x is restricted to real values. On the other hand, consider the same DE $dw/dz = w^2$ for complex $z = x + iy$ and $w = u + iv$. The formula $w = -1/z$ defines a *single* solution of the DE, in a domain D that includes every point of the complex z-plane except the isolated singular point at $z = 0$; hence, it includes both real solutions. The general solution of the same DE is the complex-valued analytic function $w = 1/(c - z)$. This function has an isolated singularity at $z = c$.

The DE $dw/dz = w^2$ is defined and analytic for all z and w, real or complex. Each particular solution $w = 1/(c - z)$ of this DE is defined in the punctured z-plane with the point $z = c$ deleted. Since this domain is connected, the solution can be continued as an analytic function from any region in it to any other. This process of *analytic continuation* is uniquely defined, for any given path of continuation.†

The (real) solution $w = u = -1/x$ of $dw/dz = w^2$ on the negative x-axis is obtained by analytic continuation in the complex plane from the solution $u = -1/x$ of $du/dx = u^2$ for $x > 0$. This is evident if we continue the solution as a complex analytic function $w = (-x + iy)/(x^2 + y^2)$ around the origin on either side. The fact that the analytic continuation of a solution of a DE is a solution of the analytic continuation of the DE is valid in general, as we shall prove in §4.

Example 1.   Consider the first-order Euler homogeneous DE

(1)
$$\frac{dw}{dz} = \frac{\gamma w}{z}, \qquad \gamma = \alpha + i\beta, \qquad \alpha, \beta \text{ real}$$

By separating variables and writing $z = re^{i\theta}$, we find the solution

(1′)
$$w = z^\gamma = e^{\gamma \ln z} = e^{(\alpha + i\beta)(\ln r + i\theta)}$$
$$= e^{(\alpha \ln r - \beta\theta)}[\cos (\beta \ln r + \alpha\theta) + i \sin (\beta \ln r + \alpha\theta)]$$

When $\beta = 0$ and $\gamma = \alpha$ is real, the analytic continuation of the real solution $u = x^\alpha$ on the positive x-axis through the upper half-plane to the negative x-axis, where $\theta = \pi$, is $(\cos \pi\alpha + i \sin \pi\alpha)|x|^\alpha$. Note that this is not equal to the real solution $|x|^\alpha$ on $x < 0$, unless α is an even integer.

The preceding example also shows that DEs involving only single-valued functions can have *multivalued* solutions in the complex plane. Unless γ is a real integer, the value of $w = z^\gamma$ changes by a factor

$$e^{2\pi i\gamma} = e^{-2\pi\beta}(\cos 2\pi\alpha + i \sin 2\pi\alpha) \neq 1$$

---

† Ahlfors, p. 182; Hille, p. 209.

when z describes a simple closed counterclockwise loop around the origin, making θ increase by 2π and $\ln z$ by $2\pi i$. This example shows that solutions of a linear DE can have *branch points* where the DE has a singular point,† even though the DE has single-valued coefficient-functions.

Example 2.   The second-order homogeneous Euler DE is

(2) $$z^2w'' + pzw' + qw = 0, \qquad p, q \text{ real constants}$$

For positive $z = x > 0$, a basis of real solutions is provided by the real and imaginary parts of the functions $z^\gamma = e^{\gamma \ln z}$, where γ is either of the roots of the *indicial equation*

(2') $$\nu^2 + (p - 1)\nu + q = 0$$

For instance, the roots of the indicial equation of the DE

$$z^2w'' + zw' + w = 0$$

are $\nu = \pm i$. A basis of complex solutions, real on the positive x-axis, is therefore provided, as in Ch. 3, §3 by the real and imaginary parts of the functions

$$w_1 = \tfrac{1}{2}(z^i + z^{-i}) = \tfrac{1}{2}[e^{i(\ln r + i\theta)} + e^{-i(\ln r + i\theta)}]$$

$$= \cosh\theta \cos(\ln r) - i \sinh\theta \sin(\ln r), \text{ and}$$

$$w_2 = \tfrac{1}{2i}(z^i - z^{-i}) = \cosh\theta \sin(\ln r) + i \sinh\theta \cos(\ln r)$$

The analytic continuation of the solution $\cos(\ln x)$, real on the positive x-axis $\theta = 0$, through the upper half-plane to the negative x-axis $\theta = \pi$ is not the solution $\cos(\ln |x|) = \cos(\ln r)$ given in Ch. 3, §3 but is the complex-valued function $\cosh\pi \cos(\ln r) - i \sinh\pi \sin(\ln r)$.

*2   MOVABLE SINGULAR POINTS

The general solution of the DE $w' = w^2$ considered in §1 is $1/(c - z)$. This function has a *pole* at the variable point c. Thus, the location of the singular point of a solution depends in this example on the particular solution. This happens for most *nonlinear* DEs; one describes the situation by saying that the "general solution" of $w' = w^2$ has a *movable* singular point.

---

† We define a *singular* point of the linear DE

$$p_0(z)w^{(n)} + p_1(z)w^{(n-1)} + \cdots + p_n(z)w = p_{n+1}(z)$$

as a point where $p_0(z) = 0$, or some $p_k(z)$ has a singular point.

A second example of movable singular points is provided by the DE $w' = z/w$. The general solution of this DE, obtained by separating variables, is the two-valued function $w = (z^2 - c^2)^{1/2}$, which has branch points at $z = \pm c$. Since c is arbitrary, the general solution has a movable branch point.

There is no significant class of nonlinear first-order normal DEs whose solutions have fixed singular points. However, the solutions of the generalized Riccati DE $w' = p_0(z) + p_1(z)w + p_2(z)w^2$ have *fixed branch points*.† This can be shown by representing $w = v'/p_2 v$ as a quotient of solutions v of the linear DE $v'' + p_1 v' + p_0 p_2 v = 0$, as in Chap. 2, §5.

A second-order nonlinear DE having a fixed singular point at $z = 0$ is

$$w'' = (w'^2/w) - (w'/z),$$

whose general solution is $w = Cz^\gamma$, with C, γ arbitrary complex constants. But nonlinear DEs with fixed singular points are highly exceptional.

It is otherwise for linear DEs, to which this chapter will be largely devoted. An nth order normal linear DE

$$L[w] = w^{(n)} + p_1(z)w^{(n-1)} + \cdots + p_{n-1}(z)w' + p_n(z)w = f(z)$$

with holomorphic coefficient functions has holomorphic solutions in any domain where the coefficient functions are holomorphic. The argument of Ch. 6, §8, can be applied to construct solutions along any path. Moreover, as in Ch. 6, §11, all the functions constructed in the Picard iteration process are holomorphic in any *simply connected* domain, say in $0 \le |z| < R$. It follows, as in Corollary 2 of Theorem 10 of Ch. 6, that the DE $L[w] = 0$ has a basis of holomorphic solutions in any such domain—and that $L[w] = f(z)$ has a solution for any choice of initial conditions compatible with the order of the DE.

It follows that the only possible singular points of the solutions of a normal linear DE occur where one or more of the coefficient-functions $p_k(z)$ has a singular point. In short, linear DEs have fixed singular points. In the remainder of this chapter, we will see how the nature of these singular points is determined by the singularities of the coefficient-functions.

3   FIRST-ORDER LINEAR EQUATIONS

The study of singular points of linear DEs in the complex domain begins with first-order DEs of the form

$$(3) \qquad\qquad w' + p(z)w = 0$$

We will treat only *isolated* singular points, assuming that p is holomorphic in some punctured disk Δ: $0 < |z| < \rho$, since any singular point can be moved to the origin by a translation of coordinates.

---

† The Riccati DE is the *only* first-order nonlinear DE with fixed branch points.

It follows that $p(z)$ can be expanded into a Laurent series,†

$$(3') \qquad p(z) = \sum_{-\infty}^{\infty} a_k z^k, \qquad 0 < |z| < \rho$$

convergent in Δ. When all a_k with $k < 0$ vanish, $p(z)$ is said to have a *removable singularity;* when there is a largest negative integer $k = -m$ for which $a_k \neq 0$, $p(z)$ is said to have a *pole of order m* there. If there are an infinite number of nonzero coefficients a_k with $k < 0$, then $p(z)$ is said to have an *essential singularity* at the origin.

In all cases, every nontrivial solution of (3) is given by Theorem 1 of Ch. 1, §3, as

$$
\begin{aligned}
(3'') \qquad w &= \exp[-\textstyle\int p(z)\, dz] \\
&= C \exp[-a_{-1} \ln z - \sum_{k=1}^{\infty}\left(\frac{a_{k-1}}{k}\right) z^k + \sum_{k=1}^{\infty}\left(\frac{a_{-k-1}}{k}\, z^{-k}\right)]
\end{aligned}
$$

As a corollary, we can represent the solution w in the form

$$(4) \qquad w = Cz^\gamma g(z), \qquad \gamma = -a_{-1}$$

where $g(z)$ is a *holomorphic* function in the domain Δ.

Caution. When γ is complex the innocent looking function $z^\gamma = e^{\gamma \ln z}$ can be quite nasty. The discussion of z^γ in Ch. 3, §2, should be reviewed in the context of Riemann surfaces. For example, writing $z = e^{i\theta}$ in the usual polar coordinate representation, but letting θ wind around the origin on an infinite-sheeted "winding surface" for $z \neq 0$, $|z^i| = e^{-\theta}$ and $\arg(z^i) = \ln r$. Hence z^i is *unbounded* for $|z| = 1$, though bounded for $|z| = r$ on any one sheet of the Riemann surface of z^i.

In general, if z^γ ($\gamma = \alpha + i\beta$), then as z tranverses any circle $|z| = r$ once counterclockwise, z^γ is multiplied by the constant

$$(4') \qquad e^{2\pi i \gamma} = e^{2\pi(-\beta + i\alpha)} = e^{-2\pi\beta}(\cos 2\pi\alpha + i \sin 2\pi\alpha)$$

Hence w is holomorphic if and only if γ is a real integer.

We now describe the basic classification of singularities for first-order linear DEs at any isolated singularity of $p(z)$. In the vicinity of any removable singularity of $p(z)$, $\gamma = a_{-1} = 0$ and so $w(z)$ can be expanded in an ordinary convergent Taylor series. Hence, if $p(z)$ has a *removable singularity* at $z = 0$, so does w.

If $p(z)$ has a pole of order one, then (4) holds, where

$$g(z) = 1 - a_0 z + \frac{(a_0^2 - a_1)z^2}{2} + \cdots$$

---

† Ahlfors, p. 182; Hille, p. 209.

is still analytic. Therefore, w has the form

(*) $$w = Cz^\gamma(1 + c_1 z + c_2 z^2 + \cdots), \qquad \gamma = -a_{-1}$$

In this case $p(z)$ is said to have a *branch pole* of order γ at $(0, 0)$, and the DE (3) to have a *regular* singular point at $z = 0$. The number γ can be real or complex, rational or irrational.

Finally, if $p(z)$ has a pole of order exeeding 1 or an essential singularity at $z = 0$, then the DE (3) is said to have an *irregular* singular point at 0. In this case, one can show that $g(z)$ in (4) has an *essential* singularity at $z = 0$. For example, the DE $w' + z^{-2}w = 0$ has the solution $w = \exp[1/\zeta] = \sum_{k=0}^\infty z^{-k}/k!$, easily found by separating variables, with an essential singularity at $z = 0$.

To prove this, suppose the contrary: that $[z^\gamma g(z)]' + p(z)z^\gamma g(z) = 0$, with $g(z) = \sum_{k=m}^\infty c_k z^k$, and $c_m \neq 0$. Solving for $p(z)$, we get

$$p(z) = \frac{zg'(z) - \gamma g(z)}{zg(z)} = -\frac{(m + \gamma) c_m z^m + \cdots}{c_m z^{m+1} + \cdots}$$

so that $p(z)$ must have a simple pole at $z = 0$ unless $M + \gamma = 0$, in which case $p(z)$ has a removable singularity.

In summary, we have proved the following result.

THEOREM 1. *Every solution of the first-order linear DE (3), with p holomorphic in Δ, has the form $w = z^\gamma g(z)$, where $\gamma = -a_{-1}$ and g is single-valued and analytic in Δ. Removable singularities, regular singular points, and irregular singular points of (3) give solutions having removable singularities, branch poles of order γ, and essential singularities at $z = 0$, respectively.*

Theorem 1 was derived by explicit calculation. To extend it to higher order linear DEs, which cannot be explicitly solved, more general arguments are needed. These require the notion of a *simple branch point* of an analytic function $w = f(z)$. This is defined as an isolated singular point z_0 near which $f(z)$ can be represented in the form $f(z) = (z - z_0)^\gamma g(z)$, where $g(z)$ is a one-valued holomorphic function in the punctured neighborhood of z_0. Expanding $g(z)$ in a Laurent series, we get the following expansion for a function f having a simple branch point at z_0:

$$f(z) = \sum_{-\infty}^{\infty} a_k(z - z_0)^{\gamma+k}$$

Not all branch points are simple; for example, the functions $\ln(z - z_0)$ and $(z - z_0)^\alpha + (z - z_0)^\beta$ have branch points at $z = z_0$ but do not have simple branch points there, unless $\alpha - \beta$ is an integer. Any branch pole (4) is a simple branch point, but a simple branch point need not be a branch pole; thus, consider $z^{1/2}e^{-1/z^2}$ at $z = 0$.

We shall now give another proof of the first result of Theorem 1, namely that every solution $w(z)$ of (3) that is not a holomorphic function has a simple branch point at $z = 0$.

Starting at a point z_0 in the given punctured disc, we continue the function w analytically counterclockwise around a closed circuit, say the circle $|z| = |z_0|$. Returning to z_0 after a complete circuit around the origin, the function $w(e^{2\pi i}z)$ $= \tilde{w}(z)$ obtained in a neighborhood of z_0 is still a solution of the DE, by Theorem 1. But $\tilde{w}$ may differ from the function w in a neighborhood of z_0, because the function w may have a branch point at $z = 0$ even though the coefficient $p(z)$ of the DE does not have a branch point there; the DE $w' = w/2z$ is a case in point.

However, the general solution of the DE (3) has the form $cw(z)$; from this it follows that

$$\tilde{w}(z) = w(e^{2\pi i}z) = cw(z), \qquad c \neq 0$$

Now, write $c = e^{2\pi i \alpha}$, where α is a suitable complex number, and consider the analytic continuation of the function

$$g(z) = z^{-\alpha}w(z)$$

around the same circuit. We obtain

$$g(ze^{2\pi i}) = (ze^{2\pi i})^{-\alpha}w(ze^{2\pi i}) = z^{-\alpha}e^{-2\pi i\alpha}w(ze^{2\pi i})$$
$$= z^{-\alpha}e^{-2\pi i\alpha}cw(z) = z^{-\alpha}w(z) = g(z)$$

This shows that g is a single-valued function in the punctured disc Δ. Thus, the function w is the product of z^α and a function without a branch point.

The idea behind this second (and deeper) proof of the first result of Theorem 1 can be applied to linear DEs of any order, as we show in the following sections.

EXERCISES A

1. Show that no solution of $w' = 1/z$ that is real on the positive x-axis can be real on the negative x-axis.

2. (a) Setting $z = re^{i\theta}$, discuss the analytic continuation to the negative x-axis of the solutions z and $z \ln z$ of $z^2w'' - zw' + w = 0$ on the positive x-axis.
   (b) Show that no nontrivial solution of $z^2w'' + 3w/8 = 0$ that is real on the positive x-axis can be real on the negative x-axis.

3. Let $w^{(n)} + a_1(z)w^{(n-1)} + \cdots + a_n(z)w = 0$ be any holomorphic linear homogeneous DE satisfied by $\ln z$. Show that $a_n(z) \equiv 0$.

*4. Show that any holomorphic linear homogeneous DE that is satisfied by $z \ln z$ is also satisfied by z.

5. Find the function g of Theorem 2 when
   (a) $p(z) = 1/z^n$ (n an integer)   (b*) $p(z) = e^{1/z}$

6. Solve the DE (4) for $p(z) = \displaystyle\sum_{k=1}^{n} \frac{1}{z - a_k}$.

7. Let $p(z)$ be holomorphic and single-valued in $|z| < \rho$ except at points a and b. Show that any solution w_1 of (3) can be written in the form

$$w_1(z) = (z - a)^{\alpha}(z - b)^{\beta}f(z)$$

where f is single-valued and holomorphic in $|z| < \rho$ except at a and b.

8. Generalize the result of the preceding exercise to the case that $p(z)$ is single-valued for $|z| < \rho$ and holomorphic at all points except $a_1, \ldots, a_n$.

9. Prove in detail that solutions of the generalized Riccati equation

$$w' = p_0(z) + p_1(z)w + p_2(z)w^2$$

can have branch points only where the $p_k(z)$ have singular points.

10. Show that, if the DE of Ex. 9 has no movable singularities, $p_2(z) \equiv 0$.

11. Show that, for analytic $p_i(z)$, the DE $dw/dz = \sum_{k=0}^{n} p_k(z)w^{-k}$ has a regular solution.

[HINT: Consider the DE satisfied by ρ/w.]

4   CONTINUATION PRINCIPLE; CIRCUIT MATRIX

A rigorous discussion of complex analytic solutions of higher order DEs involves the concept of *analytic continuation,* with which we will assume the reader to be acquainted.† It also assumes a less well-known Continuation Principle for solutions of complex analytic DEs, which may be derived as follows.

Let $F(w_1, \ldots, w_n, z)$ be an analytic complex-valued function in a domain D of $(w_1, \ldots, w_n, z)$-space. This means that F can be expanded into a convergent power series with complex coefficients in some neighborhood of each point of D. Then, as was shown in Ch. 6, §11, every solution of the nth order DE

$$(*) \qquad \frac{d^n w}{dz^n} = F(w, w', w'', \ldots, w^{(n-1)}, z)$$

is an analytic function. The function $w^{(n)}(z) - F[w(z), w'(z), \ldots, w^{(n-1)}(z), z]$ of the variable z is holomorphic and vanishes identically in the subdomain where the function w is defined. It follows that all analytic continuations of this function beyond D also vanish identically. Therefore, any analytic continuation of the function w is also a solution of the DE (*), and we have the following theorem.

THEOREM 2.   (*CONTINUATION PRINCIPLE*). *The function obtained by analytic continuation of any solution of an analytic DE, along any path in the complex plane, is a solution of the analytic continuation of the DE along the same path.*

---

† Ahlfors, p. 275; Hille, p. 184.

With this theorem in hand, let $w_1(z)$ and $w_2(z)$ be a basis of solutions of the second-order DE

(5)
$$w'' + p(z)w' + q(z)w = 0$$

where the functions p and q are single-valued and analytic in the punctured disc $\Delta: 0 < |z| < \rho$. Analytic continuation of each of these solutions counterclockwise around a circle $|z| = r < \rho$ with center at the origin yields two functions (in general different): $\tilde{w}_1(z) = w_1(ze^{2\pi i})$ and $\tilde{w}_2(z) = w_2(ze^{2\pi i})$. These are, by the Continuation Principle, also solutions of the DE (5). Since every solution of (5) is a linear combination of w_1 and w_2: the continued functions $\tilde{w}_j$ can be expressed as linear combinations of the solutions w_1 and w_2, thus:

$$\tilde{w}_1(z) = w_1(e^{2\pi i}z) = a_{11}w_1(z) + a_{12}w_2(z)$$

$$\tilde{w}_2(z) = w_2(e^{2\pi i}z) = a_{21}w_1(z) + a_{22}w_2(x)$$

The 2×2 matrix of complex constants $A = ||a_{ij}||$ is called the *circuit matrix* of the DE at the singular point $z = 0$, relative to the basis of solutions (w_1, w_2).

For instance, consider the Euler DE $z^2w'' - zw' + 3w/4 = 0$, with indicial equation $(v - \frac{1}{2})(v - \frac{3}{2}) = 0$. The functions $z^{1/2}$ and $z^{3/2}$ form a basis of solutions; hence, the circuit matrix is the diagonal (scalar) matrix

$$\begin{pmatrix} -1 & 0 \\ 0 & -1 \end{pmatrix}$$

A similar calculation shows that the DE $w'' + (2/9z^2)w = 0$ has the solution basis $z^{1/3}$, $z^{2/3}$. Relative to this basis, its circuit matrix is

$$\begin{pmatrix} \omega & 0 \\ 0 & \omega^2 \end{pmatrix}$$

where $\omega = (-1 + \sqrt{3}i)/2$ is a cube root of unity.

Higher Order DEs. A similar construction can be used to define a circuit matrix (relative to any solution basis) for the nth order linear DE

(6)
$$L[w] = \frac{d^nw}{dz^n} + p_1(z)\frac{d^{n-1}w}{dz^{n-1}} + \cdots + p_n(z)w = 0,$$

where again all the coefficient-functions $p_k(z)$ are holomorphic in the punctured disc Δ.

LEMMA. *Given a basis of solutions $w_j(z)$ of (6), analytic continuation of the $w_j(z)$ around any circle $|z| = r$ in Δ, once counterclockwise, gives a new basis of solutions of (6),*

$$\tilde{w}_j(z) = w_j(e^{2\pi i}z) = a_{j1}w_1(z) + \cdots + a_{jn}w_n(z), \qquad j = 1, 2, \ldots, n$$

Proof.   By the Continuation Principle, the $w_j(z)$ satisfy (6); since the $w_j(z)$ form a basis of solutions, the result follows.

The matrix $A = \| a_{jk} \|$ so defined is the *circuit matrix* of the DE (6) relative to the basis $w_1(z), \ldots, w_n(z)$. It represents a linear transformation of the vector space of all solutions of the nth-order linear analytic DE (6), in the following manner. If

$$w(z) = c_1 w_1(z) + c_2 w_2(z) + \cdots + c_n w_n(z)$$

is the general solution of the DE, analytic continuation of w around the same circuit γ carries $w(z)$ into the solution

$$(6') \qquad\qquad \tilde{w}(z) = w(e^{2\pi i}z) = \sum_{j,k=1}^{n} c_j a_{jk} w_k(z)$$

That is, the effect of analytic continuation around γ counterclockwise is to multiply the vector $\mathbf{c}$ by the matrix A on the right, so that $\mathbf{c} \to \mathbf{c}A$.

Using the circuit matrix A just defined, we can construct at least one "canonical" solution of (6) of the special form $z^\alpha f(z)$ with f holomorphic in the disc Δ. This is because every matrix has† at least one (complex) eigenvector (characteristic vector). Hence, for some choice of $\mathbf{c} \neq 0$, we can write $\mathbf{c}A = \lambda \mathbf{c}$, where λ is a complex number, an eigenvalue of the matrix A. Choose $\mathbf{c}$ in (6′) to be such an eigenvector, and let $f(z) = z^{-\alpha} w(z)$, where $\alpha = (\ln \lambda)/2\pi i$. It is clear that $\lambda \neq 0$, since otherwise we could retrace backward the circuit γ, continuing the solution $w = 0$ into a nonzero solution. Continuing the function $f(z) = z^{-\alpha} w(z)$ along the same circuit, we obtain, as in the proof of Theorem 2, the following result.

THEOREM 3.   *Any nth order linear* DE *(6) with coefficients holomorphic in* Δ: $0 < |z| < \rho$ *admits at least one nontrivial solution of the form*

$$(7) \qquad\qquad w(z) = z^\alpha f(z)$$

where the function f is single-valued in Δ.

5   CANONICAL BASES

A solution of a holomorphic DE (6) in Δ has a simple branch point at $z = 0$ if and only if it is carried into a constant (scalar) multiple of itself by continuation around the circuit γ. From the discussion in the preceding section, we see that a solution $w(z) = \sum_{j=1}^{n} c_j w_j(z)$ of (6) has a simple branch point at $z = 0$ if and only if the vector $\mathbf{c} = (c_1, c_2, \ldots, c_n)$ and the circuit matrix A satisfy the

---

† Birkhoff and MacLane, p. 293. As stated there, λ is a root of the characteristic equation $|A = \lambda I| = 0$.

relation $\mathbf{c}A = \lambda\mathbf{c}$, where the constant λ will then be necessarily different from zero. In other words, a linear combination $\sum_{j=1}^{n} c_j w_j$ of solutions of (6) has a simple branch point at $z = 0$ if and only if the vector $\mathbf{c}$ is an eigenvector of the circuit matrix A associated with the basis $(w_1, w_2, \ldots, w_n)$ of solutions. Thus, there are as many linearly independent solutions of (6) with simple branch points as there are linearly independent eigenvectors of the matrix A.

We shall now look for a basis of solutions with simple branch points for any second-order linear DE (5), with coefficients holomorphic in Δ.

Given two linearly independent solutions w_1 and w_2 of (5), we can construct the circuit matrix $A = \|a_{ij}\|$ as in §4. The linear combination $w = c_1 w_1(z) + c_2 w_2(z)$ then will have a simple branch point if and only if $\sum c_j a_{jk} = \lambda c_k$. By the theory of linear equations, this system of equations has a nontrivial solution if and only if the following determinant (the *characteristic equation* of the circuit matrix A) equals zero:

$$(8) \qquad |A - \lambda I| = \lambda^2 - (a_{11} + a_{22})\lambda + (a_{11}a_{22} - a_{12}a_{21}) = 0$$

Ordinarily, this characteristic equation has two distinct roots λ_1, λ_2. These roots give two linearly independent solutions $F(z) = c_1 w_1(z) + c_2 w_2(z)$ and $G(z) = d_1 w_1(z) + d_2 w_2(z)$, having simple branch points: $F(e^{2\pi i}z) = \lambda_1 F(z)$ and $G(e^{2\pi i}z) = \lambda_2 G(z)$. Relative to the canonical basis of solutions F, G, the circuit matrix is thus a *diagonal* matrix

$$\begin{pmatrix} \lambda_1 & 0 \\ 0 & \lambda_2 \end{pmatrix}$$

As in §4, $F(z) = z^\alpha f(z)$ and $G(z) = z^\beta g(z)$, where $\lambda_1 = e^{2\pi i \alpha}$, $\lambda_2 = e^{2\pi i \beta}$, and f and g are holomorphic in Δ. Such a basis is called a *canonical* basis.

When the characteristic equation has a single solution λ, the solutions may still† sometimes have a basis of the form (7). Every solution of the DE is then multiplied by the same nonzero constant $\lambda = e^{2\pi i \alpha}$ when continued around a counterclockwise circuit γ.

Otherwise, we choose a basis as follows. Let $w_1(z)$ be the solution of the form $z^\alpha f(z)$, where f is one-valued in the punctured disc whose existence was established in Theorem 2, and let $w_2(z)$ be any other linearly independent solution. Continuation of $w_2(z)$ around the circuit γ gives, as in §3, $w_2(ze^{2\pi i}) = a w_1(z) + b w_2(z)$. The circuit matrix for this basis of solutions is therefore the matrix

$$A = \begin{pmatrix} \lambda & 0 \\ a & b \end{pmatrix}$$

---

† This occurs when the matrix A is a multiple of the identity matrix. Otherwise, any 2×2 matrix A with only one eigenvalue is similar to a matrix of the form $\begin{pmatrix} \lambda & 0 \\ 1 & \lambda \end{pmatrix}$. This fact is not assumed in the present discussion.

Since the eigenvalues of such a "triangular" matrix are λ and b, and since the only eigenvalue of A was assumed to be λ, we must have $b = \lambda$ and

$$A = \begin{pmatrix} \lambda & 0 \\ a & \lambda \end{pmatrix}, \qquad a \neq 0$$

The continuation around the circuit γ of the function $h(z) = w_2(z)/w_1(z)$ is easily computed to be

$$h(ze^{2\pi i}) = \frac{a}{\lambda} + \frac{w_2(z)}{w_1(z)} = \frac{a}{\lambda} + h(z)$$

It follows that the function

$$f_1(z) = h(z) - \frac{a}{2\pi i\lambda} \ln z$$

is single-valued in $0 < |z| < \rho$ and, therefore, that the function $w_2(z)$ can be written in the form

$$w_2(z) = w_1(z)f_1(z) + \frac{a}{2\pi i\lambda} (\ln z)w_1(z)$$

In the exceptional case, since $a \neq 0$, we can make $a/2\pi i\lambda = 1$ by replacing w_2 with $2\pi i\lambda w_2/a$.

This completes the proof of the following theorem.

THEOREM 4.   *Under the hypotheses of Theorem 3, the second-order linear* DE(5) *has a basis of solutions in the neighborhood of the singular point $z = 0$, having one of the following forms:*

(9a) $$w_1(z) = z^\alpha f(z), \qquad w_2(z) = z^\beta g(z)$$

or, exceptionally,

(9b) $$w_1(z) = z^\alpha f(z), \qquad w_2(z) = w_1(z)[f_1(z) + \ln z]$$

The functions $f(z)$, $g(z)$, and $f_1(z)$ are holomorphic and single-valued in the punctured disc $0 < |z| < \rho$.

Higher Order Equations.†   The preceding discussion can be extended to nth order DEs (6). A basis of solutions $w_1, w_2, \ldots, w_n$ is called a *canonical basis* when the associated circuit matrix A is in *Jordan canonical form*‡ or, somewhat

---

† For a complete discussion see Coddington and Levinson, Ch. 4, §1.

‡ See Appendix A or Birkhoff and MacLane, p. 354.

more generally, consists of zeros except on the main diagonal and (in the case of a repeated eigenvalue) just above it. If the eigenvalues of the circuit matrix are $\lambda_1, \lambda_2, \ldots, \lambda_n$, the continuation of the solution w_j around a small circuit γ is given by one of the two formulas

$$(10a) \qquad \tilde{w}_j(z) = w_j(ze^{2\pi i}) = \lambda_j w_j(z)$$

or, when $\lambda_j = \lambda_{j-1}$,

$$(10b) \qquad \tilde{w}_j(z) = w_j(ze^{2\pi i}) = \lambda_j w_j(z) + a_{j-1}w_{j-1}(z)$$

By Theorem 3, there always is at least one solution that goes into a multiple of itself. If all the eigenvalues of the circuit matrix are distinct, then the circuit matrix can be reduced to diagonal form by suitable choice of a "canonical basis" of solutions, and the exceptional case of formula (10b) does not arise.

Example 3. Consider the nth order *Euler* DE

$$(11) \qquad L[w] = \frac{d^n w}{dz^n} + \frac{c_1}{z}\frac{d^{n-1}w}{dz^{n-1}} + \frac{c_2}{z^2}\frac{d^{n-2}w}{dz^{n-2}} + \cdots + \frac{c_n}{z^n}w = 0$$

The trial function z^ν, with unknown exponent ν, satisfies the DE if and only if the exponent ν is a root of the indicial equation of Ch. 4, §2,

$$
\begin{aligned}
(12) \quad I(\nu) = {}&\nu(\nu - 1) \cdots (\nu - n + 1) \\
&+ c_1\nu(\nu - 1) \cdots (\nu - n + 2) + \cdots + c_{n-1}\nu + c_n = 0
\end{aligned}
$$

When the roots of the indicial equation are distinct, the z^{ν_j} are a canonical basis of solutions of the Euler DE (11). The circuit matrix is the diagonal matrix with diagonal entries $\lambda_j = e \exp 2\pi i\nu_j$, where ν_j is a root of the indicial equation. When ν is a k-tuple root of the indicial equation, then the functions $z^\nu \log z$, $z^\nu(\log z)^2$, and so on, form a basis of solutions. This basis is not "canonical" when $n > 2$ (see Ex. B2), though the circuit matrix for (11) is always triangular relative to it.

EXERCISES B

1. Construct DEs (5) with circuit matrix

$$\begin{pmatrix} 1 & 0 \\ 0 & 1 \end{pmatrix}$$

   for which the functions f and g in Theorem 4 have (a) essential, and (b) removable singularities at 0.

2. (a) Compute the indicial polynomial for the homogeneous Euler DE $z^3w''' + 3z^2w''/2 + zw'/4 = w/8$.
   (b) Compute the circuit matrix of the preceding DE relative to the solution basis $z^{1/2}$, $z^{1/2}(\log z)$, and $z^{1/2}(\log z)^2$.

3. Find a DE (5) with circuit matrix $\begin{pmatrix} \lambda & 0 \\ 0 & 1/\lambda \end{pmatrix}$ for which the functions f and g in Theorem 4 have (a) poles, and (b) essential singularities. Can f and g have removable singularities?

4. Show that the requirements $0 \leqq \mathrm{Re}\{\alpha\}$, $\mathrm{Re}\{\beta\} \leqq 1$ uniquely determine the exponents α and β in formula (9a).

5. Show that, in the exceptional case of Theorem 4, the eigenvalues of the circuit matrix are equal.

6. Show that $e^{\sqrt{\lambda}}$, $e^{-\sqrt{\nu}}$ form a solution basis for the DE $w'' + \frac{1}{2z}w' - \frac{1}{4z}w = 0$. Compute the circuit matrix for the given DE relative to this basis, and also find a canonical solution basis.

7. Construct a second-order holomorphic DE (5) with a singular point at $z = 0$ whose circuit matrix has the form $\begin{pmatrix} 1 & 0 \\ 1 & 1 \end{pmatrix}$, such that f in Theorem 4 has an essential singularity at $z = 0$.

8. Show that, if $w \ln z$ satisfies a (homogeneous) linear DE (6) with holomorphic coefficients, and w is holomorphic, then w satisfies the same DE.

6   REGULAR SINGULAR POINTS

Many of the ordinary DEs of greatest interest for mathematical physics have singular points which are "regular" in the sense of the following definition.

DEFINITION.   A second-order DE

$$(13) \qquad\qquad w'' + p(z)w' + q(z)w = 0$$

analytic for $0 < |z - z_0| < \rho$, has a *regular singular point* at z_0 when $p(z)$ has at worst a simple pole at $z = z_0$, and $q(z)$ at worst a double pole there.†

In the next several sections, we shall show how to adapt the power series methods introduced in Ch. 3 to solve ordinary DEs in the neighborhood of any regular singular point. In particular, we will show that a singular point of the second-order linear DE (13) is "regular" if and only if the functions $f(z)$ and $g(z)$ of the canonical basis (9a), of solutions constructed in Theorem 4, have at worst *branch poles* there.‡ Equivalently, the condition is that a basis of solutions have the form (25) below.

---

† That is, p may either be holomorphic (have a removable singularity) or have a simple pole, and q may be holomorphic or have a pole of first or second order.

‡ Or, in the exceptional case of (9b), poles times logarithmic branch points (order of growth log r/r^n).

We now show that, near the regular singular point $z = 0$, there always exists a formal solution of the DE (13), namely, a formal power series of the form

$$(14) \qquad w = z^\nu(1 + c_1 z + c_2 z^2 + c_3 z^3 + \cdots) = z^\nu + c_1 z^{\nu+1} + c_2 z^{\nu+2} + \cdots$$

which, when substituted into (13), satisfies the DE. To calculate the coefficients c_k and the exponent ν of (14) it is convenient to rewrite (13) in the form

$$(15) \qquad L[w] = z^2 w'' + z P(z) w' + Q(z) w = 0$$

where $P(z) = \sum_{k=0}^{\infty} P_k z^k$ and $Q(z) = \sum_{k=0}^{\infty} Q_k z^k$ are convergent for $|z| < \rho$. Substituting (14) into (15) and equating to zero the coefficient of z^ν, we obtain the *indicial equation*

$$(16) \qquad I(\nu) = \nu(\nu - 1) + P_0 \nu + Q_0 = 0$$

for the exponent ν. The roots of this equation are called the *characteristic exponents* of the singular point, and $I(\nu)$ is called its *indicial polynomial*.

Equating to zero the coefficients of the higher powers of z, namely $z^{\nu+1}, \ldots, z^{\nu+n}, \ldots$, we obtain the relation

$$[(\nu + 1)\nu + P_0(\nu + 1) + Q_0]c_1 + P_1 \nu + Q_1 = 0$$

and, recursively,

$$[(\nu + n)(\nu + n - 1) + (\nu + n)P_0 + Q_0]c_n = -\sum_{k=0}^{n-1} [(\nu + k)P_{n-k} + Q_{n-k}]c_k$$

Since the left side of the preceding equation is $I(\nu + n)c_n$, the equation can be written in the form

$$(17) \qquad I(\nu + n)c_n = -\sum_{k=0}^{n-1} [(\nu + k)P_{n-k} + Q_{n-k}]c_k, \qquad n = 1, 2, 3, \ldots$$

The above equation for the coefficient c_n can be solved recursively for $c_1, c_2, c_3,$ $\ldots$, except in one case: when, for some positive integer n, both ν and $(\nu + n)$ are roots of the indicial equation. By taking a characteristic exponent having the largest real part, we can make sure that $I(\nu + n)$ does not vanish for any positive integer n, even in this case. We therefore obtain the following theorem.

THEOREM 5. *If the DE (13) has a regular singular point at $z = 0$, then at least one formal power series of the form (14) formally satisfies the DE. Unless the roots of the indicial equation differ by an integer, there are two linearly independent formal power series solutions (14) of the DE, whose exponents are the two roots of the indicial equation.*

In the special case of an *ordinary* (i.e., nonsingular) point, clearly $P_0 = Q_0 = 0$, and the indicial equation $\nu(\nu - 1) = 0$ has the roots 0, 1. In this special case, the preceding construction reduces to the methods of Ch. 4, §2. The series $z(1 + \sum_{j=1}^{\infty} c_k z^k)$ associated with the larger real root is uniquely determined but, because the roots differ by an integer and $I(1) = 0$, the series associated with the root $\nu = 0$ is not.

Example 4. A remarkable class of special functions having regular singular points at 0, 1, and ∞ (see §12), and no other singular points, is obtained by applying Theorem 5 to the *hypergeometric* DE

$$(18) \qquad z(1 - z)w'' + [\gamma - (\alpha + \beta + 1)z]w' - \alpha\beta w = 0$$

The hypergeometric DE has regular singular points at $z = 0$ and $z = 1$. The indicial equation at $z = 0$ is $\nu(\nu + \gamma - 1) = 0$, with roots $\nu_1 = 0$ and $\nu_2 = 1 - \gamma$. Unless γ is an integer, by Theorem 5 the hypergeometric DE has two formal power series solutions with exponents 0 and $1 - \gamma$. Unless γ is a negative integer or zero, one such formal power series is obtained from the easily computed recursion relations

$$(19) \qquad (n + 1)(\gamma + n)c_{n+1} = (\alpha + n)(\beta + n)c_n, \qquad n \geq 0$$

Setting $c_0 = 1$, we obtain the hypergeometric series

$$(20) \qquad F(\alpha, \beta, \gamma; z) = 1 + \frac{\alpha\beta}{\gamma} z + \frac{\alpha(\alpha + 1)\beta(\beta + 1)}{\gamma(\gamma + 1)} \frac{z^2}{2!}$$
$$+ \frac{\alpha(\alpha + 1)(\alpha + 2)\beta(\beta + 1)(\beta + 2)}{\gamma(\gamma + 1)(\gamma + 2)} \frac{z^3}{3!} + \cdots$$

From the Ratio Test, it follows that the radius of convergence of the series is at least one, and is exactly one unless α, β, or γ is a negative integer. This also may be expected from the existence theorems of Ch. 6, since the radius of convergence of a solution extends to the nearest other singular point of the coefficients of the hypergeometric DE, which is at $z = 1$. The function $F(\alpha, \beta, \gamma; z)$ defined by the power series (20) is the *hypergeometric function*, to be studied in §10 below.

7 BESSEL EQUATION

To illustrate the behavior of solutions of second-order DEs near regular singular points, we consider an example of great importance in applied mathematics. This is the Bessel DE of order n:

$$(21) \qquad z^2 w'' + zw' + (z^2 - n^2)w = 0$$

THEOREM 8. *Suppose that the roots α and $\beta = \alpha - n$ of the indicial equation of a second-order linear DE, having a regular singular point at $z = 0$, differ by a nonnegative integer n. Then there exists a canonical basis of solutions of the form*

$$(26) \qquad w_1 = z^\alpha \left(1 + \sum_{k=1}^\infty a_k z^k\right), \qquad w_2 = z^\beta \left(1 + \sum_{k=1}^\infty b_k z^k\right) + C w_1 \ln z$$

where the power series are convergent in a neighborhood of $z = 0$.

Relative to the canonical basis (26), the circuit matrix has the form

$$\begin{pmatrix} \lambda & 2\pi i C \\ 0 & \lambda \end{pmatrix}$$

with $\lambda = e^{2\pi i \alpha} = e^{2\pi i \beta}$.

*9 ALTERNATIVE PROOF OF THE FUNDAMENTAL THEOREM

Theorem 7 also can be given a more intrinsic proof by relying on the following characterization of poles of analytic functions.†

ORDER OF GROWTH THEOREM. *If $f(z)$ is holomorphic in $0 < |z| \leq R$, then $z = 0$ is a pole of $f(z)$ of order at most α, or a removable singularity, if and only if there exists a positive number C such that*

$$\sup_{0 \leq \theta \leq 2\pi} |f(re^{i\theta})| < Cr^{-\alpha}, \qquad 0 < r \leq R$$

Theorem 7 can be deduced quite easily from this and the following basic result.

LEMMA. *If the DE (5) has a regular singular point at $z = 0$, then the function $f(z)$ in Theorem 3 has at most a pole at $z = 0$.*

Proof. For any solution $w(z)$ of (5), consider the real-valued function

$$U(z) = |w(z)|^2 + |zw'(z)|^2$$

Setting $z = re^{i\theta}$, we shall majorize the derivative of this function relative to r for fixed θ; its differentiability follows by the Chain Rule.

---

† Hille, p. 213, Theorem 8.4.1.

For any differentiable complex-valued function $V(r)$ of a real variable r, as in Ch. 6, formula (3), we have

$$\left| \int_a^r V'(t) \, dt \right| \le \int_a^r |V'(t)| \, dt$$

Applying this inequality to $U(re^{i\theta})$ we obtain

$$\frac{1}{2} \left| \frac{\partial U}{\partial r} \right| \le |ww'| + \left| \frac{z^2 w'^2}{r} \right| + |z^2 w' w''|$$

where $z = re^{i\theta}$. Using the fact that $w'' = -(P(z)/z)w' - (Q(z)/z^2)w$ we obtain

$$\frac{1}{2} |\partial U/\partial r| \le |ww'| + |z^2 w'^2|/r + |P(z)||z^2 w'^2|/r + |Q(z)||ww'|$$

The functions $P(re^{i\theta})$ and $Q(re^{i\theta})$ are holomorphic in some closed disk $0 \le |z| \le R$, $R > 0$. Let M be a common upper bound for their absolute values, for $0 \le \theta \le 2\pi$ and for fixed r. This gives the inequality

$$\frac{1}{2} \left| \frac{\partial U}{\partial r} \right| \le (M + 1)|ww'| + \frac{(M + 1)|z^2 w'^2|}{r}$$

By definition of U, we have $|w|^2 \le U$, $|w'|^2 \le U/r^2$, and hence, multiplying, $|ww'| \le U/r$. We obtain, therefore,

$$\left| \frac{[\partial U(re^{i\theta})]}{\partial r} \right| \le \frac{(4M + 4)U(re^{i\theta})}{r} = \frac{KU}{r}, \qquad K > 0$$

In particular, for $0 < r \le R$, we obtain $\partial U/\partial r + KU/r \ge 0$, whence integrating between the limits r and R,

$$R^K U(Re^{i\theta}) - r^K U(re^{i\theta}) \ge 0$$

If $N = \max_{0 \le \theta \le 2\pi} U(Re^{i\theta})$, we obtain

$$U(re^{i\theta}) \le NR^K r^{-K}$$

and hence, *a fortiori*, that $|w(re^{i\theta})|^2 \le (NR^K)r^{-K}$. By the Order of Growth Theorem, with $C = NR^K$ and $\alpha = K$, the conclusion of the lemma follows.

Consequently, if the DE (5) has a regular singular point at the origin, then it has a solution given by a locally convergent power series of the form described in Theorems 5 and 7. The construction of a second solution can then be achieved as in Theorem 8.

The preceding result can be generalized to nth order linear DEs.

EXERCISES D

1. Find the exponents at $z = 0$ of the DE $w'' + (\mu/z)w' + (1/z)w = 0$. Show that this DE has a power series solution $C_\mu(z) = 1 + \sum_{k=1}^{\infty} a_k z^k$, convergent for all $|z|$. Show that $C_\mu(z) = z^{(1-\mu)/2} J_{\mu-1}(2\sqrt{z})$.

2. Show that $w'' + (n + \frac{1}{2} - z^2/4)w = 0$ has a basis of solutions

$$w_1(z) = 1 - \frac{(2n+1)z^2}{4} + \frac{(4n^2 + 4n + 3)z^4}{96} - \cdots, \text{ and}$$

$$w_2(z) = z - \frac{(2n+1)z^3}{12} + \frac{(4n^2 + 4n + 7)z^5}{480} - \cdots$$

For what values of $|z|$ do these series converge?

3. The Laguerre DE is $zw'' + (1 - z)w' + \alpha w = 0$.
   (a) Find its characteristic exponents.
   (b) Show that a nontrivial solution is given by $\sum c_k z^k$ with

$$c_{j+1} = \frac{(j - a)c_j}{(j + 1)^2}$$

4. The associated Laguerre DE is $zw'' + (k + 1 - z)w' + (n - k)w = 0$. Show that this has a polynomial solution $w = L_n^k(z)$ for any positive integers k, n.

5. Show that $e^{-z/2} z^{(k-1)/2} L_k^n(z)$ satisfies the DE

$$zw'' + 2w' + [A + Bz + C/z]w = 0$$

with $A = n - (k - 1)/2$, $B = -\frac{1}{4}$, $C = (1 - k^2)/4$.

6. Show that, if $\phi(0) \neq 0$, the substitution $w = \phi(z)v$ carries second-order linear DEs (5) having a regular singular point at the origin into DEs having the same property.

7. Generalize the result of Ex. 6 to nth order linear DEs.

8. Show that the substitution $w = z^r \phi(z)w_1$, where $\phi(0) \neq 0$ and $\phi(z)$ is regular near $z = 0$, carries a regular singular point at $z = 0$ with indicial polynomial $I(\nu)$ into one with indicial polynomial $I(\nu - r)$.

9. Do the functions $\log z$ and $(\log z)^2$ satisfy a second-order linear DE (3) with a regular singular point at $z = 0$? Do they satisfy a third-order linear DE with regular singular point at $z = 0$? Justify your answers.

*10. (a) The DE $w''' + \sum_{k=1}^{3} p_k(z)w^{(3-k)} = 0$, $p_k(z)$ holomorphic for $0 \leq |z| \leq r$, has a regular singular point at $z = 0$ if p_k has, at worst, a pole of order k. Derive an analog of the indicial equation (16) and generalize Theorem 7 to this DE.
   (b) Generalize Theorem 8 for this DE when two exponents coincide.

*10 HYPERGEOMETRIC FUNCTIONS

So far in this chapter, the behavior of solutions of DEs has been studied only near a single isolated singular point. A fascinating topic of analysis is the relation between the behavior at different singular points of analytic functions defined by DEs. This topic is beautifully illustrated by the hypergeometric functions, defined as solutions of the hypergeometric DE (18). This illustration (Example 4 of §6) is of especial interest because many common transcendental functions can be expressed in terms of the hypergeometric functions. For example, $(1 - z)^{-\alpha} = F(\alpha, \beta, \beta; z)$, $\arcsin z = zF(\tfrac{1}{2}, \tfrac{1}{2}, \tfrac{3}{2}; z^2)$, $\log (1 + z) = zF(1, 1, 2; -z)$, and so on.

According to the program laid out in Ch. 4, the properties of the hypergeometric functions can be deduced from the DE (18). For example, let us derive a formula for the derivative of the hypergeometric function $F(\alpha, \beta, \gamma; z)$. Differentiating the hypergeometric DE, we get

$$z(1 - z)w''' + [\gamma + 1 - (\alpha + 1 + \beta + 1 + 1)z]w'' - (\alpha + 1)(\beta + 1)w' = 0$$

which is again a hypergeometric DE with constants $\alpha_1 = \alpha + 1$, $\beta_1 = \beta + 1$, $\gamma_1 = \gamma + 1$. By Theorems 7 and 8, every solution of this DE holomorphic at the origin is a constant times $F(\alpha + 1, \beta + 1, \gamma + 1; z)$. This implies the formula $F'(\alpha, \beta, \gamma; z) = kF(\alpha + 1, \beta + 1, \gamma + 1; z)$. The constant k is determined by differentiating (20) at $z = 0$. This gives the differentiation formula

$$(27) \qquad F'(\alpha, \beta, \gamma; z) = \frac{\alpha\beta}{\gamma} F(\alpha + 1, \beta + 1, \gamma + 1; z)$$

The *Jacobi identity*

$$\frac{d^n}{dz^n} [z^{\alpha+n-1}F(\alpha, \beta, \gamma; z)] = \alpha(\alpha + 1) \cdots (\alpha + n - 1)z^{\alpha-1}F(\alpha + n, \beta, \gamma; z)$$

can be similarly established by multiplying both sides of the identity by $z^{1-\alpha}$, and then verifying that both sides of the resulting identity satisfy the same hypergeometric DE with constants $a_1 = \alpha + n, \beta, \gamma$.

The study of the hypergeometric DE is greatly facilitated by its *symmetry properties*. Making the substitution $w = z^{1-\gamma}u$, we obtain as a DE equivalent to (18) for the dependent variable u, a second hypergeometric DE with different constants (unless $\gamma = 1$):

$$(28) \qquad z(1 - z)u'' + [\gamma_1 - (\alpha_1 + \beta_1 + 1)z]u' - \alpha_1\beta_1 u = 0$$

where $\alpha_1 = \alpha - \gamma + 1$; $\beta_1 = \beta - \gamma + 1$, and $\gamma_1 = 2 - \gamma$. Since this DE has the solution $w_1(z) = F(\alpha_1, \beta_1, \gamma_1; z)$, we obtain at once a power series solution of (18) corresponding to the exponent $1 - \gamma$ in the form

$$(29) \qquad w_2(z) = z^{1-\gamma}F(\alpha - \gamma + 1, \beta - \gamma + 1, 2 - \gamma; z)$$

The two solutions are a *canonical basis* of solutions of (18) at the regular singular point $z = 0$.

The change of dependent variable $w = (1 - z)^{\gamma-\alpha-\beta}u$ gives another hypergeometric DE of the form (18) in the variable u with $\alpha_1 = \gamma - \alpha, \beta_1 = \gamma - \beta$,

$\gamma_1 = \gamma$. Since the solution of this DE, which is holomorphic at $z = 0$ and takes the value 1 there, is $F(\gamma - \alpha, \gamma - \beta, \gamma; z)$, we obtain the identity

(30) $$F(\alpha, \beta, \gamma; z) = (1 - z)^{\gamma - \alpha - \beta} F(\gamma - \alpha, \gamma - \beta, y; z)$$

A change of independent variable that transforms the hypergeometric DE into itself is $t = 1 - z$. This gives the DE

$$t(1 - t)w'' + [\gamma_1 - (\alpha - \beta + 1)t]w' - \alpha\beta w = 0$$

where $\gamma_1 = \alpha + \beta - \gamma + 1$. It follows that the hypergeometric DE has a second regular singular point at $z = 1$, and a basis of solutions

$$w_3(z) = F(\alpha, \beta, \alpha + \beta + 1 - \gamma; 1 - z)$$

and

$$w_4(z) = (1 - z)^{\gamma - \alpha - \beta} F(\gamma - \alpha, \gamma - \beta, \gamma - \alpha - \beta + 1; 1 - z)†$$

These functions form a *canonical basis* of solutions relative to the singular point $z = 1$. Note that the functions w_3 and w_4 are equal to linear combinations of the functions w_1 and w_2 by the uniqueness theorem for second-order linear DEs (Ch. 2, Theorem 3).

*11   THE JACOBI POLYNOMIALS

The linear transformation $z \mapsto 1 - 2z$ carries the hypergeometric DE (18) with parameters $\alpha = -n$, $\beta = n + a + b + 1$, $\gamma = a + 1$ into the *Jacobi* DE

(31) $$(1 - z^2)w'' + [a - b - (a + b + 2)z]w' + n(n + a + b + 1)w = 0$$

It carries the regular singular points 0, 1 of (18) into 1 and -1, respectively. Note that the Jacobi DE (31) goes into itself under the transformation $z \mapsto -z$, $a \rightleftharpoons b$.

Multiplying by $(1 - z)^a(1 + z)^b$, we get the self-adjoint form (Ch. 2, §5)

$$\frac{d}{dz}\left[(1 - z)^{a+1}(1 + z)^{b+1}\frac{du}{dz}\right] + n(n + a + b + 1)(1 - z)^a(1 + z)^b u = 0$$

When $a = b$, this reduces to the *ultraspherical* DE

(32) $$\frac{d}{dz}\left[(1 - z^2)^{a+1}\frac{du}{dz}\right] + n(n + 2a + 1)(1 - z^2)^a u = 0$$

---

† Assuming, of course, that the parameters α, β, γ are not chosen in such a way that the solutions coincide: thus $\gamma \neq \alpha + \beta$.

This is obtained from the partial DE $\nabla^2[r^n u(\cos\theta)] = 0$ in $(2a + 3)$-dimensional space by separation of variables; hence, its solutions play an important role in potential theory and its generalizations. Familiar special cases of the ultraspherical DE are $a = b = 0$, which gives the Legendre DE $[(1 - z^2)u']' + n(n + 1)u = 0$ (Ch. 2, §1), and $a = b = -\frac{1}{2}$, which gives the Chebyshev DE

$$(33) \qquad [(1 - z^2)^{1/2} u']' + n^2(1 - z^2)^{-1/2} u = 0$$

From the derivation of (32), it is evident that the hypergeometric functions $F(-n, n + a + b + 1, a + 1, (1 - z)/2)$ are solutions of it. It follows, inspecting the hypergeometric series (20), that if n is a nonnegative integer, this series is a *polynomial* in z unless a is a nonnegative integer $-m$, with $m < n$. Multiplying by the normalizing factor $\binom{n + a}{n}$, we get, by definition, the *Jacobi polynomials*

$$
\begin{aligned}
(34) \qquad P_n^{(a,b)}(z) &= \binom{n + a}{n} F\left(-n, n + a + b + 1, a + 1; \frac{(1 - z)}{2}\right) \\
&= (-1)^n \binom{n + b}{n} F\left(-n, n + a + b + 1; b + 1; \frac{(1 - z)}{2}\right)
\end{aligned}
$$

In turn, when $a = b$, these give the ultraspherical or Gegenbauer polynomials of index $a + \frac{1}{2}$. These are usually normalized by the formula

$$
P_n^{(a)}(z) = \frac{\Gamma(2a + n + 1)}{\Gamma(a + 1/2 + n + 1)} P_n^{(a-1/2,a-1/2)}(z)
$$

With this normalization $P_n^{(0)}(z) = T_n(z)$ is the Chebyshev polynomial of degree n: $P_n^{(1/2)}(z)$ is the Legendre polynomial of degree n, and so on.

From the differentiation formula (27) for the hypergeometric function, we infer the differentiation formula for Jacobi polynomials

$$(35) \qquad \frac{d^m}{dz^m} P_n^{(a,b)}(z) = C \cdot P_{n-m}^{(a+m,b+m)}(z)$$

$$C = 2^{-m}(n + a + b + 1)(n + a + b + 2) \cdots (n + a + b + m)$$

An expression for the Jacobi polynomials often more convenient than (34) is the *Rodrigues formula*

$$(36) \qquad P_n^{(a,b)}(z) = \frac{(-1)^n}{n!2^n} (1 - z)^{-a}(1 + z)^{-b} \frac{d^n}{dz^n} [(1 - z)^{a+n}(1 + z)^{b+n}]$$

We shall derive this formula from the identities for the hypergeometric function established in the preceding section. First, since $(1 - t)^a = F(-a, b, b, t)$, the binomial series is a special case of the hypergeometric series: $(1 - t)^{b+n} = F(a + 1, -n - b, a + 1, t)$. Using also the Jacobi identity, we justify the first two steps of

$$t^{-a}(1 - t)^{-b} \frac{d^n}{dt^n} [t^{a+n}(1 - t)^{b+n}]$$

$$= t^{-a}(1 - t)^{-b} \frac{d^n}{dt^n} [t^{a+n}F(a + 1, -n - b, a + 1; t)]$$

$$= (a + 1)(a + 2) \cdots (a + n)(1 - t)^{-b}F(a + n + 1, -n - b, a + 1; t)$$

$$= (a + 1)(a + 2) \cdots (a + n)F(-n, n + a + b + 1, a + 1; t).$$

In the last step, identity (30) for the hypergeometric function is used. The Rodrigues formula (36) follows by making the change of variable $t = (1 - z)/2$.

EXERCISES E

1. Verify the following identities:
   (a) $F(\alpha, \beta, \beta; z) = (1 - z)^{-\alpha}$
   (b) $F(\frac{1}{2}, \frac{1}{2}, \frac{3}{2}; z^2) = (\arcsin z)/z$
   (c) $F(1, 1, 2; z) = -\log (1 - z)/z$
   (d) $1 + \binom{a}{1} z + \binom{a}{2} z^2 + \cdots + \binom{a}{m} z^m = \binom{a}{m} z^m F(-m, 1, a - m + 1; -z^{-1})$
   (e) $\cos az = F[a/2, -a/2, 1/2; (\sin z)^2]$
   (f) $\log [(1 + z)/(1 - z)] = 2zF(1/2, 1, 3/2; z^2)$

2. (a) Show that (18) is equivalent to

$$\left[\frac{d}{dz} \left(z\frac{d}{dz} + \gamma - 1 \right) - z \left(z\frac{d}{dz} + \alpha \right) \left(z\frac{d}{dz} + \beta \right) \right] w = 0$$

   (b) Show that the eigenvalues of the circuit matrix of the hypergeometric DE at $z = 0$ are equal if γ is an integer.
   (c) Show that the eigenvalues of the circuit matrix for $z = 1$ are equal if $\gamma - \alpha - \beta$ is an integer.

3. (a) Show that, if α is zero or a negative integer, the hypergeometric DE (18) has a polynomial solution unless $\gamma < \alpha$ is a negative integer.
   (b) Using (34), express this solution as a Jacobi polynomial.

4. (a) Compute the characteristic exponents at $z = \pm 1$ of the Legendre DE

$$[(1 - z^2)w']' + \lambda w = 0$$

   *(b) Describe corresponding circuit matrices, taking as basic solutions an even and an odd solution.

5. (a) Show that setting $t = z^2$ in the Legendre DE gives a hypergeometric DE.
   (b) Express the Legendre polynomials as multiples of $F(\alpha, \beta, \gamma; z^2)$ for suitable α, β, γ.

6. Find the characteristic exponents at $z = \pm 1$ of the associated Legendre DE:

$$[(1 - z^2)w']' + [n(n + 1) - m^2/(1 - z^2)]w = 0$$

7. Derive from (31) the self-adjoint form of the Jacobi DE displayed in the text.

*8. Prove that $P_n^{(a,b)}(z) = k_n F\left[-n, -b - n, 1 + a; \dfrac{(z - 1)}{(z + 1)} \right]$, where

$$k_n = \binom{n + a}{n}\left(\frac{z + 1}{2}\right)^n$$

[HINT: Show that the right-hand side satisfies (31), using suitable identities for F.]

9. Find the roots of the indicial equation of the Jacobi DE (31) at $z = 1$ and $z = -1$.

10. Show that (34) defines a solution of (32) even when n is not a positive integer. What happens when n is a negative integer?

11. Using (36), show that, for $a > b > -1$

$$\int_{-1}^{1} P_m^{(a,b)}(x)P_n^{(a,b)}(x)(1 - x)^a(1 + x)^b \, dx = 0, \qquad \text{for } m \neq n$$

*12 SINGULAR POINTS AT INFINITY

Even when the coefficient-functions p and q of the second-order linear DE (5) are regular at infinity, the point at infinity may be neither a removable singularity nor a regular singular point, but an *irregular singular point*. For instance, this is true of $w'' = w$, whose solutions $e^{\pm z}$ have essential singularities at infinity.

One determines when the point at infinity is a regular singular point by making the substitution $z = 1/t$. This substitution transforms the second-order linear DE (5) into the DE

$$(37) \qquad \frac{d^2v}{dt^2} + \left[\frac{2}{t} - \frac{1}{t^2} p\left(\frac{1}{t}\right) \right] \frac{dv}{dt} + \frac{1}{t^4} q\left(\frac{1}{t}\right)v(t) = 0$$

where $v(t) = w(1/t)$. The point at infinity is said to be a *regular* singular point of the DE (5) when the origin is a regular singular point for the DE (37). This happens when the function

$$\left[\frac{2}{t} - \left(\frac{1}{t^2}\right)p\left(\frac{1}{t}\right) \right]$$

has, at worst, a pole of the first order at $t = 0$, that is, when the first coefficient in the power series expansion of $p(1/t)$ vanishes. Also, the function $t^{-4}q(1/t)$ must have, at most, a pole of the second order at $t = 0$; this happens when the first two coefficients in the power series for $q(1/t)$ vanish. This gives the following theorem.

THEOREM 9.   *The point at infinity is a regular singular point for the second-order linear DE (5) if and only if the coefficients p and q have power series expansions, convergent for sufficiently large $|z|$, of the form*

$$(38) \qquad p(z) = \frac{p_1}{z} + \frac{p_2}{z^2} + \cdots, \qquad q(z) = \frac{q_2}{z^2} + \frac{q_3}{z^3} + \cdots$$

That is, it is necessary and sufficient that the function p have a zero of at least the first order and the function q have a zero of at least the second order at infinity. In particular, the solutions of the DE are holomorphic at $z = \infty$, or $t = 0$, if and only if the coefficients

$$\left[\frac{2}{t} - \left(\frac{1}{t^2} \right) p \left(\frac{1}{t} \right) \right] \quad \text{and} \quad \left(\frac{1}{t^4} \right) q \left(\frac{1}{t} \right)$$

are regular at $t = 0$. Hence, the following corollary.

COROLLARY.   *If the coefficients $p(z)$ and $q(z)$ of the DE (5) are holomorphic for sufficiently large z, then all solutions of (5) have removable singularities at $z = \infty$ if and only if $p_1 = 2$ and $q_2 = q_3 = 0$ in (38).*

It follows from Theorem 7 that, if $z = \infty$ is a regular singular point, and if the indicial equation of (37) at $t = 0$ has roots α and β not differing by an integer, then the DE (5) has a basis of solutions of the form

$$w_j(z) = z^{-v} \left(1 + \frac{a_1}{z} + \frac{a_2}{z^2} + \cdots \right), \qquad v = \alpha, \beta$$

The indicial equation at infinity is defined, because of Theorem 9, to be the following equation for v:

$$(39) \qquad v(v - 1) + (2 - p_1)v + q_2 = 0$$

Its roots are called the *characteristic exponents* at $z = \infty$. If they differ by an integer, then there is still a solution of the form $(1/z^v)(1 + (a_1/z) + \cdots)$, but every second linearly independent solution may contain a logarithmic term.

Example 5.   The hypergeometric DE (18) has, by Theorem 9, a regular singular point at infinity with characteristic exponents α and β. In order to derive a canonical basis at infinity, it is convenient to make the substitution $u(t) = t^{-\alpha} w(1/t)$. This transforms the DE into another hypergeometric DE

$$t(1 - t)u'' + [\gamma_2 - (\alpha_2 + \beta_2 + 1)t]u' - \alpha_2 \beta_2 u = 0$$

with $\alpha_2 = \alpha$, $\beta_2 = \alpha - \gamma + 1$, $\gamma_2 = \alpha - \beta + 1$. It follows that the hypergeometric DE has the solution

$$w_5(z) = z^{-\alpha} F(\alpha, \alpha - \gamma + 1, \alpha - \beta + 1; 1/z)$$

convergent when $|z| > 1$. From the symmetry between α and β, we obtain a second solution

$$w_6(z) = z^{-\beta} F(\beta, \beta - \gamma + 1, \beta - \alpha + 1; 1/z)$$

The functions w_5 and w_6 form a canonical basis at infinity, unless $\alpha = \beta$.

*13   FUCHSIAN EQUATIONS

A homogeneous linear DE with single-valued analytic coefficients is called a *Fuchsian* DE when it has, at worst, regular singular points in the extended complex plane, including the point at infinity. Since functions whose only singular points are poles necessarily are rational functions,[†] it follows that the coefficients of any Fuchsian DE are rational functions. The most general first-order Fuchsian DE has the form (see Ex. F8)

$$w' + \left(\sum_{k=1}^{n} \frac{A_k}{z - z_k} \right) w = 0$$

The general solution of this DE is the elementary function

$$w(z) = c \prod_{k=1}^{n} (z - z_k)^{-A_k}$$

Second-order Fuchsian DEs offer much more variety; they are classified according to the number of their singular points. When the number of these is small, their study is greatly simplified by making *linear fractional transformations*[‡] of the independent variable, of the form

$$\zeta = \frac{(az + b)}{(cz + d)}, \qquad ad \neq bc$$

Any such transformation can be obtained by successive changes of variable of the forms $\zeta = z + k$, $\zeta = az$, and $\zeta = 1/z$. Each such change of variable shifts the position of the singular points of a DE, carrying branch poles of solutions into branch poles. Therefore, by Theorems 6, 7, and 8, a general linear frac-

---

† Hille, p. 217, Theorem 8.5.1.

‡ See Hille, pp. 46–50, or Ahlfors, pp. 76–89.

tional transformation transforms regular singular points into regular singular points, and the indicial equations of the transformed DE coincide with those of the original DE at corresponding points.

We first consider second-order Fuchsian DEs having at most two singular points, say at $z = z_1$ and $z = z_2$. By a linear fractional transformation of the form $\zeta = (z - z_1)/(z - z_2)$, we can send these singular points to zero and infinity. It follows from the definition of a regular singular point and from Theorem 9 that the Laurent series of $p(z)$ and $q(z)$ reduce to p_1/z and q_2/z^2, respectively. Hence, the most general Fuchsian DE of the second order with two regular singular points is equivalent to the *Euler* DE of Example 2,

$$w'' + \frac{p_1}{z} w' + \frac{q_2}{z^2} w = 0$$

after a linear fractional transformation.

The simplest Fuchsian DE of the second order whose solutions do not reduce to elementary functions is, therefore, one having three regular singular points. By a linear fractional transformation of the independent variable, we may put these singular points at $0, 1, \infty$. From the definition of a regular singular point and from Theorem 9 of §12, we can determine the coefficient-functions of a second-order Fuchsian DE with three regular singular points at $0, 1, \infty$, as follows. The coefficient $p(z)$ must have, at worst, poles of the first order at $z = 0$ and $z = 1$. It can therefore be written in the form

$$p(z) = \frac{A_1}{z} + \frac{B_1}{z - 1} + p_1(z)$$

where the function $p_1(z)$ is regular throughout the plane. However, by Theorem 9, the function $zp(z)$ has a finite limit as $|z|$ tends to infinity. Since the function $z[A_1/z) + (B_1/(z - 1))]$ is bounded as $|z|$ tends to infinity, it follows that $zp_1(z)$ is uniformly bounded. By Liouville's Theorem† it must, therefore, vanish identically.

Similarly, the coefficient $q(z)$ has at worst poles of the second order at $z = 0$ and $z = 1$, and can therefore be written in the form

$$q(z) = \frac{A_2}{z^2} + \frac{A_3}{z} + \frac{B_2}{(z - 1)^2} + \frac{B_3}{z - 1} + q_1(z)$$

where the function $q_1(z)$ is holomorphic in the finite complex plane. By Theorem 9, the function $z^2 q(z)$ remains bounded as $|z|$ tends to infinity; hence, so does the function

$$z^2 \left(\frac{A_3}{z} + \frac{B_3}{z - 1} + q_1(z) \right) = z^2 \left(\frac{(A_3 + B_3)z - A_3 + q_1(z)z(z - 1)}{z(z - 1)} \right)$$

---

† Hille, p. 204, Theorem 8.2.2.

Therefore, $A_3 = -B_3$ and, again by Liouville's Theorem, the function $q_1(z)$ vanishes identically. This completes the proof of the following theorem.

THEOREM 10.   *Any second-order Fuchsian DE with three regular singular points can be transformed by a linear fractional transformation into the form*

$$(40) \qquad w'' + \left(\frac{A_1}{z} + \frac{B_1}{z-1}\right)w' + \left[\frac{A_2}{z^2} + \frac{B_2}{(z-1)^2} - \frac{A_3}{z(z-1)}\right]w = 0$$

where the A_1 and B_1 are constants.

The differential equation (40) is called the *Riemann* DE; it evidently depends on five parameters.

With the Riemann DE are associated three pairs of characteristic exponents (λ_1, λ_2), (μ_1, μ_2), (ν_1, ν_2), belonging to the singular points 0, 1, ∞, respectively. These exponents are the roots of the indicial equations [cf. (16) and (39)]

$$\lambda(\lambda - 1) + A_1\lambda + A_2 = 0, \qquad \mu(\mu - 1) + B_1\mu + B_2 = 0$$
$$\nu^2 + (1 - A_1 - B_1)\nu + A_2 + B_2 - A_3 = 0$$

By means of these equations, we can express the parameters in the DE (40) in terms of the (characteristic) exponents:

$$A_1 = 1 - \lambda_1 - \lambda_2 \qquad\qquad\qquad A_2 = \lambda_1\lambda_2$$

$$B_1 = 1 - \mu_1 - \mu_2 \qquad\qquad\qquad B_2 = \mu_1\mu_2$$

$$A_1 + B_1 = \nu_1 + \nu_2 + 1 \qquad A_2 + B_2 - A_3 = \nu_1\nu_2$$

From the identities in the first column, we obtain the *Riemann identity*

$$(41) \qquad\qquad \lambda_1 + \lambda_2 + \mu_1 + \mu_2 + \nu_1 + \nu_2 = 1$$

Substituting into (40) we find the Riemann DE

$$(42) \qquad
\begin{aligned}
w'' &+ \left(\frac{1 - \lambda_1 - \lambda_2}{z} + \frac{1 - \mu_1 - \mu_2}{z-1}\right)w' \\
&+ \left(\frac{\lambda_1\lambda_2}{z^2} + \frac{\mu_1\mu_2}{(z-1)^2} + \frac{\nu_1\nu_2 - \lambda_1\lambda_2 - \mu_1\mu_2}{z(z-1)}\right)w = 0
\end{aligned}$$

The preceding discussion shows that the Riemann DE (40) is completely determined by the values of the exponents and the location of the singular points.

THEOREM 11.   *A Fuchsian DE of the second order with three regular singular points in the extended complex plane is uniquely determined by prescribing the two exponents at each singular point. The exponents satisfy Riemann's identity (41).*

The *hypergeometric* DE of §6 is a special case of the Riemann DE with three singular points at 0, 1, ∞. As shown in §6 and in §12, the hypergeometric DE has three regular points at 0, 1, ∞ with exponents $0, 1 - \gamma; 0, \gamma - \alpha - \beta; \alpha, \beta$ respectively.

From Theorems 6–8 and from the fact that the Riemann DE is the *unique* DE satisfying the conditions of Theorem 11, several identities for the solutions can be derived. If we make the change of dependent variable $v(z) = z^\lambda w(z)$, the function $v(z)$ has a branch pole at each of the singular points 0, 1, ∞. Therefore (cf. Theorem 9, Corollary), $v(z)$ satisfies a DE with three regular singular points at 0, 1, ∞. By Theorem 11, this must be the Riemann DE (42). The exponents of this DE are unchanged at $z = 1$, whereas they are changed to $\alpha_i + \lambda$ at $z = 0$ and to $\gamma_i - \lambda$ at infinity. A similar result holds for the more general change of dependent variable

$$v(z) = z^\lambda (z - 1)^\mu w(z)$$

Using these identities, we can prove the following fundamental

THEOREM 12. *Every Riemann* DE (40) *can be reduced to the hypergeometric* DE (18) *by a change of dependent variable of the form* $w = z^\lambda (1 - z)^\mu v(z)$.

COROLLARY. *Every second-order Fuchsian* DE *with three regular singular points can be reduced to the hypergeometric* DE *by changes of independent and dependent variable.*

Proof. The general solution $w(z)$ of the Riemann DE can be written in the form

$$w(z) = z^{\lambda_1}(1 - z)^{\mu_1} v(z)$$

where v is the general solution of a Riemann DE with exponents $0, \lambda_2 - \lambda_1$; $0, \mu_2 - \mu_1; \nu_1 + \lambda_1 + \mu_1, \nu_2 + \lambda_1 + \mu_1$. Thus, the function v is a solution of a hypergeometric DE with $\alpha = \nu_1 + \lambda_1 + \mu_1$, $\beta = \nu_2 + \lambda_1 + \mu_1$, and $\gamma = 1 - \lambda_2 + \lambda_1$, q.e.d.

EXERCISES F

1. Show that the only second-order linear DE that has just two regular singular points, at 0 and ∞, is the Euler DE $w'' + (p_0/z)w' + (q_0/z^2)w = 0$.

2. Show that no analytic linear DE (5) can have only removable singularities, if the point $z = \infty$ is included.

3. Prove in detail that any linear fractional transformation carries regular singular points into regular singular points.

4. If p and q are constant in (5), is the singular point at ∞ regular? Justify your statement.

5. Show that, unless $B = A^2/4$, the DE $(z^2 + Az + B)w'' + (Cz + D)w' + Ew = 0$ can be reduced to the hypergeometric equation by a linear substitution $z = a\zeta + b$.

6. Show that the Bessel DE has an irregular singular point at $z = \infty$.

7. Find necessary and sufficient conditions on $p(z)$ for $w' + p(z)w = 0$ to have (a) a removable singularity, and (b) a regular singular point at ∞.

8. Show that the most general first-order linear DE with $n + 1$ distinct regular singular points at $z_1, \ldots, z_n$ and ∞ is

$$w' + \left[\sum_{k=1}^{n} \frac{q_k}{(z - z_k)} \right] w = 0$$

Integrate this DE explicitly.

*9. Find the most general second-order linear DE (5) having regular singular points at $a_1, \ldots, a_n$ and ∞.

*10. Find the most general linear DE having regular singular points at $0, \infty$ and no other singular points. Show that any such DE can be integrated in terms of elementary functions.

ADDITIONAL EXERCISES

1. Show that $\int_0^{\pi/2} d\theta(1 - k^2 \sin^2 \theta)^{1/2} = (\pi/2)F(\frac{1}{2}, \frac{1}{2}, 1; k^2)$.

2. Show that the substitution $z = \zeta^m$ (m a nonzero integer) transforms DEs (5) having a regular singular point at $z = 0$ into DEs having a regular singular point at $\zeta = 0$, with the characteristic exponents multiplied by m.

3. Show that the DE $w'' + [(1 - z^2)/4z^2]w = 0$ has a basis of solutions of the form $w_1(z) = z^{1/2}[1 + z^2/16 + z^4/1024 + \cdots]$ and $w_2(z) = w_1 \log z - z^{3/2}/16 + \cdots$.

4. Find an entire function $f(z)$ and constant c for which the functions

$$w_i = f(z)^{1/2} \exp \pm \left\{ c \int_c^z \frac{dz}{[f(z)\sqrt{z(1 - z)}]} \right\}$$

are a basis of solutions of $z(1 - z)w'' + (1 - 2z)w'/2 + (az + b)w = 0$.

5. The algebraic form of the Mathieu equation is

$$4\xi(1 - \xi)u_{\xi\xi} + 2(1 - 2\xi)u_\xi + (\lambda - 16k + 32k\xi)u = 0$$

Show that this has a regular singular point at $\xi = 0$, calculate the exponents, and find a recurrence relation on the coefficients of the power series solutions.

*6. (a) If P and Q are given polynomials without common factors and if $c_{n+1}/c_n = P(n)/Q(n)$ and $\Sigma \, c_n z^n$ is convergent, show that the function $\Sigma \, c_n z^n$ satisfies the DE $zP(zd/dz) \, w - Q(zd/dz)w = 0$.

(b) Find all quadratic polynomials P and Q for which the preceding DE has regular singular points only, and express the solutions in terms of hypergeometric functions.

*7. (a) Find the eigenvalues of the circuit matrix of (18) for $z = 0$.

(b) Using the change of variable $t = 1 - z$, solve the same problem for the circuit matrix for $z = 1$.

8. Show that the function $\ln (\ln z)$ satisfies no linear DE of finite order with holomorphic coefficients.

9. Show that, if $f(0) = 0$ but $f'(0) \neq 0$, the substitution $z = f(\zeta)$ carries a regular singular point of (5) at $z = 0$ into one at $\zeta = 0$ having the same indicial equation.

10. Show that, for any nontrivial solution of the Euler DE $z^2 w'' + zw' + w = 0$ and any integer n, there exists a spiral path $\theta = h(r)$ approaching the origin, along which $\lim_{z \to 0} |z^n w| = \infty$.

*11. Let the DE $w'' + p_1(z)w' + p_2(z)w = 0$ have an isolated singular point at $z = \infty$. Show that this singular point is regular if and only if, for some $n > 0$, every solution satisfies $\lim_{z \to \infty} z^{-n} w(re^{i\theta}) = 0$ for $0 \leq \theta \leq 2\pi$.

CHAPTER 10
STURM-LIOUVILLE SYSTEMS

1 STURM-LIOUVILLE SYSTEMS

A Sturm-Liouville equation is a second-order homogeneous linear DE of the form

$$(1) \qquad \frac{d}{dx}\left[p(x)\,\frac{du}{dx} \right] + [\lambda\rho(x) - q(x)]u = 0$$

Here λ is a parameter, while p, ρ, and q are real-valued functions of x; the functions p and ρ are positive. In operational notation, with $L = D[p(x)D] - q(x)$, we can write (1) in the abbreviated form

$$(1') \qquad L[u] + \lambda\rho(x)u = 0$$

Such a DE (1) is *self-adjoint* for real λ; to ensure the existence of solutions, the functions q and ρ are assumed to be continuous and p to be continuously differentiable (of class $\mathcal{C}^1$). For a given value of λ, (1) defines a *linear operator* transforming any function $u \in \mathcal{C}^2$ into $L[u] + \lambda\rho u$. The Sturm-Liouville equation (1) is called *regular* in a *closed* finite interval $a \le x \le b$ when the functions $p(x)$ and $\rho(x)$ are positive for $a \le x \le b$. The functions p, q, and ρ, being continuous, are bounded in the interval.

For each λ, it follows from the existence theorem of Ch. 6, §8, that a regular Sturm-Liouville equation for $a \le x \le b$ has a basis of two linearly independent solutions of class $\mathcal{C}^2$.

A *Sturm-Liouville system* (or S-L system) is a Sturm-Liouville equation together with *endpoint* (or boundary) *conditions* to be satisfied by the solutions, for example $u(a) = u(b) = 0$. One type of endpoint condition we shall study is the following.

DEFINITION. A *regular* S-L *system* is a regular S-L equation (1) on a finite closed interval $a \le x \le b$, together with two *separated* endpoint conditions, of the form

$$(2) \qquad \alpha u(a) + \alpha' u'(a) = 0, \qquad \beta u(b) + \beta' u'(b) = 0$$

300

$P_n(x)$ are real eigenfunctions of this S-L system belonging to the eigenvalues $\lambda_n = n(n + 1)$.

Example 6. For fixed n, the Bessel equation of Example 2,

$$\frac{d}{dr}\left[r\frac{du}{dr}\right] + \left(k^2 r - \frac{n^2}{r}\right)u = 0, \qquad 0 < r \leq a$$

is a singular S-L equation with $p = \rho = r$, $\lambda = k^2$, and $q = n^2/r$. A singular S-L system is obtained for any $a > 0$ by imposing the endpoint conditions $u(a) = 0$, and $u(r)$ bounded as $r \to 0$.

The eigenfunctions of the preceding singular S-L systems are the Bessel functions $J_n(k_j r)$, where $k_j a$ is the jth zero of the Bessel function $J_n(x)$ of order n. It has been shown in Ch. 2, §6, that $J_n(x)$ has infinitely many zeros; it follows that the singular S-L system just defined has infinitely many eigenvalues.

The eigenfunctions of singular S-L systems are also orthogonal, provided that they are *square-integrable* relative to the weight function ρ, in the following sense.

DEFINITION. A real-valued function f is *square-integrable* on the interval I relative to a given weight function $\rho(x) > 0$ when

(10)
$$\int_I f^2(x)\rho(x)\, dx < +\infty$$

When the weight function ρ is identically equal to 1, we say simply that the function f is *square-integrable* on the interval I.

The *Schwarz inequality* holds† for square-integrable functions:

(11)
$$\left(\int_I |f(x)g(x)|\rho(x)\, dx\right)^2 \leq \int_I f^2(x)\rho(x)\, dx \int_I g^2(x)\rho(x)\, dx$$

This inequality implies in particular that the product of two such square-integrable functions is an integrable function relative to the weight function ρ, that is, the integral in parentheses on the left-hand of (11) is finite.

The right side of the boundary term in (8) vanishes in the limit, for any endpoint conditions that imply that

(12)
$$\lim_{\alpha \downarrow a, \beta \uparrow b} p(x[u(x)v'(x) - v(x)u'(x)]\Big|_{x=\alpha}^{\beta} = 0$$

---

† Birkhoff and MacLane, p. 201; see also Apostol, Vol. 2, p. 16.

The conditions $p(a) = p(b) = 0$ and $u'(x)$ bounded on the interval $[a, b]$ imply this property, for example.

When (12) holds, we obtain from (8) the identity

$$(\lambda - \mu) \int_a^b \rho(x)u(x)v(x) \, dx = 0$$

for any two square-integrable eigenfunctions u and v with eigenvalues λ and μ. The integral here may be an improper integral. If $\lambda \neq \mu$, this implies, as in the proof of the lemma of §2, that u and v are orthogonal. This proves the next theorem.

THEOREM 2. *Square-integrable eigenfunctions u and v belonging to different eigenvalues of a singular S-L system are orthogonal with weight ρ whenever the boundary term vanishes, as in* (12).

Applying this result to Example 5, we obtain the orthogonality relation for the Legendre polynomials

$$(13) \qquad \int_{-1}^1 P_m(x)P_n(x) \, dx = 0, \qquad m \neq n$$

after verifying that the boundary term vanishes. Applying it to the Bessel equation, we obtain the orthogonality relations for Bessel functions

$$(14) \qquad \int_0^a xJ_n(k_ix)J_n(k_jx) \, dx = 0, \qquad k_i \neq k_j$$

if $J_n(k_ia) = J_n(k_ja) = 0$.

Example 7.   The *Hermite* DE is

$$(15) \qquad u'' - 2xu' + \lambda u = 0, \qquad -\infty < x < +\infty$$

as in Ch. 4, §2. Using the recursion formula

$$(16) \qquad a_{k+2} = \frac{(2k - \lambda)a_k}{(k + 1)(k + 2)}, \qquad k = 0, 1, 2, \ldots$$

of Ch. 4, (9′), we obtain a polynomial solution of degree n for $\lambda = 2n$. These polynomials are commonly normalized by the condition that $a_n = 2^n$ (and $a_{n-1} = 0$); this defines the *Hermite polynomials* $H_n(x)$. For example, $H_0(x) = 1$, $H_1(x) = 2x$, $H_2(x) = 4x^2 - 2$, etc. Evidently, $H_n(x)$ is an even function for even n, and an odd function for odd n.

The Hermite DE is not an S-L equation, because it is not self-adjoint. Making the substitution $y = e^{-x^2/2}u$ in (15), we obtain the following equivalent self-adjoint S-L equation for the *Hermite functions* $y(x)$:

(17) $$y'' + [\lambda - (x^2 - 1)]y = 0, \qquad -\infty < x < \infty$$

among whose solutions for $\lambda = 2n$ are the functions $e^{-x^2/2}H_n(x)$; these functions are square-integrable and tend to zero as $x \to \pm\infty$.

That is, the functions $\phi_n(x) = e^{-x^2/2}H_n(x)$ are eigenfunctions for the singular S-L system defined by (17) and by the endpoint condition that a solution $y(x)$ must tend to zero as $x \to \pm\infty$. We shall now derive the orthogonality relations for the Hermite polynomials

(18) $$\int_{-\infty}^{\infty} H_m(x)H_n(x)e^{-x^2}\, dx = 0, \qquad m \neq n$$

For, substituting into the identity (8), we obtain

$$2(m - n) \int_{-a}^{a} H_m(x)H_n(x)e^{-x^2}\, dx = [\phi_m(x)\phi_n'(x) - \phi_m'(x)\phi_n(x)]_{x=-a}^{x=a}$$

Since the boundary terms are e^{-x^2} times a polynomial in x, and

(19) $$\lim_{x \to +\infty} x^n e^{-x^2} = 0$$

for all n, the boundary term vanishes in the limit, as in (12).

EXERCISES C

1. (a) Prove the orthogonality relations for the Bessel functions

    (*) $$\int_{0}^{L} xJ_n(\alpha x)J_n(\beta x)\, dx = 0 \qquad \text{if} \qquad J_n'(\alpha L) = J_n'(\beta L) = 0, \qquad \alpha^2 \neq \beta^2$$

    for any nonnegative integer n.
   (b) Prove the equality (*) if $\alpha J_n'(\alpha L)J_n(\beta L) = \beta J_n'(\beta L)J_n(\alpha L)$.

2. Show that the Legendre polynomials (and their constant multiples) are the only solutions of the Legendre DE that are bounded on $(-1, 1)$.

3. Show that the S-L system $[(x - a)(b - x)u']' + \lambda u = 0$, $a < b$, with $u(x)$ bounded on $a < x < b$, has the eigenvalues $\lambda = 4n(n + 1)(b - a)^2$. Describe the eigenfunctions.

4. (a) (Laguerre polynomials). Consider the singular S-L system

    $$(xe^{-x}u')' + \lambda e^{-x}u = 0 \qquad \text{on} \qquad 0 < x < +\infty$$

    with endpoint conditions that $u(0^+)$ is bounded and that $e^{-x}u(x) \to 0$ as $x \to +\infty$. Show that the values $\lambda = n$ give polynomial eigenfunctions.
   (b) Show that the preceding system has no other polynomial eigenfunctions. [HINT: Obtain the power series expansion of the general solution of the DE.]

5. Show that the eigenvalues of the singular S-L system defined by

$$\frac{d}{dx}\left[(1 - x^2)^{a+1}\frac{du}{dx}\right] + \lambda(1 - x^2)^a u = 0, \qquad a > -1$$

and the condition of being bounded on $(-1, 1)$, are $\lambda_n = n(n + 2\alpha)$.

5 PRÜFER SUBSTITUTION

We now develop a powerful method for the study of the solutions of a self-adjoint second-order linear DE

$$(20) \qquad \frac{d}{dx}\left[P(x)\frac{du}{dx}\right] + Q(x)u = 0, \qquad a < x < b$$

where $P(x) > 0$ is of class $\mathcal{C}^1$ and Q is continuous. One may want to find out how often the solutions of (20) *oscillate* on the interval under consideration, that is, the number of zeros they have for $a < x < b$. This can be done by using the Poincaré phase plane, already introduced in Ch. 2, §7. Modifying slightly the formulas used there, we make in (20) the *Prüfer substitution*

$$(21) \qquad P(x)u'(x) = r(x)\cos\theta(x); \qquad u(x) = r(x)\sin\theta(x)$$

The new dependent variables r and θ are defined by the formulas

$$(21') \qquad r^2 = u^2 + P^2 u'^2, \qquad \theta = \arctan(u/Pu')$$

r is called the *amplitude* and θ the *phase* variable. When $r \neq 0$, the correspondences $(Pu', u) \rightleftharpoons (r, \theta)$ defined by (21) are *analytic* with nonvanishing Jacobian.

For nontrivial solutions, r is always positive because, if $u(x) = u'(x) = 0$ for a given x, by the Uniqueness Theorem of Ch. 2, §4, u would be the trivial solution $u \equiv 0$.

We now derive an equivalent system of DEs for $r(x)$ and $\theta(x)$. Differentiating the relation† $\cot\theta = Pu'/u$, we get

$$-\csc^2\theta\,\frac{d\theta}{dx} = \frac{(Pu')'}{u} - \frac{Pu'^2}{u^2} = -Q(x) - \frac{1}{P}\cot^2\theta$$

If we multiply through by $-\sin^2\theta$, this expression becomes

$$(22) \qquad \frac{d\theta}{dx} = Q(x)\sin^2\theta + \frac{1}{P(x)}\cos^2\theta = F(x, \theta)$$

---

† When $\theta \equiv 0 \pmod{\pi}$, the relation is not defined. But the final equations (22)–(23) can still be derived by differentiating the relation $\tan\theta = u/Pu'$.

Differentiating $r^2 = (Pu')^2 + u^2$ and simplifying, we obtain

(23) $$\frac{dr}{dx} = \left[\frac{1}{P(x)} - Q(x)\right] r \sin\theta \cos\theta = \frac{1}{2}\left[\frac{1}{P(x)} - Q(x)\right] r \sin 2\theta$$

The system (22)–(23) is equivalent to the DE (20) in the sense that every non-trivial solution of the system defines a unique solution of the DE by the Prüfer substitution (21), and conversely. This system is called the *Prüfer system* associated with the self-adjoint DE (20).

The DE (22) of the Prüfer system is a first-order DE in θ, x alone, not containing the other dependent variable r, and it satisfies a Lipschitz condition with Lipschitz constant

$$L = \sup_{a<x<b}\left|\frac{\partial F}{\partial\theta}\right| \leq \sup_{a<x<b} |Q(x)| + \sup_{a<x<b} \frac{1}{|P(x)|}$$

The constant L is finite in any closed interval in which Q and P are continuous. Hence, the existence and uniqueness theorems of Ch. 6 are applicable, and show that the DE (22) has a unique solution $\theta(x)$ for any initial value $\theta(a) = \gamma$, provided P and Q are continuous at a.

With $\theta(x)$ known, $r(x)$ is given by (23) after a quadrature:

(23′) $$r = K \exp\left\{\frac{1}{2}\int_a^x\left[\frac{1}{P(t)} - Q(t)\right]\sin 2\theta\, dt\right\}$$

where $K = r(a)$. Each solution of the Prüfer system (22)–(23) depends on two constants: the initial *amplitude* $K = r(a)$ and the initial *phase* $\gamma = \theta(a)$. Changing the constant K just multiplies a solution $u(x)$ by a constant factor; thus, the zeros of any solution u of (20) can be located by studying only the DE (22).

6 THE STURM COMPARISON THEOREM

The *zeros* of any solution $u(x)$ of the DE (20) occur where the phase function $\theta(x)$ in the Prüfer substitution (21) assumes the values, 0, $\pm\pi$, $\pm 2\pi$, ... , that is, at all points x where $\sin\theta(x) = 0$. At each of these points $\cos^2\theta = 1$ and $d\theta/dx$ is positive, by (22) [recall that $P(x) > 0$]. Geometrically, this means that the curve $(P(x)u'(x), u(x))$ in the (Pu', u)-plane, corresponding to a solution u of the DE, can cross the Pu'-axis $\theta = n\pi$ only counterclockwise.

Now compare the DE (22) with a DE of the same form, $d\theta/dx = F_1(x, \theta)$, having coefficients $Q_1(x) \geq Q(x)$ and $P_1(x) \leq P(x)$:

$$\frac{d\theta}{dx} = Q_1(x)\sin^2\theta + \frac{1}{P_1(x)}\cos^2\theta = F_1(x, \theta)$$

If $Q_1(x) \geq Q(x)$ and $P_1(x) \leq P(x)$ in an interval I, then $F_1(x, \theta) \geq F(x, \theta)$ there. By the Comparison Theorem of Ch. 1, §11, we conclude that, if $\theta_1(x)$ is a solution of the second DE whose initial value satisfies $\theta_1(a) \geq \theta(a)$, and $\theta(x)$ is a solution of (22), then $\theta_1(x) \geq \theta(x)$ for $a \leq x \leq b$. Furthermore, we have $\theta_1(b) = \theta(b)$ only if $\theta(x) \equiv \theta_1(x)$, which implies that $u(x) \equiv cu_1(x)$, whence $F(x, \theta(x)) \equiv F_1(x, \theta_1(px))$. This implies that $Q(x) \equiv Q_1(x)$ since $d\theta/dx = 1/P_1(x) > 0$, where $\sin \theta = 0$; therefore, $\sin \theta$ can vanish only at isolated points. It also implies that $P(x) \equiv P_1(x)$, except in intervals where $\cos \theta \equiv 0$, and so $Q(x) \equiv Q_1(x) \equiv 0$ (cf. Ch. 1, §12, Corollary 1). Therefore, if $\sin \theta(a) = 0$, the number of zeros of $\sin \theta_1(x)$ for $a < x < b$ is at least the number of zeros of $\sin \theta(x)$, except when $P \equiv P_1$ and $Q \equiv Q_1$, when it is equal, and when $Q \equiv Q_1 \equiv 0$ in an interval, when it may be equal. This completes the proof of the following theorem.

THEOREM 3 (STURM COMPARISON THEOREM).   *Let $P(x) \geq P_1(x) > 0$ and $Q_1(x) \geq Q(x)$ in the* DEs

$$(24) \qquad \frac{d}{dx}\left(P(x)\frac{du}{dx}\right) + Q(x)u = 0, \qquad \frac{d}{dx}\left(P_1(x)\frac{du_1}{dx}\right) + Q_1(x)u_1 = 0$$

Then, between any two zeros of a nontrivial solution $u(x)$ of the first DE, *there lies at least one zero of every real solution of the second* DE, *except when $u(x) \equiv cu_1(x)$. This implies $P \equiv P_1$ and $Q \equiv Q_1$, except possibly in intervals where $Q \equiv Q_1 \equiv 0$.*

In the case of S-L equations, since $\rho(x) > 0$, $Q(x) \equiv Q_1(x)$ evidently implies that $\lambda = \lambda_1$.

Sturm's Separation Theorem of Ch. 2, §6 follows as a corollary, by comparing two linearly independent solutions of the same DE.

A short and easily remembered, if somewhat imprecise, summary is this: as Q increases and P decreases, the number of zeros of every solution increases.

Maxima and Minima.   For the self-adjoint DE (20), the inequality $Q(x) > 0$ implies that

$$d\theta/dx > 0 \qquad \text{if} \qquad \theta = (n + \tfrac{1}{2})\pi$$

For, in (22), $\cos \theta = 0$ and $|\sin \theta| = 1$, if $\theta = (n + \tfrac{1}{2})\pi$. Since $\cos \theta = 0$ if and only if $u' = 0$, it follows that, if $Q(x)$ is positive, any nontrivial solution of (20) has exactly one maximum or minimum between successive zeros.

7   STURM OSCILLATION THEOREM

We now consider the variation with λ in the number of zeros of the eigenfunctions of a regular S-L system (1) to (2). Setting $P(x) = p(x)$ and $Q(x) = \lambda\rho(x) - q(x)$ in (1), we obtain (20). Since $u = 0$ if and only if $\sin \theta = 0$ in (21), the zeros of any solution of (1) are the points where $\theta = 0, \pm\pi, \pm 2\pi, \ldots,$ $\pm n\pi, \ldots, \theta$ being a solution of the associated Prüfer equation

$$(25) \qquad \frac{d\theta}{dx} = [\lambda\rho(x) - q(x)] \sin^2 \theta + \frac{1}{p(x)} \cos^2 \theta, \qquad a \leq x \leq b$$

Here $p(x) > 0$, $\rho(x) > 0$ for $a \leq x \leq b$.

We now fix γ, and denote by $\theta(x, \lambda)$ the solution of (25) that satisfies an initial condition $\theta(a, \lambda) = \gamma$ for all λ, where γ is determined by the conditions†

$$(25') \qquad \tan\gamma = \frac{u(a)}{p(a)u'(a)} = \frac{\alpha'}{p(a)\alpha}, \qquad 0 \leq \gamma < \pi$$

The constants α and α' come from the initial condition $\alpha u(a) + \alpha' u'(a) = 0$. For fixed γ, the function $\theta(x, \lambda)$ is defined on the domain $a \leq x \leq b$, $-\infty < \lambda < \infty$; we shall consider its behavior there.

Applying the comparison theorem of Ch. 1, §11 (and especially Corollary 1 there) to (25), we obtain the following lemmas.

LEMMA 1. *For fixed $x > a$, $\theta(x, \lambda)$ is a strictly increasing function of the variable* λ.

LEMMA 2. *Suppose that for some $x_n > a$, $\theta(x_n, \lambda) = n\pi$, where $n \geq 0$ is an integer. Then $\theta(x, \lambda) > n\pi$ for all $x > x_n$.*

Proof. If x_n is any point where $\theta(x, \lambda) = n\pi$, then by the DE (25), we have $d\theta(x_n, \lambda)/dx_n = 1/p(x_n) > 0$. Thus, the function $\theta = \theta(x_n, \lambda)$, considered as a function of x_n, is increasing where it crosses the line $\theta = n\pi$, as shown in Figure 10.1. Hence, $\theta(x, \lambda)$ stays above this line for $x > x_n$, q.e.d.

Lemma 2, combined with the condition $0 \leq \gamma = \theta(a, \lambda) < \pi$, makes the first zero of $u(x)$ in the open interval $a < x < b$ occur where $\theta = \pi$, and the nth zero, where $\theta = n\pi$.

Our next aim is to show that, for fixed $x > a$, $\theta(x, \lambda) \to \infty$ as $\lambda \to \infty$.

In view of Lemma 2, we will have shown that $\lim_{\lambda\to\infty} \theta(x, \lambda) = \infty$ for each x, if we can show that for every integer $n > 0$, we can find a number $x_n(\lambda)$ be the smallest x such that $\theta(x, \lambda) = n\pi$. Then, all we need to show is that $x_\pi < x$ such that $\theta(x_n; \lambda) = n\pi$ for sufficiently large λ_0. Stated in different terms, let $x_n(\lambda)$ exists for large λ and that $\lim_{\lambda\to\infty} x_n(\lambda) = a$. This is done in the following lemma.

LEMMA 3. *For a given fixed positive integer n and sufficiently large λ, the function $x_n(\lambda)$ is defined and continuous. It is a decreasing function of λ, and $\lim_{\lambda\to\infty} x_n(\lambda) = a$.*

---

† We have assumed $\alpha \neq 0$. When $\alpha = 0$, set $\gamma = \pi/2$, $\tan\gamma = \infty$.

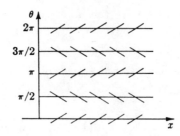

Figure 10.1   Direction field of

$$\frac{d\theta}{dx} = Q(x)\sin^2\theta + \frac{1}{P(x)}\cos^2\theta = F(x,\theta)$$

Proof.   By Theorem 3 of Ch. 6, the function $\theta(x,\lambda)$ is a continuous function of both variables x and λ for $a \le x \le b$ and $-\infty < \lambda < \infty$. We shall first prove that, if the function $x_n(\lambda)$ is well-defined, that is, if $\theta(x,\lambda) = n\pi$ for some x, then $x_n(\lambda)$ is a monotonic decreasing function of λ. To prove this result, it suffices to prove that $\theta(x,\lambda)$ is an increasing function of λ. But this is the conclusion of Lemma 1.

We now show that, for fixed n, the function $x_n(\lambda)$ is well-defined for large enough λ. This amounts to saying that, for large enough λ, there is an x in the inverval $a < x < b$ for which $\theta(x,\lambda) = n\pi$. We can translate this statement into an equivalent statement for the solutions of the DE (1), using (21). It is equivalent to saying that every nontrivial solution of (1) has at least n zeros in the interval $a < x < b$, since $\theta(x,\lambda)$, being a continuous function of x, must take all values between $\theta(a,\lambda) = \gamma < \pi$ and $n\pi$.

Now, let q_M and p_M be the maxima of $q(x)$ and $p(x)$, respectively, and let ρ_m be the minimum of $\rho(x)$ for $a \le x \le b$. A solution of the DE

$$(26) \qquad\qquad p_M u'' + (\lambda\rho_m - q_M)u = 0, \qquad \lambda > \frac{q_M}{\rho_m}$$

is the function $u_1(x) = \sin k(x - a)$, where $k^2 = (\lambda\rho_m - q_M)/p_M$. The successive zeros of this function are spaced at a distance $\pi\sqrt{p_M/(\lambda\rho_m - q_M)}$ apart. By the Sturm Comparison Theorem (Theorem 3 above), any nontrivial solution $u(x)$ of the Sturm-Liouville equation (1) must have at least one zero between any two zeros of the function $u_1(x)$. Since $u_1(x)$ has n zeros on (a,b) when λ is sufficiently large, it follows that $u(x)$ has at least n zeros and, therefore, that $\theta(x,\lambda)$ takes the value $n\pi$ for sufficiently large λ, as we wanted to show.

The number $x_n(\lambda)$ falls between the $(n-1)$st and the nth zero of $u_1(x)$, and both these zeros tend to a as $\lambda \to \infty$. Therefore, we have $x_n(\lambda) \to a$ as $\lambda \to \infty$, q.e.d.

We are now ready to prove the following result.

THEOREM 4 (OSCILLATION THEOREM). *The solution $\theta(x; \lambda)$ of DE (25) satisfying the initial condition $\theta(a, \lambda) = \gamma$, $0 \leq \gamma < \pi$ for each λ, is a continuous and strictly increasing function of λ for fixed x on $a < x \leq b$. Moreover,*

$$(27) \qquad \lim_{\lambda \to \infty} \theta(x; \lambda) = \infty, \qquad \lim_{\lambda \to -\infty} \theta(x; \lambda) = 0$$

for $a < x \leq b$.

The first sentence was proved in Lemmas 1–3. The first formula of (27) was proved in Lemma 1.

We shall now prove the second formula of (27). Choose numbers $\gamma < \gamma_1 < \pi$ and $\epsilon > 0$. The slope of the segment in the $x\theta$-plane joining the points (a, γ_1) and (x_1, ϵ) where $a < x_1 \leq b$, equals $(\epsilon - \gamma_1)/(x_1 - a)$. For a point (x, θ) on this segment, the slope of $\theta(x, \lambda)$, as given by (25), will be less than the slope of the segment for large negative λ. Therefore, the function $\theta(x, \lambda)$ will lie below the segment for $a \leq x \leq x_1$, for all sufficiently large negative λ. We conclude that $\theta(x_1, \lambda) < \epsilon$ for sufficiently large negative λ. Since, by the argument used to prove Lemma 2, $\theta(x_1, \lambda) > 0$, it follows that $|\theta(x_1, \lambda)| < \epsilon$. And since ϵ and x_1 are arbitrary, the proof is complete.

We now derive an estimate for the positions of the zeros of a solution of a regular S-L equation (1), by comparing it with equation (26) and with

$$(28) \qquad p_m u'' + (\lambda \rho_M - q_m) u = 0$$

where p_m and q_m are the minima of $p(x)$ and $q(x)$, and ρ_M the maximum of $\rho(x)$ for $a \leq x \leq b$.

Consider solutions of (26) and (28) for which $u(a)/p(a)u'(a) = \tan \gamma$. The zeros of these solutions can be determined by inspection. They are $a + (n\pi - \gamma)/\sqrt{(\lambda\rho_m - q_M)/\rho_M}$ and $a + (n\pi - \gamma)/\sqrt{(\lambda\rho_M - q_m)\rho_M}$, respectively. Applying the Sturm Comparison Theorem, we obtain the following Corollary.

COROLLARY. *Let x_n be the nth zero of a nontrivial solution of the S-L equation (1). Then*

$$(29) \qquad \sqrt{\frac{p_m}{\lambda\rho_M - q_m}} \leq \frac{x_n - a}{n\pi - \gamma} \leq \sqrt{\frac{p_M}{\lambda\rho_m - q_M}}$$

The preceding results have been proved under the assumption that $\alpha \neq 0$ in (2). If $\alpha = 0$, we can use the same argument when $\beta \neq 0$, by changing the independent variable to $t = a + b - x$. If $\alpha = \beta = 0$, we can still prove the foregoing results with $\gamma = \pi/2$.

EXERCISES D

1. For $u'' + [\lambda - q(x)]u = 0$, with separated endpoint conditions (2), show that all eigenvalues are positive if $q(x) > 0$, $\alpha\alpha' < 0$, and $\beta\beta' > 0$.

2. Show that the number of negative eigenvalues of a regular S-L system is always finite and is at most 1 if $q(x) > 0$.

3. Show that any finite sequence of eigenvalues of a regular S-L system is unbounded.

4. Find all solutions of the DE $\theta' = A \sin^2 \theta + B \cos^2 \theta$, where A and B are positive constants. [HINT: Relate this to a Prüfer system.]

5. Show that, at all points x where a solution $u(x)$ of $(Pu')' + Qu = 0$ has a minimum or a maximum, $d\theta/dx = Q(x)$.

6. Extend the Sturm Oscillation Theorem to the case where $\alpha = \beta = 0$ in (2).

7. Derive Theorem 3 from the Sturm Comparison Theorem of Ch. 2 by introducing the new dependent variables $t = \int_a^x ds/P(s)$ and $t_1 = \int_a^x ds/P_1(s)$.

8. For any solution of $u'' + q(x)u = 0$, $q(x) < 0$, show that the product $u(x)u'(x)$ is an increasing function. Infer that a nontrivial solution can have at most one zero.

9. Show that, if $q(x) < 0$ in $u'' + p(x)u' + q(x)u = 0$, no nontrivial solution of the DE can have more than one zero.

10. (a) Show that $J_n(x)$ is increasing for $0 < x \leq |n|$. [HINT: Use the identity $x(xJ_n')' \equiv (n_2 - x^2)J_n$.]
    (b) Prove that, if x_0 is the first positive zero of J_n and y_0 is that of J_n', then

$$|n| \leq y_0 < x_0$$

*11. (a) Let $u(x)$ be a solution of $(Pu')' + Qu = 0$, where $P > 0$, $P' > 0$, $Q > 0$, and $(P'/Q)' > 1$. Show that the zeros of u, u', u'' follow one another cyclically.
    (b) Infer that the zeros of J_n, J_{n+1}, J_{n+2} follow one another cyclically.

12. (Sturm Convexity Theorem). In $u'' + Q(x)u = 0$, let $Q(x)$ be increasing. Show that $x_n - x_{n-1} < x_{n+1} - x_n$, where $\{x_n\}$ is the sequence of successive zeros of a nontrivial solution u.

13. For the modified Bessel function $I_0(y) = J_0(iy)$, without considering its Taylor series, show that $I_0'(y) > 0$ and $1 < I_0(y) < \cosh y$ for all $y > 0$.

14. Show that, in the Sturm Oscillation Theorem, $\theta(x, \lambda) \to \infty$ as $\lambda \to \infty$, uniformly in any subinterval $a' \leq x \leq b$, $a' > a$.

15. Show that $\theta(x, \lambda) \to 0$ as $\lambda \to -\infty$, uniformly in any subinterval $a' \leq x \leq b$, $a' > a$.

8   THE SEQUENCE OF EIGENFUNCTIONS

The existence of an infinite sequence of eigenfunctions of a regular S-L system that consists of the DE (1), together with the separated endpoint conditions (2), that is, the conditions

$$(30) \qquad A[u] = \alpha u(a) + \alpha' u'(a) = 0, \qquad B[u] = \beta u(b) + \beta' u'(b) = 0$$

will now be proved.

We first transform these endpoint conditions into equivalent endpoint conditions for the phase function $\theta(x, \lambda)$ of the Prüfer system (22)–(23) associated with the DE (1). If $\alpha \neq 0$, then the function $\theta(x, \lambda)$ must satisfy the initial condition $\theta(a, \lambda) = \gamma$, where γ is the smallest positive number $0 \leq \gamma < \pi$ such that $p(a) \tan \gamma = -\alpha'/\alpha$. When $\alpha = 0$, we chose $\gamma = \pi/2$. Similarly, we choose $0 < \delta \leq \pi$ so that $\tan \delta = -\beta'/\beta p(b)$.

A solution $u(x)$ of the DE (1) for $a \leq x \leq b$ is an eigenfunction of the regular S-L problem obtained by imposing the endpoint conditions (3) if and only if, for the corresponding phase function defined by (21'),

$$(31) \qquad \begin{aligned} \theta(a, \lambda) &= \gamma, \qquad \theta(b, \lambda) = \delta + n\pi, \qquad n = 0, 1, 2, \dots \\ 0 &\leq \gamma < \pi, \qquad 0 < \delta \leq \pi \end{aligned}$$

Clearly, any value of λ for which conditions (31) are satisfied is an eigenvalue of the given regular S-L system, and conversely. Let $\theta(x, \lambda)$ be the solution of (25) for the initial condition $\theta(a, \lambda) = \gamma$. Figure 10.2 shows graphs of the function $\theta = \theta(x, \lambda)$ for various values of the parameter λ. The waviness of the lines expresses the fact that $1/P(x)$ in (23) is independent of λ, whereas $Q(x) = \lambda \rho - q$ tends to infinity with λ. As a result, the slope of the graph is $1/p(x)$ for all λ when $\theta \equiv 0 \pmod{\pi}$, although it tends to infinity with λ for all other θ.

Since $\theta(b, \lambda)$ is an increasing function of λ, and $\theta(b, \lambda) > 0$ by Lemma 2 of § 7, as λ increases from $-\infty$, there is a first value λ_0 for which the second of the conditions (31) is satisfied. For this eigenvalue, we have $\theta(b, \lambda_0) = \delta$. As λ increases, there is an infinite sequence of λ_n for which the second boundary condition is satisfied, namely those for which $\theta(b, \lambda_n) = \delta + n\pi$, for some nonnegative integer n. Each of these values gives an eigenfunction

$$(32) \qquad u_n(x) = r_n(x) \sin \theta(x, \lambda_n)$$

of the S-L system. Furthermore, the eigenfunction belonging to λ_n has exactly n zeros in the interval $a < x < b$, by Theorem 4. This proves all but the last statement of the following theorem.

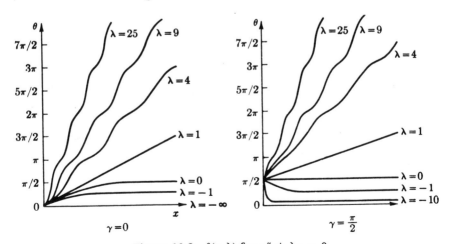

Figure 10.2 $\theta(x, \lambda)$ for $u'' + \lambda u = 0$.

THEOREM 5. *Any regular S-L system has an infinite sequence of real eigenvalues* $\lambda_0 < \lambda_1 < \lambda_2 < \cdots$ *with* $\lim_{n \to \infty} \lambda_n = \infty$. *The eigenfunction* $u_n(x)$ *belonging to the eigenvalue* λ_n *has exactly n zeros in the interval* $a < x < b$ *and is uniquely determined up to a constant factor.*

Only the last assertion wants verification. Any two solutions of (1) that satisfy the same initial condition $\alpha u(a) + \alpha' u'(a) = 0$ are linearly dependent, by the Uniqueness Theorem of Ch. 2, §4.

EXERCISES E

1. Show that for a regular S-L system, if $q(x)$ is increased to $q_1(x) > q(x)$, each nth eigenvalue of the new system is larger than that of the old.

2. Show that for a regular S-L system, if $\rho(x)$ is increased to $\rho_1(x) > \rho(x)$, all positive eigenvalues decrease and any negative eigenvalue increases.

3. Discuss the asymptotic behavior, as $n \to \infty$ of the nth eigenvalue of the S-L systems defined by $u'' + \lambda u = 0$, and the endpoint conditions:
   (a) $u(0) = 0$, $u(\pi) + u'(\pi) = 0$.
   (b) $u(0) = 0$, $u(\pi) = u'(\pi)$.

   That is, find constants a_0, a_1 such that $\sqrt{\lambda_n} = n + a_0 + a_1/n + 0(1/n^2)$.

4. For regular S-L systems with two sets of endpoint conditions, (30) and

$$\alpha_1 u(a) + \alpha_1' u'(a) = 0, \qquad \beta u(b) + \beta' u'(b) = 0$$

   show that, if $\alpha_1'/\alpha_1 < \alpha'/\alpha$, the eigenvalues of the second system are smaller than the corresponding eigenvalues of the first.

5. (a) Given $(Pu')' + Qu = 0$, and $(P_1v')' + Q_1v = 0$, $P_1(x) > 0$, $Q_1(x)$ continuous, establish *Picone's identity*.

$$\int_a^b [Q_1(x) - Q(x)]u(x)^2 \, dx + \int_a^b [P(x) - P_1(x)]u'(x)^2 \, dx$$

$$+ \int_a^b P_1(x)\left[u'(x) - \frac{u(x)v'(x)}{v(x)} \right]^2 dx = 0$$

   where $u(a) = u(b) = 0$ and $v(x) \neq 0$ in $[a, b]$.
   (b) Infer the Sturm Comparison Theorem from Picone's identity.

*6. (Szegö's Comparison Theorem). Under the hypothesis of the Sturm Comparison Theorem for $a < x < b$, $P \equiv P_1$, $Q \not\equiv Q_1$, let $u(x) > 0$, $u_1(x) > 0$ for $a < x < b$, and $\lim_{x \to a} P(x)[u'u_1 - uu_1'] = 0$. Show that, if $u(b) = 0$, there is an x_2 in (a, b) such that $u_1(x_2) = 0$. .

9   THE LIOUVILLE NORMAL FORM

By changes of dependent and independent variables of the form

$$(33) \qquad u = y(x)w, \qquad t = \int^b h(x) \, dx; \qquad y > 0, \qquad h > 0$$

we can simplify the S-L equations (1) considerably. If the functions y and h are positive and continuous in the given interval, the first substitution leaves the location of zeros unchanged, while the second one distorts the range of the independent variable, preserving the order, and leaves the number of zeros of a solution in corresponding intervals unchanged. The equivalent DE in w and t is obtained from the identity $d/dx = h(x)\, d/dt$, which is obtained from the second of equations (33). When substituted into the S-L equation (1), this identity gives

$$0 = h[hp(yw)_t]_t + (\lambda\rho - q)yw$$

$$= h\{pyhw_{tt} + [(hp)_ty + 2hpy_t]w_t + (hpy_t)_tw\} + (\lambda\rho - q)yw$$

Dividing through by the coefficient pyh^2 of w_{tt}, we obtain the equivalent DE (for $h, y \in \mathcal{C}^2$),

$$w_{tt} + (pyh)^{-1}[(hp)_ty + 2hpy_t]w_t + [(pyh)^{-1}(hpy_t)_t + h^{-2}p^{-1}(\lambda\rho - q)]w = 0$$

The term $\lambda(\rho/ph^2)w$ reduces to λw if and only if $h^2 = \rho/p$. The coefficient of w_t vanishes if and only if $(hp)_t/hp = -2y_t/y$, which can be achieved by choosing $y^2 = (hp)^{-1}$. Therefore, a simplified equivalent DE in w and t is obtained by choosing

(34)
$$u = w/\sqrt[4]{p(x)\rho(x)}, \qquad t = \int \sqrt{\rho(x)/p(x)}\, dx$$

This substitution reduces (1) to *Liouville normal form*. Since p and ρ are positive throughout the interval of definition (cf. §1), this change of variables makes $h(x)$ and $y(x)$ positive and of class $\mathcal{C}^2$ whenever p and ρ are of class $\mathcal{C}^2$.

THEOREM 6. *Liouville's substitution (34) transforms the S-L equation (1) with coefficient functions p, $\rho \in \mathcal{C}^2$ and $q \in \mathcal{C}$ into the Liouville normal form*

(35)
$$\frac{d^2w}{dt^2} + [\lambda - \hat{q}(t)]w = 0$$

where

(36)
$$\hat{q} = \frac{q}{\rho} + (p\rho)^{-1/4}\frac{d^2}{dt^2}[(p\rho)^{1/4}]$$

Evaluating the second derivative in (36) and using the identity $d/dt = (p/\rho)^{1/2}\, d/dx$, we get the alternative rational form

(36')
$$\hat{q} = \frac{q}{\rho} + \frac{p}{4\rho}\left[\left(\frac{p'}{p}\right)' + \left(\frac{\rho'}{\rho}\right)' + \frac{3}{4}\left(\frac{p'}{p}\right)^2 + \frac{1}{2}\left(\frac{p'}{p}\right)\left(\frac{\rho'}{\rho}\right) - \frac{1}{4}\left(\frac{\rho'}{\rho}\right)^2\right]$$

If the DE (1) is defined in $a \le x < b$, and t is the definite integral $t = \int_a^x \sqrt{\rho(s)/p(s)}\, ds$, then the equivalent DE (35) is defined in the interval $[0, c)$, where $c = \int_a^b \sqrt{\rho(x)/p(x)}\, dx$. An S-L equation (1) with p, $\rho \in \mathcal{C}^2$ and $q \in \mathcal{C}$ is transformed by Liouville's substitution into an S-L equation (35) with $\hat{q} \in \mathcal{C}$, since the denominator in (36) remains bounded away from 0.

COROLLARY 1. *Liouville's reduction (34) transforms regular S-L systems into regular S-L systems, separated and periodic boundary conditions into separated and periodic boundary conditions. The transformed system has the same eigenvalues as the original system.*

Let $u(x)$ and $v(x)$ be transformed into the functions $f(t)$ and $g(t)$ by Liouville's reduction (34). From the identity

$$(37) \qquad \int_0^c f(t)g(t)\, dt = \int_a^b u(x)v(x)\sqrt{p(x)\rho(x)}\,\sqrt{\frac{\rho(x)}{p(x)}}\, dx = \int_a^b u(x)v(x)\rho(x)\, dx$$

we infer the following result

COROLLARY 2. *Liouville's reduction (34) transforms functions orthogonal with weight ρ into orthogonal functions with unit weight.*

The Bessel equation (3) of Example 2, §1,

$$(38) \qquad\qquad (xu')' + \left(k^2 x - \frac{n^2}{x}\right)u = 0$$

is the special case $p = \rho = x$, $q = n^2/x$ of the DE (1). Hence, Liouville's reduction (34) is $u = w/\sqrt{x}$ and $x = t$, which leads to the equivalent DE

$$\frac{d^2 w}{dx^2} + \left[k^2 - \frac{n^2 - \frac{1}{4}}{x^2}\right]w = 0, \qquad w = x^{1/2}u$$

If $n = \frac{1}{2}$, this is the trigonometric DE $w'' + k^2 w = 0$, having a basis of solutions $\cos kx$ and $\sin kx$ ($k = 1, 2, 3, \ldots$). Since $J_{1/2}(0) = 0$, it follows that $J_{1/2}(x)$ is a constant multiple of $(\sin x)/\sqrt{x}$.

EXERCISES F

1. (a) Show that the self-adjoint form of the Hermite DE (15) is the S-L equation

$$[e^{-x^2}u']' + \lambda e^{-x^2}u = 0$$

   (b) Show that the Liouville normal form of this is the S-L equation (17) for the Hermite functions.

2. Show that the Liouville normal form of the self-adjoint form of the Jacobi DE is, for $x = \cos t$

$$w_{tt} + \left[\frac{(\frac{1}{4} - \alpha^2)}{4 \sin^2 (t/2)} + \frac{(\frac{1}{4} - \beta^2)}{4 \cos^2 (t/2)} + \left(n + \frac{(\alpha + \beta + 1)}{2} \right)^2 \right] w = 0$$

3. Show that the self-adjoint form of the hypergeometric DE is the singular S-L equation

$$[x^y(1 - x)^{\alpha+\beta+1-y}u']' - [\alpha\beta x^{y-1}(1 - x)^{\alpha+\beta-y}]u = 0$$

What is the Liouville normal form for this DE?

4. Compute the Liouville normal form for the Legendre DE, setting $x = -\cos t$, $-\pi < t < 0$.

*5. Show that every solution of the Legendre DE is square-integrable on $[-1, 1]$ and satisfies the endpoint conditions $\lim_{x \to \pm 1} (1 - x^2)u(x) = 0$.

*6. The Laguerre DE is $xu'' + (1 - x)u' + \lambda u = 0$. Show that its self-adjoint form is the S-L equation $[xe^{-x}u']' + \lambda e^{-x}u = 0$. What is its Liouville normal form?

*7. Show that the Legendre polynomial $P_n(x)$ has exactly n zeros. [HINT: Reduce the Legendre DE to Liouville normal form and apply Ex. E6.]

*8. If $x_1 = \cos t_1, \ldots, x_n = \cos t_n$ are the zeros of $P_n(x)$, $x_j < x_{j+1}$, show j that $2\pi(-1)/(2n + 1) < t_j < 2\pi j/(2n + 1)$, for $2 \le j \le n$. [HINT: Use the Liouville normal form and Ex. E6.]

10 MODIFIED PRÜFER SUBSTITUTION

By applying a modification of the Prüfer substitution to the Liouville normal form of an S-L system, we can obtain asymptotic formulas for the nth eigenfunction $u_n(x)$, valid for large n.

Using the Liouville substitution, any regular S-L system can be transformed into a regular S-L system consisting of the equation

(39) $$u'' + [\lambda - q(x)]u = u'' + Q(x)u = 0, \qquad Q(x) = \lambda - q(x)$$

and separated boundary conditions of the same form

(40) $$\alpha u(a) + \alpha'u'(a) = 0, \qquad \beta u(b) + \beta'u'(b) = 0$$

The constants α, α', β, β' are usually changed, but we still have $\alpha^2 + \alpha'^2 \ne 0$ and $\beta^2 + \beta'^2 \ne 0$. By Theorem 6, Corollary 1, the eigenvalues of this system are the same as those of the original system, and the eigenfunctions are obtained from those for the Liouville normal form through the Liouville substitution. To study the distribution of eigenvalues and magnitude of the eigenfunctions, it, therefore, suffices to treat the system (39)–(40). In §§10–11, we shall use mainly (40).

We shall assume from now on that $Q(x) > 0$ for $a \leq x \leq b$, that is, that $\lambda > q(x)$ and $Q \in \mathcal{C}^1$. We introduce the functions $R(x, \lambda)$ and $\phi(x, \lambda)$, the *modified amplitude* and *modified phase*, which are defined in the terms of a given solution $u(x, \lambda)$ of (39) by the equations

$$(41) \qquad u = \frac{R}{\sqrt[4]{Q}} \sin \phi, \qquad u' = R \sqrt[4]{Q} \cos \phi$$

These equations constitute the *modified Prüfer* system for the DE (39).

We shall now derive a pair of DEs for R and ϕ that are equivalent to (39). We have†

$$(42) \qquad \cot \phi = \frac{1}{\sqrt{Q}} \frac{u'}{u}, \qquad R^2 = \sqrt{Q} u^2 + \frac{1}{\sqrt{Q}} u'^2$$

Differentiating the first of these equations, we obtain (using $u'' = -Qu$)

$$(\csc^2 \phi)\phi' = \frac{Qu^2 + u'^2}{Q^{1/2}u^2} + \frac{1}{2} \frac{Q'}{Q^{3/2}} \frac{u'}{u}$$

Using the second equation, this simplifies to

$$(\csc^2 \phi)\phi' = \frac{R^2}{u^2} + \frac{1}{2} \frac{Q'}{Q} \cot \phi$$

and, multiplying by $\sin^2 \phi$ and simplifying,

$$(43) \qquad \phi' = Q^{1/2} + \frac{1}{4} \frac{Q'}{Q} \sin 2\phi$$

To derive the DE satisfied by R, differentiate the second equation in (42), obtaining the identity

$$2RR' = 2Q^{-1/2}(Quu' + u'u'') + \left(\frac{Q'}{2Q}\right)(Q^{1/2}u^2 - Q^{-1/2}u'^2)$$

The first term vanishes since $u'' = -Qu$, leaving the DE

$$(44) \qquad \frac{R'}{R} = \frac{Q'}{4Q} (\sin^2 \phi - \cos^2 \phi) = \frac{-Q'}{4Q} \cos 2\phi$$

---

† When $u \neq 0$, these equations are valid. When $u = 0$, set $\tan \phi = \sqrt{Q}u/u'$ and proceed similarly.

In terms of λ and q, the *modified Prüfer system* is

(45a) $$\phi' = \sqrt{\lambda - q} - \frac{q'}{4(\lambda - q)} \sin 2\phi$$

(45b) $$\frac{R'}{R} = \frac{q'}{4(\lambda - q)} \cos 2\phi$$

Clearly, to every nontrivial solution of (39) there corresponds a solution of the modified Prüfer system, and conversely. Furthermore, we know that $R > 0$, unless R vanishes identically.

Equations (45a) and (45b) determine the asymptotic behavior of the solutions of (39) as $\lambda \to \infty$. The fundamental result is the following.

THEOREM 7. *Let $\phi(x, \lambda)$ and $R(x, \lambda)$ be solutions of the system (45a) and (45b), where $q(x) \in \mathcal{C}^1$ is bounded. Then, as $\lambda \to \infty$,*

(46) $$\phi(x, \lambda) = \phi(a, \lambda) + \sqrt{\lambda}(x - a) + \frac{O(1)}{\sqrt{\lambda}}$$

and

(47) $$R(x, \lambda) = R(a, \lambda) + \frac{O(1)}{\lambda}$$

Intuitively, Theorem 7 states that for large λ the modified phase ϕ is approximately a linear function of $\sqrt{\lambda}$, and the modified amplitude function R is approximately constant.

The Symbol $O(1)$. The symbol $O(1)$ used here and later signifies a function $f(x, \lambda)$ of x and λ, defined for all sufficiently large λ, which is uniformly bounded for $a \leq x \leq b$ as $\lambda \to \infty$. Hence, $O(1)/\lambda^s$ signifies a function $f(x, \lambda)$ such that $\lambda^s f(x, \lambda)$ is uniformly bounded. The symbol $O(1)/\lambda^s$ is also often written $O(\lambda^{-s})$, as has been done in analogous contexts in Chs. 7 and 8.

The formula $f(x, \lambda) = O(1)$, where f is a given function, is not an ordinary equation. Thus, to write $O(1) = f(x, \lambda)$ would be meaningless, since $O(1)$ is not a function. The formula means simply that f remains uniformly *bounded* for all x as $\lambda \to \infty$, and that no other property of the function f is needed for the purpose at hand. Using this definition, the following important properties of the symbol $O(1)$ can be easily verified:

$$O(1) + O(1) = O(1); \qquad O(1)O(1) = O(1); \qquad \int_a^b O(1) \, dx = O(1)$$

for any finite a, b. Again, if α and β are real numbers with $\alpha \leq \beta$, then $O(1)/\lambda^\alpha + O(1)/\lambda^\beta = O(1)/\lambda^\alpha$. Finally, if $q(x)$ is any bounded function of x, then by

Taylor's formula we have, as $\lambda \to \infty$

$$[\lambda - q(x)]^\alpha = \lambda^\alpha \left[\frac{1 - q(x)}{\lambda}\right]^\alpha = \lambda^\alpha - \alpha q(x)\lambda^{\alpha-1} + O(1)\lambda^{\alpha-2}$$

The preceding formulas will be used freely in subsequent computations.

Proof.   For all λ for which $|q(x)| < \lambda$ on $[a, b]$, we have as before

$$\frac{q'}{\lambda - q} = \frac{q'}{\lambda}\left(1 + \frac{O(1)}{\lambda}\right) = \frac{q'}{\lambda} + \frac{O(1)}{\lambda^2}$$

$$\sqrt{\lambda - q} = \sqrt{\lambda}\left(1 - \frac{q}{\lambda}\right)^{1/2} = \sqrt{\lambda} - \frac{q}{2\sqrt{\lambda}} + \frac{O(1)}{\lambda^{3/2}}$$

We now compare the solutions of the DEs (45a) and (45b) with the solutions $\phi_1(x, \lambda) = \phi(a, \lambda) + \sqrt{\lambda}(x - a)$ and $R_1(x, \lambda) \equiv R_1(a)$ of

$$\phi' = \sqrt{\lambda} \qquad \text{and} \qquad (\log R)' = 0$$

using Theorem 3 of Ch. 6. In making this comparison, we set $\epsilon = O(1)/\sqrt{\lambda}$, and replace **x** and **y** with the functions $\phi(x, \lambda)$ and $\phi_1(x, \lambda)$, respectively. If $\phi_1(a, \lambda) = \phi(a, \lambda)$, the inequality (7) of Ch. 6 gives $|\phi(x, \lambda) - \phi_1(x, \lambda)| \leq O(1)/\sqrt{\lambda}$, and since $\phi_1(x, \lambda) = \phi(a, \lambda) + \sqrt{\lambda}(x - a)$, equation (46) follows.

Similary, to derive (47), compare $R(x, \lambda)$ with $R_1(x, \lambda)$, using the identity $e^{O(1)/\lambda} = 1 + O(1)/\lambda$ obtained from Taylor's formula.

*11   THE ASYMPTOTIC BEHAVIOR OF BESSEL FUNCTIONS

We shall now use the modified Prüfer substitution to study the asymptotic behavior of solutions of the Bessel DE (3) as $x \to \infty$. The substitution $u = w/\sqrt{x}$ reduces (3) to the Liouville normal form

$$(48) \quad w'' + [1 - (M/x^2)]w = 0, \qquad 0 < x < \infty, \qquad M = n^2 - \tfrac{1}{4}$$

whose solutions are $w(x) = \sqrt{x}Z_n(x)$, where $Z_n(x)$ is a solution of the Bessel DE [see (38)]. The modified Prüfer system for (48) is then obtained by setting $Q(x) = 1 - M/x^2$ in (43) and (44). This gives

$$(49a) \qquad \phi'(x) = \sqrt{1 - \frac{M}{x^2}} + \frac{M \sin 2\phi}{2(x^3 - Mx)}$$

$$(49b) \qquad \frac{R'(x)}{R(x)} = \frac{-M \cos 2\phi}{2(x^3 - Mx)}$$

Expanding the right sides of these equations, we have as $x \to \infty$, since $(1 - M/x^2)^{1/2} = 1 - M/2x^2 + O(1)/x^4$,

$$\phi'(x) = 1 - \frac{1}{2}\frac{M}{x^2} + \frac{O(1)}{x^3}, \qquad \frac{R'(x)}{R(x)} = \frac{O(1)}{x^3}$$

Here $O(1)$ denotes a function of x that remains bounded as $x \to \infty$. Integrating the first of these equations between any $x > \sqrt{M}$ and $y > x$, we obtain

$$\phi(x) - \phi(y) = x - y - \frac{M}{2x} - \frac{M}{2y} + \frac{O(1)}{x^2}$$

Keeping x fixed and letting $y \to \infty$, we find that $\phi_\infty = \lim_{y \to \infty} [y - \phi(y)]$ is finite. This gives $\phi(x) = \phi_\infty + x - M/(2x) + O(1)/x^2$.

If $\sqrt{M} < x < y$, integration of the second equation gives, similarly, $\log R(x) - \log R(y) = O(1)/x^2$. Taking exponentials and letting $y \to \infty$, we get

$$R(x) = R_\infty \exp [O(1)/x^2] = R_\infty + O(1)/x^2$$

where $R_\infty = \lim_{y \to \infty} R(y)$.

It follows that every solution of the Bessel DE (48) has the asymptotic form

$$Z_n(x) = x^{-1/2} \left[R_\infty + \frac{O(1)}{x^2} \right] \sin \left(\phi_\infty + x - \frac{M}{2x} + \frac{O(1)}{x^2} \right)$$

Since $\sin (A + O(1)/x^2) = \sin A + O(1)/x^2$, the preceding display can be rewritten as

$$Z_n(x) = R_\infty x^{-1/2} \sin \left(x + \phi_\infty - \frac{M}{2x} \right) + \frac{O(1)}{x^{5/2}}$$

The solution Z_n is uniquely determined by the constants R_∞ and ϕ_∞ above. For, if two solutions had the same asymptotic amplitude R_∞ and phase ϕ_∞, their difference would be a solution having modified amplitude $R(x) = O(1)/x^{5/2}$. Since

$$R(x) = R_\infty \exp \left[\frac{O(1)}{x^2} \right]$$

this would imply $R \equiv u \equiv 0$. Setting $x_\infty = \pi/2 + \phi_\infty$, this proves the following theorem.

THEOREM 8.   *To every nontrivial solution of the Bessel DE (3), there corresponds an asymptotic phase constant x_∞ and a limiting modified amplitude R_∞. The solution is uniquely determined by x_∞ and R_∞; every solution $Z_n(x)$ of Bessel's DE can be* *expressed as $x \to \infty$ in the form*

$$(50) \qquad Z_n(x) = \frac{R_\infty}{\sqrt{x}} \cos\left(x + x_\infty - \frac{(n^2 - 1/4)}{2x}\right) + \frac{O(1)}{x^{5/2}}$$

For the Bessel function $J_n(x)$, it can be shown that $x_\infty = n\pi/2 + \pi/4$ and that $R_\infty = \sqrt{2/\pi}$. The Neumann function $Y_n(x)$ is defined likewise by the conditions $x_\infty = n\pi/2 + 3\pi/4$ and $R_\infty = \sqrt{2/\pi}$. Thus, the Neumann function $Y_n(x)$ is defined by the condition that it has the same asymptotic amplitude as $J_n(x)$, with an asymptotic phase lag of $\pi/2$ radians.

That is, the asymptotic relation between $J_n(x)$ and $Y_n(x)$ is, for large positive x, the same as that between $\cos x$ and $\sin x$. The *Hankel function* $H_n(x) = J_n(x) + iY_n(x)$ is, therefore, analogous to the complex exponential function $e^{ix} = \cos x + i \sin x$.

12   DISTRIBUTION OF EIGENVALUES

We shall next show that the asymptotic distribution of the eigenvalues of all regular S-L systems is the same: the trigonometric DE $u'' + \lambda u = 0$ is typical. We shall treat in detail the case of separated endpoint conditions (2), also assuming $\alpha'\beta' \neq 0$ for uniformity. We can assume the given S-L system reduced to Liouville normal form (39)–(40), because this does not change the eigenvalues or the condition $\alpha'\beta' \neq 0$.

For the trigonometric DE and the boundary conditions $u(a) = \dot{u}(b) = 0$, the nth eigenfunction is $\sin[n\pi(x - a)/(b - a)]$ and the nth eigenvalue is $\lambda_n = n^2\pi^2/(b - a)^2$, $n = 1, 2, 3, \ldots$. For $u(a) = u'(b) = 0$, $u_n(x) \sin \sqrt{\lambda_n}(x - a)$, where $\lambda_n = (n + \frac{1}{2})^2\pi^2/(b - a)^2$. For $u'(a) = u'(b) = 0$, the $(n + 1)$st eigenfunction is $\cos \sqrt{\lambda_n}(x - a)$, where $\lambda_n = n^2\pi^2/(b - a)^2$ and $n = 0, 1, 2, \ldots$.

We will treat in detail, here and in §13, regular S-L systems satisfying separated endpoint conditions (2) with $\alpha'\beta' \neq 0$. We will show that $\sqrt{\lambda_n} = [n\pi/(b - a)] + O(1)/n$ in this case, $n = 0, 1, 2, \ldots$. That is, unless $\alpha' = 0$ or $\beta' = 0$ in (40), the asymptotic behavior of the eigenvalues and eigenfunctions is similar to that of $u'' + \lambda u = 0$, with the endpoint conditions $\alpha = \beta = 0$.

THEOREM 9. *For the regular S-L system (39)–(40), let $\alpha'\beta' \neq 0$. Then the eigenvalues λ_n are given, as $n \to \infty$, by the asymptotic formula*

$$(51) \qquad \sqrt{\lambda_n} = \frac{n\pi}{b - a} + \frac{O(1)}{n}$$

Here $O(1)$ denotes a function of n that is uniformly bounded for all integers $n \geq 0$.

Proof. Let $A = -\alpha/\alpha'$ and $B = -\beta/\beta'$. By assumption, A and B are finite. Choose a solution $\phi(x, \lambda)$ of (39) satisfying the initial condition

$$(52) \qquad \cot \phi(a, \lambda) = \frac{A}{\sqrt{\lambda - q(a)}}, \qquad 0 \leq \phi(a, \lambda) < \pi$$

According to (41), the solution $u(x, \lambda)$ corresponding to ϕ will be an eigenfunction if and only if

$$
(53) \qquad \cot \phi(b, \lambda) = \frac{B}{\sqrt{\lambda - q(b)}}
$$

Condition (52) can be simplified by expanding arccot x around $x = \pi/2$ to a first-order approximation in $1/\sqrt{\lambda}$. This gives, as $\lambda \to \infty$,

$$
(54) \qquad \phi(a, \lambda) = \frac{\pi}{2} + \frac{A}{\sqrt{\lambda}} + \frac{O(1)}{\lambda^{3/2}}
$$

Condition (53) can be simplified by a similar expansion. For the $(n + 1)$st eigenvalue, the modified phase function changes asymptotically to $n\pi + O(1)/\sqrt{\lambda_n}$. This gives, for $x = A/\sqrt{\lambda - q(a)}$

$$
(54') \qquad \phi(b, \lambda_n) = \frac{\pi}{2} + n\pi + \frac{O(1)}{\sqrt{\lambda_n}}
$$

Subtracting (54) from (54'), and comparing with (46) of Theorem 7, we obtain the equation

$$
(55) \qquad \phi(b, \lambda_n) = \phi(a, \lambda_n) = n\pi + \frac{O(1)}{\sqrt{\lambda_n}} = \sqrt{\lambda_n}(b - a) + \frac{O(1)}{\sqrt{\lambda_n}}
$$

Letting $\lambda_n \to \infty$ we obtain $\lim_{n \to \infty} n\pi \lambda_n^{-1/2} = (b - a)$, or $\sqrt{\lambda_n} = K_n n$, where the K_n tend to $\pi/(b - a)$. Substituting into (55), we obtain

$$
\sqrt{\lambda_n} = \frac{n\pi}{b - a} + \frac{O(1)}{\sqrt{\lambda_n}} = \frac{n\pi}{b - a} + \frac{O(1)}{n}, \qquad \text{q.e.d.}
$$

COROLLARY. *If* λ_n *is the sequence of nonzero eigenvalues of a regular* S-L *system, then* $\sum_{n=0}^{\infty} \lambda_n^{-2} < \infty$.

13   NORMALIZED EIGENFUNCTIONS

A square-integrable function u on an interval $a < x < b$ is *normalized* relative to a weight function ρ when

$$
\int_a^b u^2(x)\rho(x) \, dx = 1.
$$

In the case of the eigenfunctions of (39), $\rho(x) \equiv 1$. Our aim is to show that the normalized eigenfunctions of (39) and (40) behave approximately like cosine functions, provided that $\alpha'\beta' \neq 0$. [The cases $\alpha = 0$ and $\beta = 0$ are similar, after phase-shifts of $\pi/2$ in (54) and (54').]

THEOREM 10.   *Let $u_n(x)$ $(n = 0, 1, 2, \ldots)$ be the sequence of normalized eigenfunctions of the regular S-L system* (39)–(40), *with $\alpha'\beta' \neq 0$. Then*

$$
(56) \qquad\qquad u_n(x) = \sqrt{\frac{2}{b-a}} \cos \frac{n\pi(x-a)}{b-a} + \frac{O(1)}{n}
$$

The proof of this theorem will be carried out in three steps. For an eigenfunction $u_n(x)$, with eigenvalue λ_n, we have by (41)

$$
(57) \qquad\qquad u_n(x) = \frac{R(x, \lambda_n)}{\sqrt[4]{\lambda n - q(x)}} \sin \phi(x, \lambda_n), \qquad a \leq x \leq b
$$

In order to obtain formula (56), we obtain asymptotic expressions separately in terms of n for each of the three factors appearing in (57). This is done in the following three lemmas.

LEMMA 1.   *Let $\phi(x, \lambda)$ be as in the proof of Theorem 9. Then as $\lambda \to \infty$,*

$$
(58) \qquad\qquad \int_a^b \sin^2 \phi(x, \lambda)\, dx = \frac{b-a}{2} + \frac{O(1)}{\lambda^{1/2}}
$$

Proof.   Using $\phi(x, \lambda)$ as the variable of integration in (55), and recalling from (46) that $dx/d\phi = (d\phi/dx)^{-1} = -\lambda^{-1/2} + O(1)\lambda^{-3/2}$, we have

$$
\int_a^b \sin^2 \phi(x, \lambda)\, dx = \int_{\phi(a,\lambda)}^{\phi(b,\lambda)} \sin^2 \phi\, \frac{dx}{d\phi}\, d\phi
$$

$$
= (\lambda^{-1/2} + O(1)\lambda^{-3/2}) \int_{\phi(a,\lambda)}^{\phi(b,\lambda)} \sin^2 \phi\, d\phi
$$

The last integral can be evaluated explicitly. Apply Theorem 7, to obtain

$$
\int_{\phi(a,\lambda)}^{\phi(b,\lambda)} \sin^2 \phi\, d\phi = \left[\frac{\phi}{2} - \frac{\sin 2\phi}{4} \right]_{\phi(a,\lambda)}^{\phi(b,\lambda)} = \frac{\lambda^{1/2}(b-a)}{2} + O(1)
$$

Substituting into the previous displayed formula and simplifying, we obtain (58).

A second step toward our result is the following lemma.

LEMMA 2.   *Let $u(x, \lambda)$ be a solution of* (39). *Then, as $\lambda \to \infty$*

$$
(59) \qquad \left(\int_a^b u^2(x)\, dx \right)^{1/2} = R(a, \lambda)\lambda^{-1/4} \sqrt{\frac{b-a}{2}} \left(1 + \frac{O(1)}{\lambda^{1/2}} \right) + \frac{O(1)}{\lambda^{5/4}}
$$

Proof. Expressing u in terms of R by (41), and then expanding R as in Theorem 7, formula (47), we have

$$\int_a^b u^2(x)\, dx = \left[R(a, \lambda) + \frac{O(1)}{\lambda} \right]^2 \int_a^b [\lambda - q(x)]^{-1/2} \sin^2 \phi \, dx$$

Since $(\lambda - q)^{-1/2} = \lambda^{-1/2} + O(1)\lambda^{-3/2}$, we get after simplifying and using (58)

$$\int_a^b u^2(x)\, dx = \left[R(a, \lambda) + \frac{O(1)}{\lambda} \right]^2 (\lambda^{-1/2} + O(1)\lambda^{-3/2}) \left(\frac{b-a}{2} + O(1)\lambda^{-1/2} \right)$$

$$= \left[R(a, \lambda) + \frac{O(1)}{\lambda} \right]^2 \left(\frac{b-a}{2\lambda^{1/2}} + \frac{O(1)}{\lambda} \right)$$

Hence, taking square roots

$$\left(\int_a^b u^2(x)\, dx \right)^{1/2} = \left(R(a, \lambda) + \frac{O(1)}{\lambda} \right) \left(\frac{b-a}{2\lambda^{1/2}} + \frac{O(1)}{\lambda} \right)^{1/2}$$

$$= \frac{R(a, \lambda)}{\lambda^{1/4}} \sqrt{\frac{b-a}{2}} \left(1 + \frac{O(1)}{\lambda^{1/2}} \right) + \frac{O(1)}{\lambda^{5/4}}, \qquad \text{q.e.d.}$$

COROLLARY. *If, in Lemma 2, $\int_a^b u^2(x, \lambda)\, dx = 1$, then*

(60) $$R(a, \lambda) = \sqrt{\frac{2}{b-a}}\, \lambda^{1/4}[1 + O(1)\lambda^{-1/2}]$$

Proof. Formula (59) gives the following condition on the amplitude function of a normalized solution:

$$1 - \frac{O(1)}{\lambda^{5/4}} = \frac{R(a, \lambda)}{\lambda^{1/4}} \sqrt{\frac{b-a}{2}} \left(1 + \frac{O(1)}{\lambda^{1/2}} \right)$$

Solving for R, and taking the asymptotic form of the quotient, we get (60), q.e.d.

LEMMA 3. *Let λ_n be the nth eigenvalue ($\lambda_0 < \lambda_1 < \lambda_2 < \cdots$) of the S-L system* (39)–(40). *Then, as $n \to \infty$, unless $\alpha'\beta = 0$*

(61) $$\sin \phi(x, \lambda_n) = \cos \frac{n\pi(x-a)}{b-a} + O(1)\lambda_n^{-1/2}$$

Proof. By Theorem 7, (46), we have

$$\phi(x, \lambda_n) = \phi(a, \lambda_n) + \sqrt{\lambda_n}(x - a) + \frac{O(1)}{\sqrt{\lambda_n}}$$

Moreover by (54), $\phi(a, \lambda_n) = \pi/2 + O(1)/\sqrt{\lambda_n}$. Substituting back into the preceding formula, we get

(62)
$$\sin \phi(x, \lambda_n) = \sin [\sqrt{\lambda_n} (x - a) + \pi/2] + O(1)/\sqrt{\lambda_n}$$
$$= \cos [\sqrt{\lambda_n} (x - a)] + O(1)/n$$

We now apply Theorem 9 to this formula. By formula (51) and the mean value theorem, we have

$$\cos [\sqrt{\lambda_n} (x - a)] - \cos \left[\frac{n\pi(x - a)}{(b - a)} \right] = O(1)n^{-1} = O(1)\lambda_n^{-1/2}$$

Substituting into the right-hand side of (62), we obtain (61), q.e.d.

The proof of Theorem 10 can now be completed as follows. Of the three factors in equality (57), $1/\sqrt[4]{\lambda n - q(x)}$ can be replaced by the first-order approximation $(\lambda - q)^{-1/4} = \lambda^{-1/4} + O(1)\lambda^{-5/4}$. The factor $R(x, \lambda_n)$ is estimated by the Corollary to Lemma 2, and the factor $\sin \phi(x, \lambda_n)$ is estimated by Lemma 3. Substituting all these expressions into (57) and simplifying, we obtain

$$u_n(x) = \sqrt{\frac{2}{b - a}} \cos \left[\frac{n\pi(x - a)}{b - a} \right] + O(1)\lambda_n^{-1/2}$$

Since $\lambda_n^{-1/2} = O(1)n^{-1}$, this gives Theorem 10.

EXERCISES G

1. For any DE $u'' + u + \rho(x)u = 0$ with $\rho(x) = O(x^{-2})$ as $x \to +\infty$, show that, for every solution $u(x)$, constants A and x_1 can be found for which

$$u(x) = A \cos (x - x_1) + O(x^{-1}) \qquad \text{as } x \to \infty$$

*2. Establish the following formula for Legendre polynomials:

$$P_n(\cos \theta) = \frac{A}{(\sin \theta)^{1/2}} \cos \left[\left(n + \frac{1}{2}\right) \theta - \frac{\pi}{4} \right] + O(n^{-1/2}) \qquad \text{for} \qquad 0 < \theta < \pi$$

for some constant A. [HINT: Find a DE satisfied by $P_n(\cos \theta)$.]

3. Show that the relative maxima of $x^{1/2}|J_n(x)|$ form an increasing sequence if $0 < n < \frac{1}{2}$ and a decreasing sequence if $n > \frac{1}{2}$.

*4. (Sonin-Polya Theorem). Show that if in $(Pu')' + Qu = 0$, $P, Q \in \mathcal{C}^1[a, b]$, $Q(x) \neq 0$, and $P(x)Q(x)$ are nondecreasing, the successive maxima of $|u(x)|$ form a nonincreasing sequence, and that equality occurs if and only if $Q(x) = 1/P(x)$. [HINT: Show that the derivative of $\phi(x) = u(x)^2 + P(x)u'(x)^2/Q(x)$ is nonpositive.]

*5. Show that the values of $|P(x)Q(x)|^{1/2}|u(x)|$ at those points where $u'(x) = 0$ are a monotonic increasing or decreasing sequence, according as the values of $P(x)Q(x)$ are decreasing or increasing. [HINT: Consider $v(x) = P(x)Q(x)\phi(x)$, ϕ as in Ex. 4.]

14 INHOMOGENEOUS EQUATIONS

Inhomogeneous second-order linear equations, of the form

$$(63) \qquad L[u] = p_0(x)u'' + p_1(x)u' + p_2(x)u = f(x), \qquad p_0(x) \neq 0, \quad p_0(x) \in \mathcal{C}^1$$

subject to *homogeneous* separated endpoint conditions (30), can be solved by use of *Green's functions*. The method of solution generalizes that for *two endpoint* problems described in Ch. 2, §§9, 11. The discussion given there, which covers the case $\alpha' = \beta' = 0$ of (30), can now be reviewed to advantage.

Before introducing Green's functions, we first analyze the problem with *inhomogeneous* separated endpoint conditions:

$$(64) \qquad A[u] = \alpha u(a) + \alpha' u'(a) = \alpha_1, \qquad B[u] = \beta u(b) + \beta' u'(b) = \beta_1$$

Let $U(x)$ be the solution of $L[u] = 0$ satisfying the initial conditions $U(a) = \alpha'$, $U'(a) = -\alpha$; let $V(x)$ be the solution of $L[u] = 0$ satisfying $V(b) = \beta'$, $V'(b) = -\beta$; let $F(x)$ be the solution of $L[u] = f(x)$ satisfying $F(a) = F'(a) = 0$. The existence and uniqueness of these functions follow from Theorem 7, Corollary 2, of Ch. 6, §8. For any constants, c, d, the function

$$w(x) = cU(x) + dV(x) + F(x)$$

satisfies the inhomogeneous DE (63). Moreover, we have

$$A[w] = d(\alpha V(a) + \alpha' V'(a)) = dA[V]$$

$$B[w] = c(\beta U(b) + \beta' U'(b)) + B[F] = cB[U] + B[F]$$

If U and V are linearly independent, their Wronskian $W = UV' - VU'$ never vanishes. Hence,

$$A[V] = \alpha V(a) + \alpha' V'(a) = -U'(a)V(a) + U(a)V'(a) \neq 0$$

Similarly, $B[U] = -W(b) \neq 0$. Therefore, equations (65) for the unknowns c and d have a unique solution for any values given to $A[w]$ and $B[w]$.

On the other hand, if U and V are linearly dependent, their Wronskian vanishes identically. Hence, $U(x)$ satisfies $A[U] = \alpha\alpha' + \alpha'(-\alpha) = 0$ and $B[U] = 0$. This proves the following theorem.

THEOREM 11. *Either* DE *(63) has a solution w satisfying the boundary conditions $A[w] = \alpha_1$ and $B[w] = \beta_1$, for any given constants α_1 and β_1, or else the homogeneous* DE $L[u] = 0$ *has an eigenfunction with eigenvalue 0, satisfying the homogeneous conditions $A[u] = 0$ and $B[u] = 0$.*

15 GREEN'S FUNCTIONS

We now show that, in the first case of the preceding theorem, there exists a *Green's function* $G(x, \xi)$ defined for $a \le x, \xi \le b$, such that the solution of (63) subject to the boundary conditions (30) is given by

$$(66) \qquad u(x) = \int_a^b G(x, \xi) f(\xi)\, d\xi = \mathcal{G}[f]$$

Note that $\mathcal{G}$ is an *integral operator* (Ch. 2, §9) whose kernel is the Green's function $G(x, \xi)$.

This result has already been established in Ch. 2, §11, for the endpoint conditions $u(a) = u(b) = 0$; it will now be generalized to arbitrary *homogeneous* separated endpoint conditions (30): $A[u] = B[u] = 0$.

In this general case, $G(x, \xi)$ can be constructed by the method used in Ch. 2. For each fixed ξ, $G(x, \xi)$ is a solution of the homogeneous DE $L[G] = 0$ on the intervals $[a, \xi]$ and $[\xi, b]$, satisfying the homogeneous endpoint conditions $A[u] = 0$ and $B[u] = 0$, respectively. It is continuous across $x = \xi$ (i.e., across the principal diagonal of the square $a \le x, \xi \le b$), and its derivative $\partial G/\partial x$ jumps by $1/p_0(x)$ across this diagonal. In other words, we have

$$G(x, \xi) = \begin{cases} \partial(\xi) U(x) V(\xi), & a \le x \le \xi \\ \epsilon(\xi) V(x) U(\xi), & \xi \le x \le b \end{cases}$$

where the factor $\epsilon(\xi)$ above is chosen to give $\partial G/\partial x$ a jump of $1/p_0(\xi)$ across $x = \xi$. Thus

$$\frac{\partial G}{\partial x}(\xi^+, \xi) - \frac{\partial G}{\partial x}(\xi^-, \xi) = \epsilon(\xi)\{U(\xi) V'(\xi^+) - V(\xi) U'(\xi^-)\} = \frac{1}{p_0(\xi)}$$

We are therefore led to try the kernel

$$(67) \qquad G(x, \xi) = \begin{cases} \dfrac{U(x) V(\xi)}{p_0(\xi) W(\xi)}, & a \le x \le \xi \\[2ex] \dfrac{U(\xi) V(x)}{p_0(\xi) W(\xi)}, & \xi \le x \le b \end{cases}$$

where $W = UV' - VU'$ is the Wronskian of U and V.

THEOREM 12. *Unless $W \equiv 0$, equations (66) and (67) yield for any continuous function f on $[a, b]$ a solution $u(x)$ of the DE $L[u] = f(x)$ that satisfies the boundary conditions $A[u] = B[u] = 0$.*

That is, unless the homogeneous linear boundary-value problem $L[u] = A[u] = B[u] = 0$ admits an eigenfunction, the function defined by (67) is a Green's function for the system $L[u] = f$, $A[u] = B[u] = 0$.

The proof is like that given in Ch. 2, §11. Rewriting (66) as

$$u(x) = \int_a^x G(x, \xi)f(\xi) \, d\xi + \int_x^b G(x, \xi)f(\xi) \, d\xi$$

and differentiating, we have, by Leibniz' rule,

$$u'(x) = \int_a^x G_x(x, \xi)f(\xi) \, d\xi + \int_x^b G_x(x, \xi)f(\xi) \, d\xi$$

The endpoint contributions give $G(x, x^-)f(x^-) - G(x, x^+)f(x^+) = 0$; they cancel since $G(x, \xi)$ and f are continuous for $x = \xi$. Differentiating again, we have, by Leibniz' rule

$$u''(x) = \int_a^x G_{xx}(x, \xi)f(\xi) \, d\xi + G_x(x, x^-)f(x^-)$$
$$+ \int_x^b G_{xx}(x, \xi)f(\xi) \, d\xi - G_x(x, x^+)f(x^+)$$

The two terms corresponding to the contributions from the endpoints come from the sides $x > \xi$ and $x < \xi$ of the diagonal; since f is continuous, their difference is $[G_x(x^+, x) - G_x(x^-, x)]f(x) = f(x)/p_0(x)$. Simplifying, we obtain

$$u''(x) = \int_a^b G_{xx}(x, \xi)f(\xi) \, d\xi + \frac{f(x)}{p_0(x)}$$

From the foregoing identities, we can calculate $L[u]$. It is

$$L[u] = \int_a^b L_x[G(x, \xi)]f(\xi) \, d\xi + f(x) = f(x)$$

where $L_x[G(x, \xi)]$ stands for the sum $p_0 G_{xx} + p_1 G_x + p_2 G$. This sum is zero except on the diagonal $x = \xi$, where it is undefined. This gives the identity (63).

Since $G(x, \xi)$, as a function of x, satisfies the boundary conditions (30) for all ξ, it follows from (66), by differentiating under the integral sign and using Leibniz' Rule again, that u satisfies the same boundary conditions. This completes the proof of the theorem.

In operator language (cf. Ch. 2, §3), we have shown that the operator $f \to \mathcal{G}[f]$ transforms the space $\mathcal{C}[a, b]$ of continuous functions on the interval $[a, b]$ into the space $\mathcal{C}^2[a, b]$ of functions of class $\mathcal{C}^2$, and that this operator is a *right inverse* of the operator L. In other words, we have $L[\mathcal{G}[f]] = f$ for all continuous f. In operator notation, we can write $\mathcal{G} = L^{-1}$.

EXERCISES H

In Exs. 1–5 show that Green's function is as specified.

1. $u'' = -f$, $u(0) = u'(1) = 0$; $G(x, \xi) = \begin{cases} x & \text{for} & x \leq \xi \\ \xi & \text{for} & x > \xi \end{cases}$

2. $u'' = f$, $u(-1) = u(1) = 0$; $G(x, \xi) = -[|x - \xi| + x\xi - 1]/2$

3. $xu'' + u' = f$, $u(x)$ bounded as $x \to 0$, $u(1) = 0$;

$$G(x, \xi) = \begin{cases} \log \xi & \text{for} & x \leq \xi \\ \log x & \text{for} & x > \xi \end{cases}$$

4. $u'' - u = f$, $u(x)$ bounded as $|x| \to \infty$; $G(x, \xi) = -\exp(|x - \xi|)]/2$.

5. $u'' - u' = f(x)$, $u(0) = u'(1) = 0$.

$$G(x, \xi) = \begin{cases} e^{-\xi}(1 - e^x), & x \leq \xi \\ (e^{-\xi} - 1), & x \geq \xi \end{cases}$$

6. Find Green's function for $u'' - u = f$ with $u(-a) = u(a) = 0$. Show that, as $a \to \infty$, it approaches that of Ex. 4.

7. Show that $L[u] + \lambda u = 0$ for nontrivial u, $\lambda \neq 0$ and given homogeneous endpoint conditions (30), if and only if $\mathcal{G}[u] = \mu u$ for $\mu = 1/\lambda$.

*8. Show that the Green's function G of a regular S-L system is a symmetric function of x and ξ, in the sense that $G(x, \xi) = G(\xi, x)$.

*16   THE SCHROEDINGER EQUATION

The *Schroedinger equation* of quantum mechanics in one space dimension is the DE

(68) $$\psi'' + \left(\frac{2m}{\hbar^2}\right) [E - V(x)]\psi = 0$$

Physically, the function $V(x)$ has the significance of potential energy; the constant m stands for the mass of the particle; the constant E is an energy parameter; $\hbar = h/2\pi$ is a universal constant, whose numerical value depends on the units used. The "wave function" $\psi(x)$ may be real or complex; $\psi\psi^* \, dx = |\psi|^2 \, dx$ is the probability that the particle under consideration will be "observed" in the interval $(x, x + dx)$. The *eigenvalues* of (68) for varying E are the *energy levels* of the associated physical system.

The DE (68) is precisely the Liouville normal form

(69) $$u'' + [\lambda - q(x)]u = 0$$

of a general S-L equation, with $\lambda = 2mE/\hbar^2$ and $q = 2mV/\hbar^2$. But, in most physical applications, one is concerned with the *infinite interval* $(-\infty, \infty)$. On this interval, the "endpoint" condition that a solution remain bounded as $x \to \pm\infty$ defines a *singular* S-L system (cf. §4). In problems involving the Schroedinger equation, it is customary among physicists to define the *spectrum* of this S-L system as the set of all eigenvalues for which eigenfunctions exist. The set of isolated points (if any) in this spectrum is called the *discrete spectrum;* the part (if any) that consists of entire intervals is called the *continuous spectrum*. We shall adopt this suggestive terminology here; unfortunately, its logical extension to boundary value problems generally is very technical, even for ordinary DEs.†

For regular S-L systems, we have proved that the spectrum is always discrete, and the eigenfunctions are (trivially) square-integrable. We now describe a simple singular S-L system whose spectrum is continuous and whose eigenfunctions are not square-integrable.

Example 8. The S-L system of a *free particle* is

$$(70) \qquad\qquad u'' + \lambda u = 0, \qquad -\infty < x < +\infty$$

For every positive number $\lambda > 0$, this DE has two linearly independent bounded solutions $\sin(\sqrt{\lambda}x)$ and $\cos(\sqrt{\lambda}x)$. For $\lambda = 0$, it has the bounded solution $u = 1$, and no other linearly independent eigenfunction. For $\lambda < 0$, it has the linearly independent unbounded solutions $\sinh(\sqrt{\lambda}\,x)$ and $\cosh(\sqrt{\lambda}\,x)$, and no nontrivial bounded solution. Hence, the *spectrum* of the free particle is *continuous:* it consists of the half-line $\lambda \geq 0$.

Example 9. In the case of a *harmonic oscillator,* the potential energy $V(x)$ is a constant multiple of x^2. By a change of unit $x \to kx$, we can reduce the resulting Schroedinger DE to the normal form

$$(71) \qquad\qquad u'' + (\lambda - x^2)u = 0$$

Comparing with Example 7 of §4, we see that this has the eigenfunctions $e^{-x^2/2}H_n(x)$, for $\lambda = 2n + 1$ ($n = 0, 1, 2, \ldots$). These eigenfunctions are even *square-integrable*.

For any value of λ not an odd positive integer, the recurrence relation $a_{k+2} = (2k - \lambda + 1)a_k/(k + 1)(k + 2)$ satisfied by $H_\lambda(x)$ may be compared with that for the Taylor series $e^{\beta x^2} = \Sigma\, \beta^r x^{2r}/(r!)$, namely $c_{2r+2} = \beta c_{2r}/(r + 1)$. Setting $2r = k$, we see that, for all sufficiently large k,

$$\frac{a_{k+2}}{a_k} > \frac{c_{k+2}}{c_k} > 0 \qquad \text{if} \qquad \beta < 1$$

Hence $|H_\lambda(x)| > Be^{\beta x^2} - p_\lambda(x)$, where $B > 0$, $p_\lambda(x)$ is a polynomial and (say) $\beta = \frac{3}{4}$. It follows that $|e^{-x^2/2}H_\lambda(x)| > Be^{x^2/4} - O(1)$ is *unbounded,* unless λ is an

---

† See, for example, Coddington and Levinson, pp. 252–269.

odd positive integer. Finally, if $u(x)$ is any bounded nontrivial solution of (71), the same is true of $u(-x)$ and of $[u(x) + u(-x)]/2$, $[u(x) - u(-x)]/2$. This shows that, if (71) has an eigenfunction, it must have an odd eigenfunction or an even eigenfunction. Since either of these would be defined up to a constant factor by the relation

$$a_{k+2} = \frac{(2k - \lambda + 1)a_k}{(k + 1)(k + 2)}$$

on its coefficients, we see that the *Hermite functions* $e^{-x^2/2}H_n(x)$ are the *only eigenfunctions* of the harmonic oscillator.

*17   THE SQUARE-WELL POTENTIAL

In Example 8, the spectrum is continuous; in Example 9, it is discrete. We now describe a Schroedinger equation whose spectrum is partly continuous and partly discrete.

Example 10. A *square-well* potential is one satisfying $V(x) = -C^2$ on $|x| < a$, and $V(x) = 0$ when $|x| > a$. This leads to the Schroedinger DE with *discontinuous* $q(x)$:

$$(72) \qquad u'' + \lambda u = \begin{cases} 0 & \text{on} & |x| > a \\ -C^2 u & \text{on} & |x| < a \end{cases}$$

The eigenfunctions can again be determined explicitly.†

If $u(x)$ is any eigenfunction, then so is $u(-x)$, and so are the even part $[u(x) + u(-x)]/2$ and the odd part $[u(x) - u(-x)]/2$ of $u(x)$. Hence, (72) has a basis of eigenfunctions consisting exclusively of *even* and *odd* eigenfunctions, that is, satisfying $u'(0) = 0$ or $u(0) = 0$.

For $\lambda > 0$, every solution of (72) has the form $A \cos \sqrt{\lambda}x + B \sin \sqrt{\lambda}x$ for $|x| > a$. Hence, every nontrivial solution of (72) is an eigenfunction and, as in Example 8, the spectrum includes the entire half line $\lambda \geq 0$.

For $\lambda < -C^2$, on the other hand, the continuation to $|x| > a$ of both the even solution $\cosh (\sqrt{-\lambda - C^2}x)$ and the odd solution $\sinh (\sqrt{-\lambda - C^2}x)$, from the interval $|x| < a$, can be shown (see Theorem 14 below) to satisfy $u(x) > 0$, $u'(x) > 0$, and $u''(x) > 0$ for all positive x. Hence, they are both unbounded. In summary, the spectrum contains no points on $\lambda < -C^2$: there is no bounded solution with eigenvalue $\lambda < -C^2$.

In the interval $-C^2 < \lambda < 0$, one can show that the spectrum is discrete by working out the implications of the Sturm Oscillation Theorem. The solutions

---

† This is done with the usual understanding (Ch. 2, Ex. A12) that a "solution" of (72) is a function $u \in \mathcal{C}^1$ which satisfies (72), and so is of class $\mathcal{C}^2$ where $q(x)$ is continuous.

bounded for $x > a$ are the functions $A \exp(-\sqrt{-\lambda}x)$ which satisfy $u'(a)/u(a)$ $= -\sqrt{-\lambda}$. Writing $\mu = \sqrt{\lambda + C^2}$, we see that the even solutions $A \cos \mu x$ of (72) satisfy the same boundary condition $u'(a)/u(a) = -\sqrt{-\lambda}$ if and only if $\mu \tan \mu a = \sqrt{-\lambda}$. The odd solutions $B \sin \mu x$ satisfy it if and only if $\mu \cot \mu a = -\sqrt{-\lambda}$. Solving the preceding transcendental equations graphically, we see that the number of even eigenfunctions and the number of odd eigenfunctions belonging to the discrete spectrum are both approximately equal to aC/π. Moreover, every eigenfunction that corresponds to the discrete spectrum is square-integrable, and conversely.

*18   MIXED SPECTRUM

The preceding example is typical of a wide class of Schroedinger equations—namely, all those having a "potential well" dying out at infinity. We first treat the continuous portion of the spectrum.

LEMMA 1.   *In the normalized Schroedinger equation (69), let q be continuous and satisfy $q(x) = B/x + 0(1/x^2)$ as $x \to \infty$, for some constant B. Then, for $\lambda > 0$, every solution has infinitely many zeros and is bounded.*

Proof.   The first statement follows from the Sturm Comparison Theorem, comparing with the DE $u'' + \lambda u/2 = 0$.

To prove the second statement, first change the independent variable to $t = \sqrt{\lambda}x$, giving the DE $u_{tt} + Q(t)u = 0$, with $Q(t) = 1 - q(t/\sqrt{\lambda})/\lambda$. Applied to the new DE, the Prüfer substitution (21) gives, by (22),

$$(73) \qquad \frac{d\theta}{dt} = Q(t) \sin^2 \theta + \cos^2 \theta = 1 + \frac{A}{t} \sin^2 \theta + \frac{O(1)}{t^2}, \qquad A = \frac{-B}{\sqrt{\lambda}}$$

Moreover, $r^2 = u^2 + u_t^2$ is given by (23'), as

$$r^2 = K^2 \exp \left\{ \int_a^t \left[\frac{A}{s} + O\left(\frac{1}{s^2}\right) \right] (\sin 2\theta) \frac{ds}{d\theta} d\theta \right\} = K^2 \exp \left\{ \int_a^t F(s) \, ds \right\}$$

where $F(s) = (A/s)(\sin 2\theta) + O(1)/s^2$, and the limits of integration refer to s. Using the expression $dt/d\theta = 1 - (A/t) \sin^2 \theta + O(1)/t^2$, derivable from the preceding display, we have

$$\int_a^t F(s) \, ds = \int_a^t \left[\frac{A}{s} + O\left(\frac{1}{s^2}\right) \right] \sin 2\theta \, d\theta$$

The first term on the right side above can be integrated by parts:

$$\int_a^t F(s) \, ds = \left[-\frac{A}{2s} \cos 2\theta \right]_a^t - \int_a^t \frac{A}{2s^2} \cos 2\theta \, ds + \int_a^t O\left(\frac{1}{s^2}\right) \sin 2\theta \, d\theta$$

The boundedness of the first two terms on the right side of this equation is evident; the last term is bounded because $d\theta/ds = 1 + O(1)/s$. Hence, u^2 is bounded because $u^2 \leq r^2 \leq K^2 \exp\{\int_a^t F(s)\,ds\}$.

Combining Lemma 1 with the analogous result for negative x, we obtain the following result.

THEOREM 13.   *If $q \in \mathcal{C}$ satisfies $q(x) = A/x + O(1/x^2)$ as $x \to +\infty$ and $q(x) = B/x + O(1/x^2)$ as $x \to -\infty$, then the spectrum (69) includes the half line $\lambda > 0$.*

As regards the discrete portion of the spectrum, the key result is the following lemma, which characterizes the asymptotic behavior for large x of a wide class of DEs that have nonoscillatory solutions, such as the modified Bessel equation of Ch. 9, §7.

LEMMA 2.   *In the Schroedinger equation (69), let q be continuous, let $\lim_{x \to +\infty} q(x) = 0$, and let $\lambda = -k^2 < 0$. For any ϵ, $0 < \epsilon < k$, there exist two solutions $u_1(x)$ and $u_2(x)$ of (69) such that, for all sufficiently large x,*

(74)
$$e^{(k-\epsilon)x} \leq u_1(x) \leq e^{(k+\epsilon)x}, \qquad e^{(-k-\epsilon)x} \leq u_2(x) \leq e^{(-k+\epsilon)x}$$

Proof.   Choose a so large that $(k - \epsilon)^2 < q(x) - \lambda < (k + \epsilon)^2$ for all $x \geq a$, and let $u_1(x)$ be the solution defined by the initial conditions $u_1(a) = e^{ka}$, $u_1'(a) = ke^{ka}$. Then $\tau(x) = u_1'/u_1$ satisfies $\tau(a) = k$ and the Riccati equation $\tau' = G(x, \tau) = q(x) - \lambda - \tau^2$. For the DEs

$$\rho' = F(x, \rho) = (k - \epsilon)^2 - \rho^2, \qquad \sigma' = H(x, \sigma) = (k + \epsilon)^2 - \sigma^2$$

it is clear that $F(x, \tau) \leq G(x, \tau) \leq H(x, \tau)$ on the domain $\tau \geq k - \epsilon > 0$. Moreover, the solutions $\rho(x) = k - \epsilon$ and $\sigma(x) = k + \epsilon$ of the displayed DEs satisfy $\rho(a) < \tau(a) < \sigma(a)$. Hence, by the Comparison Theorem of Ch. 1, §11, we have $k - \epsilon = \rho(x) \leq \tau(x) \leq \sigma(x) = k + \epsilon$. Integrating, we get the first inequality of (74).

We now derive the second inequality. As in Ch. 2, §5, a linearly independent solution of the DE (69) is given by

$$u_2(x) = 2ku_1(x) \int_x^\infty \frac{ds}{u_1^2(s)}$$

The first inequality of (74), applied to the integral on the right, gives the inequalities

$$\frac{1}{2(k - \epsilon)} e^{-2(k-\epsilon)x} \geq \int_x^\infty \frac{ds}{u_1(s)^2} \geq \frac{1}{2(k + \epsilon)} e^{-2(k+\epsilon)x}$$

Multiplying through by $2ku_1(x)$, and using (74) again, we get

(75) $$\frac{e^{2\epsilon x}}{1 - \epsilon/k} e^{-(k-\epsilon)x} \geq u_2(x) \geq \frac{e^{-2\epsilon x}}{1 + \epsilon/k} e^{-(k+\epsilon)x}$$

But, for any η such that $0 < 3\epsilon < \eta < k$ we have, for sufficiently large x,

$$e^{-(k-\eta)x} \geq \frac{e^{2\epsilon x}}{1 - \epsilon/k} e^{-(k-\epsilon)x} \quad \text{and} \quad \frac{e^{-2\epsilon x}}{1 + \epsilon/k} e^{-(k+\epsilon)x} \geq e^{-(k+\eta)x}$$

Applying these inequalities to (75), we obtain the second formula of (74) with η in place of ϵ. Since, for any η with $0 < \eta < k$, we can find $\epsilon = \eta/6$ with $0 < 3\epsilon < \eta < k$, and the proof is complete.

COROLLARY 1. *On $(0, \infty)$, let $q(x)$ be continuous and satisfy $\lim_{x\to\infty} q(x) = q_0$. Then every solution of the Schroedinger equation with $\lambda < q_0$ that is bounded on the interval $(0, \infty)$ is square-integrable.*

COROLLARY 2. *Let $q(x) \in \mathcal{C}$ on the line $(-\infty, \infty)$, and let $q(x)$ tend to limits q_0 and q_1, respectively, as $x \to \pm\infty$. Then every eigenfunction with eigenvalue $\lambda < \min (q_0, q_1)$ is square-integrable.*

The final conclusions can be summarized in a single theorem.

THEOREM 14. *Let $q(x)$ be as in Theorem 13. Then, for $\lambda > 0$, the spectrum is continuous. For $\lambda < 0$, the eigenfunctions are square-integrable.*

It can also be shown that, for $\lambda > 0$, the eigenfunctions are not square-integrable, and that for $\lambda < 0$, the spectrum is discrete.

EXERCISES I

1. Show that the S-L system: $u'' + \lambda u = 0$, $0 \leq x < \infty$, $\alpha u(0) + \alpha' u'(0) = 0$, $u(x)$ bounded as $x \to \infty$, has a continuous spectrum $0 < \lambda < \infty$ of $\alpha\alpha' \neq 0$.

2. Show that, if $u'' - q(x)u = 0$, $0 \leq x < \infty$, with $q(x)$ bounded, the DE cannot have two square-integrable linearly independent solutions. [HINT: Use the Wronskian.]

3. In $u'' + [\lambda - q(x)]u = 0$, $0 \leq x < \infty$, if $q(x) \to +\infty$ as $x \to \infty$, show that, for any λ, the DE has exactly one square-integrable solution up to a constant factor.

*4. Under the assumptions of Ex. 3, show that the S-L system corresponding to the boundary condition $u(0) = 0$, $u(x)$ square-integrable in $[0, \infty)$, has an infinite sequence of eigenvalues.

5. Show that, if the DE $u'' + q(x)u = 0$, $0 \leq x < \infty$, $q \in \mathcal{C}$ has a solution $u_1(x)$ with $\lim_{x\to\infty} u_1(x) = 1$, it also has a solution $u_2(x)$ such that $\lim_{x\to\infty} u_2(x)/x = 1$.

6. In $u'' + q(x)u = 0$, $0 \leq x < \infty$, suppose that $\int_0^\infty x|q(x)|\, dx < \infty$. Show that the DE

has a solution with $\lim_{x\to\infty} u(x) = 1$. [HINT: Show, by successive approximations, that the integral equation $u(x) = 1 - \int_x^\infty (t - x)q(t)u(t)\,dt$ has a solution.]

7. Suppose that all solutions of the DE $u'' + q(x)u = 0$ are bounded as $x \to \infty$ and that $\int_0^\infty \rho(x)\,dx < \infty$, $\rho(x) > 0$. Show that, for all λ, all solutions of the DE $u'' + (q(x) + \lambda\rho(x))u = 0$ are also bounded as $x \to \infty$. [HINT: Consider the inhomogeneous DE $u'' + qu = -\lambda\rho u$, and show that the integral equation obtained by variation of parameters has a bounded solution.]

8. Show that, if $k^2 > 0$ and $\int_0^\infty |q(x) - k^2|\,dx < \infty$, all solutions of the DE $u'' + q(x)u = 0$ are bounded as $x \to \infty$.

*9. Show that solutions of the generalized Laguerre DE

$$u'' + \frac{2}{x}u' + \left[\frac{\lambda}{x} - \left(\frac{1}{4} + \frac{\alpha}{x^2}\right)\right]u = 0$$

are $u = e^{-x/2}x^{(k-1)/2}L_n^{(k)}(x)$, where $L_n^{(k)}(x) = d^k[L_n(x)]/dx^k$, for $\alpha = (k^2 - 1)/4$ and $\lambda = n - (k - 1)/2$, n, k any nonnegative integers.

ADDITIONAL EXERCISES

1. Show that, if $a, b > 0$, the singular S-L system

$$\frac{d}{dx}\left[(1 + x)^{b+1}(1 - x)^{a+1}\frac{du}{du}\right] + \lambda(1 + x)^b(1 - x)^a u = 0, \qquad -1 < x < 1$$

with the endpoint condition that u remain bounded as $x \to \pm 1$, has the eigenvalues $\lambda_n = n(n + a + b + 1)$ and eigenfunctions $u_n(x) = P_n^{(a,b)}(x)$ (Jacobi polynomials).

2. Obtain orthogonality relations for the Jacobi polynomials.

3. Using Rodrigues' formula, show that between any two zeros of $P_n^{(a,b)}$ there is exactly one zero of $P_{n+1}^{(a,b)}$, if $a, b > -1$.

*4. Derive the following identities for Legendre polynomials:
   (a) $\int_{-1}^1 P_n^2(x)\,dx = 2/(2n + 1)$ (b) $\int_{-1}^1 xP_n(x)P_n'(x)\,dx = 2n/(4n^2 - 1)$
   [HINT: Use Rodrigues' formula and integrate by parts.]

*5. Show that the Legendre DE, with the endpoint condition

$$\lim_{x\to\pm 1} [(1 - x^2)u'(x)] = 0$$

has the Legendre polynomials as eigenfunctions, and no other eigenfunctions.

6. Show that there exists a bounded differentiable function g on $a < x < b$, satisfying the inequality $g' + g^2/P(x) + Q(x) \le 0$, if and only if no solution of $(Pu')' + Qu = 0$ has more than one zero on $a \le x \le b$.

*7. Show that, if $\int_a^b |Q(x)|\,dx \le 4/(b - a)$, no nontrivial solution of $u'' + Q(x)u = 0$ can have more than one zero in $a \le x \le b$. [HINT: By Theorem 3, it can be assumed that $Q \ge 0$. Changing coordinates so that $a = 0$, $b = 1$, use Ex. 6 with

$$g(x) = \int_x^1 Q(t)\,dt + \begin{cases} (1/x) - 4, & 0 < x \le \frac{1}{2} \\ 1/(x - 1), & \frac{1}{2} \le x < 1 \end{cases}$$

*8. (Fubini). Show that if, for $a \le x \le b$,

$$p'(x) + p(x)^2 - q(x) \le p_1'(x) + p_1(x)^2 - q_1(x)$$

then, between any two zeros of a solution of $u'' + 2p_1u' + q_1u = 0$, there is at least one zero of $u'' + 2pu' + qu = 0$. [HINT: See Ch. 2, Ex. B4.]

9. For a regular S-L system with $\alpha\alpha' < 0$, and $\beta\beta' < 0$, and λ less than the smallest eigenvalue, show that the Green's function is negative.

CHAPTER 11

EXPANSIONS IN EIGENFUNCTIONS

1 FOURIER SERIES

One of the major mathematical achievements of the nineteenth century was the proof that all sufficiently smooth functions can be *expanded* into infinite series, whose terms are constant multiples of the *eigenfunctions* of any S-L system with discrete spectrum. The present chapter will be devoted to proving this result for regular S-L systems and explaining some of its applications.

The most familiar example of such an expansion into eigenfunctions is the expansion into Fourier series. We begin by recalling† from the advanced calculus two basic results about Fourier series. The first of these is the following.

FOURIER'S CONVERGENCE THEOREM. *Let $f(x)$ be any continuously differentiable periodic function of period 2π, and let*

$$(1) \qquad a_k = \frac{1}{\pi} \int_{-\pi}^{\pi} f(x) \cos kx \, dx, \qquad b_k = \frac{1}{\pi} \int_{-\pi}^{\pi} f(x) \sin kx \, dx$$

Then the infinite series

$$(2) \qquad a_0/2 + a_1 \cos x + b_1 \sin x + a_2 \cos 2x + b_2 \sin 2x + \cdots$$

converges uniformly to $f(x)$.

Note that the nonzero terms $a_1 \cos x$, $b_1 \sin x$, ... in (2) are actually themselves eigenfunctions of the periodic Sturm-Liouville system in question (Example 3 of Ch. 10); hence $f(x)$ is represented as a *sum* of eigenfunctions in (2). However, we shall adopt the usual convention of referring to the normalized $\cos kx$ and $\sin kx$ in (2) as *the* eigenfunctions of the system.

Though there exist continuous functions whose Fourier series are not convergent, the following sharpened form of Fourier's Convergence Theorem applies to all continuous periodic functions.

---

† Fourier's theorem is proved in Courant and John, p. 594 ff; Fejér's Theorem is proved in Widder, p. 423.

FEJÉR'S CONVERGENCE THEOREM. *Let $f(x)$ be any continuous periodic function of period 2π, and let*

$$\sigma_N(x) = \frac{1}{N}\left\{\sum_{n=0}^{N-1}\left[\frac{a_0}{2} + \sum_{k=1}^{n}(a_k\cos kx + b_k\sin kx)\right]\right\}$$

$$= \frac{a_0}{2} + \sum_{k=1}^{N-1}(\alpha_k^N\cos kx + \beta_k^N\sin kx)$$

where $\alpha_k^N = (1 - (k/N))a_k$, $\beta_k^N = (1 - (k/N))b_k$, be the arithmetic mean of the first N partial sums of the Fourier series of $f(x)$. Then the sequence of functions $\sigma_N(x)$ converges uniformly to $f(x)$.

The preceding results, which we will assume as known, yield as corollaries the following statements about cosine series and about sine series. Let $f(x)$ be continuous on $0 \le x \le \pi$; define a function $g(x)$ for $-\pi \le x \le \pi$ by the equation $g(x) = f(|x|)$. Since $g(-\pi) = g(\pi)$, $g(x)$ can be extended to an even periodic function of period 2π, which is defined and continuous for all real x. By symmetry, all coefficients b_k are zero in the Fourier series of $g(x)$. Applying Fejér's and Fourier's Convergence Theorems, we have the following corollary.

COROLLARY 1. *Any continuous function on $0 \le x \le \pi$ can be approximated uniformly and arbitrarily closely by linear combinations of cosine functions. If the function is of class $\mathcal{C}^1$ and $f'(0) = f'(\pi) = 0$, then it can be expanded into a uniformly convergent series of cosine functions:*

(2')
$$f(x) = (a_0/2) + a_1\cos x + a_2\cos 2x + \cdots$$

By a linear transformation of the independent variable, the preceding result can be extended to any closed interval $[a, b]$; the required cosine functions are the functions $\cos[k\pi(x - a)/(b - a)]$.

Similarly, if $f(0) = f(\pi) = 0$, define $h(x)$ as $f(x)$ on $0 \le x \le \pi$, and as $-f(-x)$ on $-\pi \le x \le 0$. This gives an odd continuous periodic function of period 2π, in whose Fourier series all a_k vanish.

COROLLARY 2. *Any function of class $\mathcal{C}^1$ on $0 \le x \le \pi$ that satisfies $f(0) = f(\pi) = 0$ can be expanded into a uniformly convergent series of sine functions.*

The preceding corollaries are examples of expansions into the eigenfunctions of the regular S-L systems defined by the DE $u'' + \lambda u = 0$ and the two separated endpoint conditions $u'(0) = u'(\pi) = 0$ and $u(0) = u(\pi) = 0$, respectively. We will prove below that analogous expansions are possible into the eigenfunctions of *any* regular S-L system.

2 ORTHOGONAL EXPANSIONS

Let $\phi_1(x)$, $\phi_2(x)$, $\phi_3(x)$, . . . be any bounded, square-integrable functions on an interval I: $a < x < b$, orthogonal with respect to a positive weight function $\rho(x)$, so that

$$(3) \qquad \int_I \phi_h(x)\phi_k(x)\rho(x)\,dx = 0 \qquad \text{if} \qquad h \neq k$$

Suppose that a given function $f(x)$ can be expressed as the limit of a *uniformly* convergent series of multiples of the ϕ_k, so that

$$(4) \qquad f(x) = c_1\phi_1(x) + c_2\phi_2(x) + c_3\phi_3(x) + \cdots = \sum_{h=1}^{\infty} c_h\phi_h(x)$$

Multiplying both sides of (4) by $\phi_k(x)\rho(x)$, and integrating term-by-term over the interval—as is possible for uniformly convergent series—we get from the orthogonality relations (3) the equation

$$\int_I f(x)\phi_k(x)\rho(x)\,dx = \sum_{h=1}^{\infty} \int_I c_h\phi_h(x)\phi_k(x)\rho(x)\,dx = c_k \int_I \phi_k{}^2(x)\rho(x)\,dx$$

Hence, the coefficients c_h in (4) must satisfy the equation

$$(5) \qquad c_h = \left\{ \int_I f(x)\phi_h(x)\rho(x)\,dx \right\} \Big/ \left\{ \int_I \phi_h{}^2(x)\rho(x)\,dx \right\}$$

When the ϕ_k are the trigonometric functions, from this identity we obtain, as a special case, the coefficients $c_1 = a_0/2$, $c_2 = a_1$, $c_3 = b_1$, . . . of the Fourier series (1)–(2) with $\rho = 1$, using the familiar integrals

$$\int_{-\pi}^{\pi} dx = 2\pi, \qquad \int_{-\pi}^{\pi} \cos^2 kx\,dx = \int_{-\pi}^{\pi} \sin^2 kx\,dx = \pi$$

for any nonzero integer k.

We can summarize the preceding result as follows.

THEOREM 1. *If a function $f(x)$ is the limit $f(x) = \Sigma c_k\phi_k(x)$ of a uniformly convergent series of constant multiples of bounded square-integrable functions $\phi_k(x)$ that are orthogonal with respect to a weight function $\rho(x)$, the coefficients c_h are given by (5).*

The preceding conclusion holds provided that one can integrate the series $\Sigma c_h\phi_h(x)\phi_k(x)\rho(x)$ term-by-term on the interval I. This holds much more generally than for uniform convergence, e.g., for mean-square convergence as defined in §3.

The preceding conclusion was justified by using the fact that uniformly convergent series can be integrated term-by-term on any finite interval *I*. Many other series of orthogonal functions also can be integrated term-by-term, and formula (5), therefore, also holds for them, as we shall prove in later sections.

3 MEAN-SQUARE APPROXIMATION

So far, we have considered only uniformly convergent series, because these can be integrated term-by-term. The notion of convergence most appropriate for orthogonal expansions is, however, not uniform convergence but *mean-square convergence*, which we now define.

DEFINITION. Let *f* and the terms of the sequence $\{f_n\}$ $(n = 1, 2, 3, \ldots)$ be square-integrable real functions. The sequence $\{f_n\}$ is said to converge to *f* in the *mean* square on *I*, with respect to the positive weight function $\rho(x)$, when

$$(6) \qquad \int_I [f_n(x) - f(x)]^2 \rho(x)\, dx \to 0, \qquad \text{as } n \to \infty$$

Now, suppose that $\phi_1, \phi_2, \phi_3, \ldots$ form an infinite sequence of *square*-integrable functions on the interval *I*, *orthogonal* with respect to the weight function ρ, and let $f_n(x) = \gamma_1 \phi_1(x) + \cdots + \gamma_n \phi_n(x)$ be the *n*th partial sum of the series $\sum_{k=1}^{\infty} \gamma_k \phi_k(x)$. To make the partial sums f_n converge in the mean square to *f* as rapidly as possible, we choose the coefficients γ_k so as to minimize the expression:

$$(7) \qquad E = E(\gamma_1, \ldots, \gamma_n) = \int_I \left[f(x) - \sum_{k=1}^{n} \gamma_k \phi_k(x) \right]^2 \rho(x)\, dx$$

Expanding (7), and using the orthogonality relations (3), the function *E* of the variables $\gamma_1, \gamma_2, \ldots, \gamma_n$ is given by the expression

$$(7') \qquad E = \int_I f^2 \rho\, dx - 2\sum_{k=1}^{n} \gamma_k \int_I f\phi_k \rho\, dx + \sum_{k=1}^{n} \gamma_k^2 \int_I \phi_k^2 \rho\, dx$$

Now, consider the numbers $\gamma_1, \gamma_2, \ldots, \gamma_n$ that minimize the function *E*. Since *E* is differentiable in each of its variables, the minimum can be attained only by setting every $\partial E/\partial \gamma_k = 0$. That is, a necessary condition for a minimum is that the γ_k satisfy the equations

$$0 = -2\int_I f\phi_k \rho\, dx + 2\gamma_k \int_I \phi_k^2 \rho\, dx$$

Solving for γ_k, we get $\gamma_k = \{\int f\phi_k \rho\, dx\}/\{\int \phi^2 \rho\, dx_k\}$, which is the same as equation (5) for the c_k, in another notation.

We now show that the choice $\gamma_k = c_k$, where, as in (5),

(8) $$c_k = \left\{ \int_I f(x)\phi_k(x)\rho(x)\,dx \right\} \bigg/ \left\{ \int_I \phi_k{}^2(x)\rho(x)\,dx \right\}$$

does indeed give a minimum for E. A simple calculation, completing the square, gives for E the expression

(8') $$E = \int_I \left[f - \sum \gamma_k \phi_k \right]^2 \rho\,dx$$
$$= \int_I f^2\rho\,dx + \sum_{k=1}^{n} \left[-c_k{}^2 + (\gamma_k - c_k)^2 \right] \int_I \phi_k{}^2\rho\,dx$$

The right side shows that the minimum is attained if and only if $\gamma_k = c_k$. This proves the following result, for any interval I.

THEOREM 2. *Let $\{\phi_k(x)\}$ be a sequence of orthogonal square-integrable functions, and let f be square-integrable. Then, among all possible choices of $\gamma_1, \ldots, \gamma_n$, the integral (7) is minimized by selecting $\gamma_k = c_k$, where c_k is defined by (8).*

The coefficients c_k are called the *Fourier coefficients* of f relative to the orthogonal sequence ϕ_k.

The partial sum $c_1\phi_2(x) + \cdots + c_n\phi_n(x)$ in Theorem 1 is thus, for each n, the *best mean-square approximation to $f(x)$* among all possible sums $\gamma_1\phi_1(x) + \cdots + \gamma_n\phi_n(x)$; it is often called the *least square approximation* to $f(x)$ because it minimizes the mean square difference (7). The remarkable feature of least-square approximation by orthogonal functions is that the kth coefficient γ_k in the list $(\gamma_1, \ldots, \gamma_k)$ which gives the best mean-square approximation to f is *the same for all $n \geq k$.* This "finality property" does not hold, for example, in the case of least-squares approximation by nonorthogonal functions, or of the approximations in Fejér's Theorem, or of best *uniform* approximation minimizing the functional $\sup_{a<x<b}| f(x) - \Sigma_{k=1}^{n} c_k\phi_k(x)|$.

Orthonormal Functions. The preceding formulas become much simpler when the orthogonal functions ϕ_k are *orthonormal*, in the sense that $\int \phi_k{}^2\rho\,dx = 1$. For a sequence of orthonormal functions, the formula for the Fourier coefficients is $c_k = \int_I f\phi_k\rho\,dx$. We can easily construct, from any sequence ϕ_k of orthogonal functions, an orthonormal sequence ψ_k by setting $\psi_k = \phi_k/[\int_I \phi_k{}^2\rho\,dx]^{1/2}$. For example, the functions

$$\frac{1}{\sqrt{2\pi}}, \qquad \frac{1}{\sqrt{\pi}} \cos kx, \qquad \frac{1}{\sqrt{\pi}} \sin kx$$

are orthonormal on $-\pi \leq x \leq \pi$ with respect to the weight function $\rho(x) \equiv 1$.

Substituting the condition $\int \phi_k^2 \rho \, dx = 1$ into (8′) and remarking that E is nonnegative, we obtain the following important corollary.

COROLLARY 1. *Let $\sum_1^n c_k \phi_k$ be the least-square approximation to f by a linear combination of orthonormal functions ϕ_k. Then*

$$(9) \qquad \sum_1^n c_k^2 \leq \int_I f^2(x)\rho(x) \, dx$$

For the right member of (9) to be finite, it is necessary that $f^2 \rho$ be integrable, that is, that f be *square-integrable* with respect to the weight function ρ. When this is the case, the integrals (8) are also well-defined by the Schwarz inequality. Under these circumstances, since the right side of (9) is independent of n, if we let n tend to infinity, we will still have

$$(10) \qquad \sum_1^\infty c_k^2 \leq \int_I f^2(x)\rho(x) \, dx < +\infty \qquad \text{(Bessel inequality)}$$

That is, *the Fourier coefficients of any square-integrable function f form a square-summable sequence of numbers if the ϕ_k are orthonormal.*

EXERCISES A

1. Show that $a_k \cos kx + b_k \sin kx = (1/\pi) \int_{-\pi}^{\pi} f(t) \cos [k(t-x)] \, dt$.

2. Show that $\frac{1}{2} + \sum_{k=1}^n \cos kx = \sin [(2n+1)x/2]/[2 \sin (x/2)]$.

3. Using Ex. 2, infer that

$$\frac{a_0}{2} + \sum_{k=1}^n (a_k \cos kx + b_k \sin kx) = \frac{1}{\pi} \int_{-\pi}^{\pi} f(t) \frac{\sin [(2n+1)(t-x)/2]}{2 \sin [(t-x)/2]} \, dt$$

4. (a) Prove in detail Corollaries 1 and 2 of Fejér's Theorem, discussing with care the differentiability at 0 and π of the periodic functions constructed.
   (b) Find necessary and sufficient conditions for a continuous function on $[0, \pi]$ to be uniformly approximable by a linear combination of functions $\sin kx$.

5. Show that, in Fejér's convergence theorem,

$$\sigma_n(x) = \frac{1}{2\pi n} \int_{-\pi}^{\pi} f(x+t) \left[\frac{\sin (nt/2)}{\sin (t/2)} \right]^2 \, dt$$

*6. Prove Fejér's theorem, assuming that Ex. 5 holds.

For the regular S-L systems in Exs. 7 and 8, (a) find the eigenvalues and eigenfunctions, (b) obtain an expansion formula for a function $f \in \mathcal{C}^1$ into a series of eigenfunctions.

7. $u'' + \lambda u = 0$, $u(0) = 0$, $u'(\pi) = 0$, $0 \leq x \leq \pi$.

8. The same DE with $u'(0) = 0$, $u(\pi) = 0$.

9. Show that the trigonometric functions are orthogonal, for any a, in

$$-\pi - a \leq x \leq \pi - a$$

4   COMPLETENESS

The most important question about a sequence of continuous functions ϕ_k ($k = 1, 2, 3, \ldots$), orthogonal and square-integrable with respect to a weight function ρ, is the following: Can *every* square-integrable function f be expanded into an infinite series† $f = \sum_1^\infty c_k\phi_k$ of the ϕ_k? When this is possible for every continuous f,‡ the sequence of orthogonal functions ϕ_k is said to be *complete*.

Using the fundamental equation (8') on mean-square approximation, we can reformulate the definition of completeness as follows. In order that

$$\lim_{n\to\infty} \int_I \left[f(x) - \sum_{k=1}^n \gamma_k\phi_k \right]^2 \rho(x)\, dx = 0$$

it is necessary and sufficient that

$$\lim_{n\to\infty} \left\{ \left[\int_I f^2\rho\, dx - \sum_1^n c_k^2 \int_I \phi_k^2\rho\, dx \right] + \sum_{k=1}^n (\gamma_k - c_k)^2 \int_I \phi_k^2\rho\, dx \right\} = 0$$

Since the term in square brackets is nonnegative by the Bessel inequality (10), and since $\int \phi_k^2\rho\, dx > 0$ for any nontrivial ϕ_k, the limit is zero if and only if $\gamma_k = c_k$ for all k, and equality holds in the Bessel inequality (10). This proves the following results.

THEOREM 3.   *A sequence $\{\phi_k\}$ of functions $\phi_k(x)$, orthogonal and square-integrable with positive weight $\rho(x)$ on an interval I, is complete if and only if*

$$\int_I f^2(x)\rho(x)\, dx = \sum_{k=1}^\infty \left\{ \left[\int_I f(x)\phi_k(x)\rho(x)\, dx \right]^2 \bigg/ \int_I \phi_k^2(x)\rho(x)\, dx \right\}$$

for all continuous square-integrable functions f.

---

† Here and below, the equation $f = \sum_1^\infty c_k\phi_k$ is to be interpreted in the sense of mean-square convergence, namely, that the partial sums $\sum_{k=1}^n c_k\phi_k$ converge in the mean square to the function f with respect to ρ.

‡ If every continuous function can be expanded into a series $\sum_1^\infty c_k\phi_k$, then many discontinuous functions also have such an expansion, convergent in the mean square. The class of all such functions is that of all Lebesgue square-integrable functions (see §11). We are here considering only continuous functions in order to avoid assuming a knowledge of the Lebesgue integral.

COROLLARY 1.   *If the $\phi_k(x)$ are orthonormal, a necessary and sufficient condition for completeness is the validity of the* Parseval *equality*

$$(11) \qquad \int_I f^2(x)\rho(x)\,dx = \sum_{k=1}^{\infty} \left[\int_I f(x)\phi_k(x)\rho(x)\,dx \right]^2$$

for all continuous square-integrable functions f.

For example, take the case of Fourier series. In the notation of (1), the condition for the completeness of the functions 1, cos kx, sin kx on $-\pi \le x \le \pi$ is that, for all continuous functions f,

$$(12) \qquad \pi \left[\frac{a_0^2}{2} + \sum_{k=1}^{\infty} (a_k^2 + b_k^2) \right] = \int_{-\pi}^{\pi} f^2(x)\,dx$$

It follows from Fourier's Convergence Theorem, integrating the squares of the partial sums of (2), that the identity (12) holds if f is a continuously differentiable periodic function.

We shall now prove that the identity (12) holds for all continuous periodic functions f. By Fejér's Convergence Theorem, the sums

$$(13) \qquad \sigma_N(x) = \frac{a_0}{2} + \sum_{k=1}^{N-1} \left(1 - \frac{k}{N}\right) a_k \cos kx + \sum_{k=1}^{N-1} \left(1 - \frac{k}{N}\right) b_k \sin kx$$

converge uniformly for $-\pi \le x \le \pi$ to a continuous periodic function $f(x)$. Therefore, $\int_{-\pi}^{\pi} \sigma_N^2\,dx$ converges as $N \to \infty$ to $\int_{-\pi}^{\pi} f^2\,dx$. Evaluating the integral by (13), we find that

$$(14) \qquad \lim_{N\to\infty} \pi \left[\frac{a_0^2}{2} + \sum_{k=1}^{N-1} \left(1 - \frac{k}{N}\right)^2 (a_k^2 + b_k^2) \right] = \int_{-\pi}^{\pi} f^2(x)\,dx$$

Now, by the Bessel inequality, we have

$$(14') \qquad \pi \left[\frac{a_0^2}{2} + \sum_{k=1}^{\infty} (a_k^2 + b_k^2) \right] \le \int_{-\pi}^{\pi} f^2\,dx < \infty$$

Since $[1 - (k/N)]^2 \le 1$, it follows that, if we replace the sum in square brackets on the left side of (14) by $a_0^2/2 + \sum_{k=1}^{N} (a_k^2 + b_k^2)$, we will get an increasing sequence whose limit is at least equal to $\int f^2\,dx$. But, by (14'), this limit is at most equal to $\int f^2\,dx$. Hence, the limit is exactly $\int f^2\,dx$, and (12) is proved. Since any continuous function on $-\pi \le x \le \pi$ can be given an arbitrarily close mean-square approximation by a continuous function satisfying $f(-\pi) = f(\pi)$, this proves

COROLLARY 2. *The trigonometric functions* 1, cos *kx,* sin *kx* (*k* = 1, 2, . . .) *are a complete orthogonal sequence in the interval* $-\pi \leq x \leq \pi$.

Using the method of Corollary 2 of §1 and changing variables, we obtain another corollary.

COROLLARY 3. *The functions* cos [*kπ*(*x* − *a*)/(*b* − *a*)], (*k* = 0, 1, 2, . . .), *form a complete orthogonal sequence in the interval* $a \leq x \leq b$.

We conclude this section with the following criterion for completeness of a sequence of orthogonal functions, which relates the notion of completeness to that of *approximation* in the sense of mean-square convergence.

THEOREM 4. *Let* $\{\phi_k\}$ (*k* = 1, 2, . . .) *be any sequence of orthogonal square-integrable functions on an interval I, relative to a weight function* $\rho > 0$. *The sequence is complete if and only if every continuous square-integrable function can be approximated arbitrarily closely in the mean square by a linear combination of the* ϕ_k.

Proof. The condition is clearly necessary. Conversely, suppose that, given $\epsilon > 0$, we can find a linear combination $\Sigma_{k=1}^{n} \gamma_k \phi_k$ such that

$$\int_I \left(f - \sum_{k=1}^{n} \gamma_k \phi_k \right)^2 \rho \, dx < \epsilon$$

If we replace each of the γ_k by the Fourier coefficients c_k of f relative to ϕ_k—as given by formula (5)—then by Theorem 2 the square integral on the left decreases:

$$\int_I \left(f - \sum_{k=1}^{n} c_k \phi_k \right)^2 \rho \, dx < \epsilon$$

But this is precisely what we had set out to prove.

5 ORTHOGONAL POLYNOMIALS

We shall now prove the completeness of the eigenfunctions of some of the singular S-L systems studied in Ch. 10. These are the S-L systems on a *finite* interval whose eigenfunctions are polynomials, such as the Legendre polynomials.

We can use any positive *weight function* $\rho(x)$ on an interval (*a, b*) with the property that $\int_a^b x^n \rho(x) \, dx$ is convergent for all $n \geq 0$, to construct an infinite sequence of polynomial functions $P_0(x), P_1(x), P_2(x), \ldots$ with $P_n(x)$ of degree n, which are orthogonal on (*a, b*) with respect to this weight function, so that

(15) $$\int_a^b P_m(x) P_n(x) \rho(x) \, dx = 0, \qquad m \neq n$$

Equations (15) define $P_n(x)$ uniquely up to an arbitrary factor of proportionality, the normalization constant.

Given a weight function $\rho(x)$, one can compute the $P_n(x)$ explicitly from (15); the computations will not be described here.† Instead, we shall derive some interesting general properties of orthogonal polynomials.

We shall first establish the fact that, on any *finite* interval, such sequences of orthogonal polynomials are *complete*. To prove this, we will need the following result.

LEMMA. *Every uniformly convergent sequence of continuous functions is mean-square convergent on any interval I, with respect to any integrable positive weight function* ($\int_I \rho\, dx < \infty$).

This follows immediately from the inequality

$$(16) \qquad \int_I [f_n(x) - f(x)]^2 \rho(x)\, dx \leq \max\, [(f_n(x) - f(x))^2] \int_I \rho(x)\, dx$$

valid when I is any finite or infinite interval. On an infinite interval, however, we must carefully check the integrability of the weight function. For instance, the functions $f_n(x) = n^{-1/2} \exp(-x^2/n^2)$ converge uniformly to the zero function on the interval $-\infty < x < \infty$, but the integrals $\int_{-\infty}^{\infty} f_n^2(x)\, dx$ do not converge to zero.

Using this lemma, it is easy to prove the completeness of a sequence of *orthogonal polynomials* defined on a *finite* interval I, relative to any continuous integrable weight function $\rho(x)$ from the fundamental

WEIERSTRASS APPROXIMATION THEOREM. *Let $f(x)$ be any function continuous on a finite closed interval $a \leq x \leq b$, and let $\epsilon > 0$ be any positive number. Then there exists a polynomial $p(x)$, such that $|p(x) - f(x)| \leq \epsilon$, for all x on $a \leq x \leq b$.*‡

From this theorem, and the inequality (16), we infer

THEOREM 5. *Let $P_n(x)$ ($n = 0, 1, 2, \ldots$) be a polynomial function of degree n. For a fixed interval $I: a \leq x \leq b$, let*

$$\int_a^b P_m(x)P_n(x)\rho(x)\, dx = 0, \qquad \text{if} \qquad m \neq n$$

where $\rho(x)$ is a continuous integrable positive weight function. Then the orthogonal polynomials $P_n(x)$ are complete on I.

Proof. Let $p(x)$ be any polynomial of degree n. We can find c_n such that $p(x) - c_n P_n(x)$ is a polynomial of degree $n - 1$ or less. Hence, by induction on n, we

---

† It is the Gram-Schmidt orthogonalization process applied to the vectors $1, x, x^2, \ldots$. This process can be applied in any Euclidean vector space (Birkhoff and MacLane, p. 204).

‡ See Widder, p. 426, or Courant-Hilbert, Vol. 1, p. 65.

can express $p(x)$ as a finite linear combination of $P_0(x), \ldots, P_n(x)$. By the Weierstrass Approximation Theorem, we can approximate uniformly *any* continuous function arbitrarily closely by a suitable polynomial $p(x)$. By the preceding lemma, every continuous function can, therefore, be approximated arbitrarily closely in the mean square by a linear combination of the P_k. The result now follows from Theorem 4.

The completeness of Legendre, Chebyshev, Gegenbauer (or ultraspherical), and other Jacobi polynomials (see Ch. 9, §11) follows as a corollary. But it is harder to prove the completeness of polynomials orthogonal on semi-infinite and infinite intervals, such as the Hermite polynomials and the Laguerre polynomials introduced in the next section.

EXERCISES B

1. Using Fourier's Convergence Theorem, show that the eigenfunctions of $u'' + \lambda u = 0$ for the separated boundary conditions $u(0) = u'(\pi) = 0$ are complete on $(0, \pi)$.

2. Show that, if $f_n \to f$ in the mean square, and c_k, $c_k^{(n)}$ are the Fourier coefficients of f, f_n relative to a given orthonormal sequence ϕ_k, then $c_k^{(n)} \to c_k$ uniformly in k.

3. Using expansions into Legendre polynomials, obtain a formula for the best mean-square approximation in $|x| \leq 1$ of a square-integrable function by polynomials of degree $\leq n$.

*4. Using the Liouville substitution, obtain from Fourier's theorem an expansion theorem for functions $f \in \mathcal{C}^2[-1, 1]$ into series of Chebyshev polynomials.

5. Show that, given square-integrable functions $f_1, \ldots, f_n$, a sequence $\phi_1, \ldots, \phi_m$ of orthonormal square-integrable functions can be found for which f_k is a linear combination of $\phi_1, \ldots, \phi_k$, $1 \leq k \leq m$.

*6   PROPERTIES OF ORTHOGONAL POLYNOMIALS

We shall now develop some of the properties of orthogonal polynomials which depend only on the fact that they are orthogonal, irrespective of completeness. These properties apply to the polynomials whose completeness was proved in §5 (see last paragraph). They apply also to the Hermite and Laguerre polynomials, whose completeness will not be proved in this book. All these polynomials have been met before, except the Laguerre polynomials, which we now define.

Example 1.   Consider the singular S-L system consisting of the DE

$$(17) \qquad (xu')' + \left[\alpha + \frac{(2 - x)}{4} \right] u = 0, \qquad 0 < x < \infty$$

with the endpoint conditions that $u(x)$ is bounded as $x \to \infty$ and as $x \to 0$. Setting $u(x) = v(x)e^{-x/2}$, we get the Laguerre DE

$$(18) \qquad xv'' + (1 - x)v' + \alpha v = 0$$

Trying $v = \sum_{k=0}^{\infty} a_k x^k$, the Method of Undetermined Coefficients (Ch. 4, §2) gives the recurrence relation $a_{k+1} = (k - \alpha)a_k/(k + 1)^2$. Hence, we have

$$(19) \qquad a_k = (-1)^k \alpha(\alpha - 1)(\alpha - 2) \cdots \frac{(\alpha - k + 1)a_0}{(k!)^2}$$

The series is a polynomial if and only if $\alpha = n$, a nonnegative integer; otherwise, it represents a function that grows exponentially at infinity. Normalizing the polynomial (for $\alpha = n$) by the condition $a_n = (-1)^n/(n!)$, we get the *Laguerre polynomials*†:

$$(19') \qquad L_n(x) = \sum_{k=0}^{n} (-1)^k \binom{n}{k} \frac{x^k}{k!}$$

For example, $L_0(x) = 1$, $L_1(x) = 1 - x$, $L_2(x) = 1 - 2x + \frac{1}{2}x^2$,

$$L_3(x) = 1 - 3x + \tfrac{3}{2}x^2 - \tfrac{1}{6}x^3, \ldots .$$

Thus, the functions $L_n(x)e^{-x/2}$ are eigenfunctions of a singular S-L system. These functions are certainly square-integrable, together with their derivatives, hence Theorem 2 of Ch. 10 applies, giving the orthogonality relations

$$\int_0^{\infty} e^{-x} L_m(x)L_n(x)\, dx = 0, \qquad m \neq n$$

We shall now consider some of the fundamental properties of an arbitrary sequence of *orthogonal polynomials* (not necessarily solutions of a DE) $P_0(x)$, $\ldots, P_n(x), \ldots$, where P_n is of degree n. As in §5, we assume that the weight function $\rho(x)$ is such that all products $x^n \rho(x)$ are integrable on I. We do not assume that the interval I is finite.

We first derive a result similar to the Sturm Oscillation Theorem.

THEOREM 6. *Let $\{P_n\}$ ($n = 0, 1, 2, \ldots$) be any sequence of polynomials orthogonal on a given interval (a, b) where $P_n(x)$ has degree n. Then P_n has n distinct zeros, all contained in the interval (a, b).*

Proof. Suppose that $P_n(x)$ has fewer than n zeros in (a, b). Let $x_1, \ldots, x_m$ ($m < n$) be those zeros at which $P_n(x)$ changes sign. Then the polynomial $(x - x_1)(x - x_2) \cdots (x - x_m)P_n(x)$ would be of constant sign. Hence

$$\int_a^b (x - x_1) \cdots (x - x_m)P_n(x)\rho(x)\, dx \neq 0, \qquad m < n$$

---

† The normalizing condition $a_n = 1$ is also often used, and makes some formulas simpler.

But $(x - x_1) \cdots (x - x_m)$ is a polynomial of degree lower than n. Therefore, it can be written as a linear combination of the polynomials $P_0, \ldots, P_m$, say $\sum_{k=0}^{m} c_k P_k(x)$, $m < n$. Hence, we have

$$\int_a^b (x - x_1) \cdots (x - x_m) P_n(x) \rho(x) \, dx = \int_a^b \sum_{k=0}^{m} c_k P_k(x) P_n(x) \rho(x) \, dx$$

$$= \sum_{k=0}^{m} c_k \int_a^b P_k P_n \rho \, dx = 0$$

a patent contradiction, since the integrand is positive except at the x_i. Hence, $P_n(x)$ has at least n zeros on (a, b). Since a polynomial of degree n has, at most, n zeros, the proof is complete.

Next, we shall establish a *recursion formula* for an arbitrary system of orthogonal polynomials.

THEOREM 7. *Any three orthogonal polynomials of consecutive degree satisfy a linear relation*

(20) $$P_{n+1}(x) = (A_n x + B_n) P_n(x) + C_n P_{n-1}(x)$$

for suitable constants A_n, B_n, C_n.

Proof. First choose A_n such that $P_{n+1}(x) - x A_n P_n(x)$ is a polynomial of degree n or less, so that

$$P_{n+1}(x) - x A_n P_n(x) = \gamma_0 P_n(x) + \gamma_1 P_{n-1}(x) + \cdots + \gamma_n P_0$$

Multiplying both sides by $P_k(x)\rho(x)$, integrating from a to b, and using the orthogonality relation, we find that $\gamma_k = 0$ for $k = 2, 3, \ldots, n$. Hence

$$P_{n+1}(x) - x A_n P_n(x) = \gamma_0 P_n(x) + \gamma_1 P_{n-1}(x)$$

Therefore set $\gamma_0 = B_n$ and $\gamma_1 = C_n$, q.e.d.

The numerical values of the constants A_n, B_n, C_n in Theorem 7 depend on the normalizing factors used to define the orthogonal polynomials considered. For convenience, we have listed in Table 1 the recursion coefficients for some common polynomials.

EXERCISES C

Establish the following formulas for Hermite polynomials (see Ch. 4, §2):

1. $H_{n+1}(x) = 2x H_n(x) - 2n H_{n-1}(x)$.

2. $H_n'(x) = 2n H_{n-1}(x)$.

Table 1.   Recursion Coefficients

Polynomial	A_n	B_n	C_n
Legendre	$\dfrac{2n+1}{n+1}$	0	$\dfrac{-n}{n+1}$
Chebyshev	2	0	-1
Gegenbauer	$\dfrac{2n+\lambda}{n+1}$	0	$\dfrac{1-n-2\lambda}{n+1}$
Hermite	2	0	$-2n$
Laguerre	$\dfrac{-1}{n+1}$	$\dfrac{2n+1}{n+1}$	$\dfrac{-n}{n+1}$

*3. $\sum_{k=0}^{\infty} H_k(x)t^k/k! = e^{xt-t^2/2}$ (generating function).

4. Establish the recursion formula for the Laguerre polynomials:

$$L_n(x) = L_n'(x) - L_{n+1}'(x)$$

[HINT: Differentiate the recursion formula for L_n.]

5. Show that for the functions $\phi_n(x) = e^{x/2}\, d^n/dx^n(e^{-x}x^n)$ are orthogonal in $0 < x < \infty$.

6. Infer from Ex. 5 that $\phi_n(x) = (n!)\, e^{-x/2}L_n(x)$, where L_n is the nth Laguerre polynomial.

7. Show that, if $L_n(x)$ is the Laguerre polynomial of degree n, then $d^k[L_n(x)]/dx^k$ satisfies the DE

$$xv'' + (k+1-x)v' + (n-k)v = 0$$

8. Prove the recursion formula of Table 1:

$$(n+1)L_{n+1}(x) = (2n+1-x)L_n(x) - nL_{n-1}(x)$$

9. Let $P_k(x)$ be a sequence of orthonormal polynomials with weight function ρ in $a < x < b$, and let $c_k = \int_a^b f(x)P_k(x)\rho(x)\, dx$. Show that the partial sum

$$\sigma_n(x) = \sum_{k=0}^{n} c_k P_k(x)$$

coincides with $f(x)$ in at least $n+1$ points of the interval. [HINT: Use a method similar to the proof of Theorem 6.]

*10. Show that, in Theorem 6, between any two zeros of P_n, there is exactly one zero of P_{n+1}.

*11. Show that the Legendre polynomials are the only Gegenbauer polynomials for which the maximum of $|P_n^a(x)|$, namely $P_n^a(1)$, is independent of n.

*12. (a) Expand the function $(1 - 2xh + h^2)^{-1/2}(|x| < 1)$ into a series of Legendre polynomials, and show that the nth Fourier coefficient is $2h^n/(2n+1)$.

(b) Obtain from (a) the formula

$$(1 - 2xh + h^2)^{-1/2} = \sum_{k=0}^{\infty} h^k P_k(x) \qquad \text{for} \qquad |h| < \sqrt{2} - 1$$

*13. Show that the generating function for the Laguerre polynomials is,

$$\frac{1}{1-t} \exp\left[\frac{i-xt}{1-t}\right] = \sum_{n=0}^{\infty} t^n L_n(x)/n!$$

In Exs. 14–17, $D = d/dx$ and $\rho(x)$ is positive. The method of proof is to find by induction a S-L equation satisfied by the expressions given.

*14. Show that the only orthogonal polynomials of the form

$$p_n(x) = K_n(\rho(x))^{-1} D^n[\rho(x)] \qquad \text{for} \qquad \rho(x) \in \mathcal{C}^{\infty}$$

are the Hermite polynomials.

*15. Show that the only orthogonal polynomials of the form

$$p_n(x) = K_n(\rho(x))^{-1} D^n[\rho(x)(ax + b)]$$

are the Laguerre polynomials, after a change of independent variable.

*16. Show that the only system of orthogonal polynomials of the form

$$K_n(\rho(x))^{-1} D^n[\rho(x)(ax^2 + bx + c)]$$

are the Jacobi polynomials, after a change of independent variable.

*17. Show that the only sequences of orthogonal polynomials that satisfy a Rodrigues formula $p_n(x) = K_n[\rho(x)]^{-1} D^n[\rho(x)p(x)]$, where p is a given polynomial, are the Jacobi, Laguerre, and Hermite polynomials.

*7   CHEBYSHEV POLYNOMIALS

The Chebyshev polynomials $T_n(x)$ were introduced in Ch. 9, §11, as solutions of the self-adjoint DE

$$(21) \qquad [(1 - x^2)^{1/2} u']' + \lambda(1 - x^2)^{-1/2} u = 0, \qquad -1 < x < 1$$

The Liouville normal form of this DE is $u_{\theta\theta} + \lambda u = 0$, which is obtained by setting $w = u$ and $\theta = \int_{-1}^{x} d\xi/\sqrt{1 - \xi^2}$, or $x = -\cos\theta$, $0 < \theta < \pi$. For integral n and $\lambda = n^2$, two linearly independent solutions of this equation are

$$(21a) \qquad \cos n\theta = T_n(\cos\theta) = T_n(x)$$

and $S_n(x)$ given by the formula

(21b) $S_n(x) = \sin n\theta = \sin \theta U_{n-1}(x)$

The $U_m(x)$ defined by (21b) are also polynomials of degree m, called *Chebyshev polynomials of the second kind*. Their theory is parallel to that of the $T_n(x)$ and is developed in Exs. D3–D7.

From the forms of the preceding explicit solutions, we see that the functions $T_n(x)$ are eigenfunctions of the singular S-L system defined from (21) by the boundary conditions that $u'(-1)$ and $u'(1)$ be finite. All solutions of (21) are bounded at the singular points $x = \pm 1$, as is apparent from inspection of explicit solutions (21a) to (21b) and also from a calculation of the roots $\nu = 0$, $\frac{1}{2}$ of the indicial equation $2\nu^2 - \nu = 0$ of the normal form of (21). But only multiples of the $T_n(x)$ have bounded derivatives at the endpoints.

Minimax Property.   The most striking property of the Chebyshev polynomials is contained in the following result.

THEOREM 8. *Among all monic polynomials $P(x) = x^n + \sum_{k=0}^{n-1} a_k x^k$ of degree n, $2^{1-n}T_n(x)$ minimizes $\max_{-1 \le x \le 1} |P(x)|$ (Minimax property).*

Proof.   For $n = 1$, the result follows by inspection. For $n \ge 2$, it follows by induction from the recursion formula

$$T_n(x) = 2xT_{n-1}(x) - T_{n-2}(x)$$

which is equivalent to the trigonometric identity

$$\cos(m\theta + \theta) = 2\cos\theta \cos m\theta - \cos(m\theta - \theta), \qquad n = m + 1$$

Next, since $T_n(\cos \theta) = \cos n\theta$, we have that

$$\max_{-1 \le x \le 1} |2^{1-n}T_n(x)| = 2^{1-n}$$

In order to establish the statement, it therefore suffices to show that for any monic polynomial of degree n we have

$$\max_{-1 \le x \le 1} |x^n + a_{n-1}x^{n-1} + \cdots + a_0| \ge 2^{1-n}$$

Suppose this were not so. Then, we could find a monic polynomial $p(x)$ of degree n such that $\max_{-1 \le x \le 1} |p(x)| < 2^{1-n}$. Now, the polynomial $2^{1-n}T_n(x) - p(x)$ is of degree $n - 1$. We shall reach a contradiction by showing that this polynomial has n distinct zeros.

To see this, notice that the polynomial $2^{1-n}T_n(x)$ takes alternately the values $\pm 2^{1-n}$ at $n + 1$ points $x_0 = -1 < x_1 < \cdots < x_n = 1$, which immediately results from $T_n(\cos \theta) = \cos n\theta$. Since $|p(x_k)| < 2^{1-n}|T_n(x_k)|$, it follows that the polynomial $2^{1-n}T_n(x) = p(x)$ takes alternately positive and negative values at $n + 1$ points. Therefore, this polynomial of degree $n - 1$ must have at least n distinct zeros, and so, must vanish identically, q.e.d.

Chebyshev Equioscillation Principle. An important partial generalization of Theorem 8 is the Chebyshev Equioscillation Principle. This states that, if $f(x) \in C[a, b]$, then there is a unique polynomial $p(x)$ of degree $n - 1$ that minimizes max $|p(x) - f(x)|$ on $[a, b]$. The difference $P(x) = p(x) - f(x)$ vanishes at $n - 1$ points on $[a, b]$, and $|P(x)|$ assumes its maximum value at n points. Setting $f(x) = x^n$, we have $P(x) = T_n(x)$.

EXERCISES D

1. Show that the DE $[(1 - x^2)^{3/2}u']' + \lambda(1 - x^2)^{1/2}u = 0$ can be reduced to a DE with constant coefficients by setting $v(\theta) = (\sin \theta)u(\cos \theta)$.

2. Show that the endpoint conditions $\lim_{x \to \pm 1} \sqrt{1 - x^2} \, u(x) = 0$ give an S-L system with eigenvalues $\lambda_n = n(n + 2)$, from the DE of Ex. 1.

3. Show that the eigenfunction belonging to the eigenvalue λ_n is a Chebyshev polynomial of the second kind.

4. Using Ex. 3, obtain an expansion theorem of a smooth function into a series of Chebyshev polynomials of the second kind.

5. Show that $T_n(x) = U_n(x) - xU_{n-1}(x)$.

6. Show that $(1 - x^2)U_{n-1}(x) = xT_n(x) - T_{n+1}(x)$.

7. Express $U_n(x)$ in terms of the hypergeometric function.

8. Expand the function arccos x on $(-1, 1)$ into a series of Chebyshev polynomials of the first kind.

*9. Infer the Weierstrass Approximation Theorem from Fejér's Theorem.

8   EUCLIDEAN VECTOR SPACES

The concepts of mean-square convergence and completeness have suggestive geometric interpretations. These interpretations are based on the properties of inner products.

Consider the set of all real functions f, g, h, . . . , continuous and square-integrable on an interval I, with respect to a fixed positive weight function ρ. The interval I may be open or closed, finite, semi-infinite, or infinite. Define the *inner product* of two such functions f, g as the integral

$$(22) \qquad (f, g) = \int_I f(x)g(x)\rho(x) \, dx, \qquad \rho(x) > 0$$

The following formulas are immediate:

$$(f + g, h) = (f, h) + (g, h), \qquad (f, g) = (g, f)$$
$$(cf, g) = c(f, g); \qquad (f, f) > 0, \quad \text{unless} \quad f \equiv 0$$

Hence, with respect to the inner product (f, g), this set of functions is a *Euclidean vector space*† (or "inner product space").

For real functions, the integral (6) in the definition of mean-square convergence, is the inner product $(f_n - f, f_n - f)$; hence it is the square of the *distance* $\|f_n - f\| = (f_n - f, f_n - f)^{1/2}$ between f_n and f in the Euclidean vector space E. Therefore, $f_n \to f$ in the *mean square*, relative to ρ, means that the Euclidean *distance* from f_n to f in E tends to zero.

This distance enjoys the properties of distance in ordinary space, including the *triangle inequality* and the *Schwarz inequality*

$$-\|f\| \cdot \|g\| \le (f, g) = \|f\| \cdot \|g\| \cos \angle(f, g) \le \|f\| \cdot \|g\|$$

The Schwarz inequality shows that f and g are *orthogonal* if and only if the angle

$$\theta = \angle(f, g) = \arccos [(f, g)/\|f\| \cdot \|g\|], \qquad 0 \le \theta \le \pi$$

is 90°; it gives a geometrical interpretation to the definition of orthogonal functions.

We shall now generalize Theorems 1, 2, 3, and 4 to an arbitrary Euclidean vector space E.

If $\{\phi_k\}$ is a sequence of orthogonal vectors in E, and $f \in E$ is given, consider the squared distance

$$E(\gamma_1, \ldots, \gamma_n) = \left\| f - \sum_{k=1}^{n} \gamma_k \phi_k \right\|^2 = \left(f - \sum_{k=1}^{n} \gamma_k \phi_k, f - \sum_{k=1}^{n} \gamma_k \phi_k \right)$$

Defining $c_k = (f, \phi_k)/(\phi_k, \phi_k)$, we obtain, as in §3,

$$\left\| f - \sum_{k=1}^{n} \gamma_k \phi_k \right\|^2 = (f, f) - \sum_{k=1}^{n} [c_k^2 (\phi_k, \phi_k)] + \sum_{k=1}^{n} [(c_k - \gamma_k)^2 (\phi_k, \phi_k)]$$

Geometrically, the *least-square* approximation $\sum_{k=1}^{n} c_k \phi_k$ to f appears as the *orthogonal projection* of the vector f onto the *subspace* S of all linear combinations $\gamma_1 \phi_1 + \cdots + \gamma_n \phi_n$ of $\phi_1, \ldots, \phi_n$. This is because, in the orthogonal projection onto a subspace S of a vector $\mathbf{c}$ issuing from the origin, the component of $\mathbf{c}$ perpendicular to S is the *shortest* vector from S to $\mathbf{c}$. The coefficients γ_k are given by the direction cosine formulas of analytic geometry.

Completeness of the ϕ_k is defined as in §4, as the property that

$$\lim_{n \to \infty} \left\| f - \sum_{k=1}^{n} c_k \phi_k \right\|^2 = 0, \qquad \text{that is,} \qquad \sum_{k=1}^{\infty} c_k \phi_k = f$$

---

† The reader should familiarize himself with this notion; the space is *not* assumed to be finite-dimensional.

for every f in E. It has a simple geometric interpretation in any Euclidean vector space E. The relation $f = \Sigma_1^\infty c_k\phi_k$ holds if and only if the distance $\|f - \Sigma_1^n c_k\phi_k\|$ tends to zero as $n \to \infty$. That is, as in the proof of Theorem 3, the condition for completeness is that we can approximate any f arbitrarily closely by finite linear combinations $c_1\phi_1 + \cdots + c_n\phi_n$ of the orthogonal vectors ϕ_k whose completeness is in question. This idea is most vividly expressed in terms of the concept of a dense subset of a Euclidean vector space.

DEFINITION.   A subset S of a Euclidean vector space E is *dense* in E if and only if, for any f in E and positive numbers $\delta > 0$, an element s can be found in S such that $\|s - f\| < \delta$.

As in Theorem 4, a set $\{\phi_n\}$ of orthogonal elements of E is *complete* if and only if the set S of all finite linear combinations $\Sigma_1^n \gamma_k\phi_k$ of the ϕ_k is dense in E.

The Parseval equality is easily derived in any Euclidean vector space E. Consider the sequence of best mean-square approximations

$$f_n = \sum_{k=1}^n c_k\phi_k, \qquad c_k = (f, \phi_k)/(\phi_k, \phi_k)$$

to a given vector f in E. If the finite linear combinations $\Sigma_1^n c_k\phi_k$ are dense in E, then the square distance

$$\left\| \sum_{k=1}^n c_k\phi_k - f \right\|^2 = (f, f) - \sum_{k=1}^n \left[\frac{(f, \phi_k)^2}{(\phi_k, \phi_k)} \right] \geq 0$$

must tend to zero as $n \to \infty$. Hence, if the sequence $\{\phi_n\}$ is *complete*, then

(23) $$\sum_{k=1}^\infty [(f, \phi_k)^2/(\phi_k, \phi_k)] = (f, f) \qquad \text{(Parseval equality)}$$

Applying Parseval's equality to the vector $f + g$ and then expanding and simplifying, we obtain more generally

(23') $$\sum_{k=1}^\infty [(f, \phi_k)(g, \phi_k)/(\phi_k, \phi_k)] = (f, g)$$

valid whenever $\{\phi_k\}$ is complete and f, g are square-integrable. Even if Parseval's equality fails, we still get

(24) $$\sum_{k=1}^\infty \frac{(f, \phi_k)^2}{(\phi_k, \phi_k)} \leq (f, f) \qquad \text{(Bessel inequality)}$$

If Parseval's equality fails, then strict inequality will occur in (24) for some f in E. For such an f, we have by Theorem 1

$$\left\| f - \sum_{k=1}^{n} \gamma_k \phi_k \right\| \geq \left\| f - \sum_{k=1}^{n} c_k \phi_k \right\| \geq \sqrt{\delta} > 0$$

for any choice of γ_k. Since δ is independent of n, this shows that the ϕ_k cannot be a complete set of orthogonal vectors. This gives another proof of Theorem 3, which we now restate for an arbitrary Euclidean vector space.

THEOREM 9. *A sequence $\{\phi_k\}$ of orthogonal vectors of a Euclidean vector space E is* complete *if and only if the Parseval equality (23) holds for all f in E.*

9 COMPLETENESS OF EIGENFUNCTIONS

The completeness of the eigenfunctions of a regular Sturm-Liouville system is a consequence of the asymptotic formulas of Ch. 10, and a geometric property of sets of orthonormal vectors in Euclidean vector spaces. This property is stated in the following theorem of N. Bary.

THEOREM 10. *Let $\{\phi_n\}$ be any* complete *sequence of orthonormal vectors in a Euclidean vector space E, and let $\{\psi_n\}$ be any sequence of orthonormal vectors in E that satisfies the inequality*

(25) $$\sum_{n=1}^{\infty} \|\psi_n - \phi_n\|^2 < +\infty$$

Then the ψ_n are complete in E.

This result will be proved in §§10, 11. It is plausible intuitively, because it asserts that completeness is preserved in passing from a set of orthonormal vectors ϕ_n to any nearby system.

Assuming Theorem 10 provisionally, we can establish the completeness of the eigenfunctions of a regular S-L system as follows.

Consider the asymptotic formula (56) of Ch. 10,

$$u_n(x) = \sqrt{\frac{2}{(b-a)}} \cos\left[\frac{n\pi(x-a)}{(b-a)}\right] + \frac{O(1)}{n}$$

If $u_n(x)$ is the nth normalized eigenfunction of a regular S-L system in Liouville normal form, and if $\phi_n(x) = \sqrt{2/(b-a)} \cos(n\pi(x-a)/(b-a))$, this gives $|u_n(x) - \phi_n(x)| = O(1)/n$. Squaring and integrating, we obtain

$$\|u_n - \phi_n\|^2 = \int_I [u_n(x) - \phi_n(x)]^2 \, dx = \frac{O(1)}{n^2}$$

Since the series $1 + \frac{1}{4} + \frac{1}{9} + \cdots + 1/n^2 + \cdots$ converges (to $\pi^2/6$), this implies the following lemma.

LEMMA. *Let $u_n(x)$ be the nth normalized eigenfunction of any regular S-L system in Liouville normal form, with $\alpha'\beta' \neq 0$, and let*

$$\phi_n(x) = \sqrt{\frac{2}{(b-a)}} \cos\left[\frac{n\pi(x-a)}{(b-a)}\right]$$

Then the ϕ_n are an orthonormal sequence, and

$$(26) \qquad \sum_{n=1}^{\infty} \|u_n - \phi_n\|^2 < +\infty$$

Since the cosine functions are complete (by Corollary 3 of Theorem 3), it follows from this lemma and Theorem 10 that the eigenfunctions of any regular S-L system in Liouville normal form with $\alpha'\beta' \neq 0$ are a complete set of orthonormal functions.

As shown in Ch. 10, §9, the transformation to Liouville normal form, applied to the (normalized) eigenfunctions, carries the inner product

$$(\phi, \psi) = \int_I \phi(x)\psi(x)\rho(x)\,dx$$

into the inner product

$$(u, v) = \int_a^b u(x)v(x)\,dx$$

Therefore,† the change of variable that leads to a Liouville normal form carries complete orthonormal sequences, relative to a weight function ρ, into complete orthonormal sequences. Hence, the eigenfunctions of regular S-L systems not in Liouville normal form are also complete.

Finally, since similar arguments cover the case $\alpha'\beta' = 0$, we have the following result.

THEOREM 11. *The eigenfunctions of any regular S-L system are complete in the Euclidean vector space of square-integrable continuous functions, on the interval $a \leq x \leq b$, relative to the weight function ρ.*

---

† Since distance and convergence are defined in terms of inner products in any Euclidean vector space.

*10 HILBERT SPACE

The set of real numbers differs from the set of rational numbers by the *completeness* property that every Cauchy sequence of real numbers is convergent.†
This property of *completeness* has an analog for Euclidean vector spaces (and, more generally, for metric spaces).

DEFINITION. In a Euclidean vector space E, a *Cauchy sequence* is an infinite sequence of vectors f_n such that

(27) $\|f_m - f_n\| \to 0$ as $m, n \to \infty$

The space E is called *complete* when, given any Cauchy sequence $\{f_n\}$, there exists a vector f in E such that $\|f_n - f\| \to 0$ as $n \to \infty$. A complete Euclidean vector space is called a Hilbert space.

Any finite-dimensional Euclidean vector space is complete, but the Euclidean vector space of continuous square-integrable functions defined in §8 is not complete, as will appear presently.

Example 2. Let (ℓ_2) denote the Euclidean vector space of all infinite sequences $\mathbf{a} = \{a_k\} = (a_1, a_2, a_3, \ldots)$ of real numbers which are square-summable, that is, which satisfy $\Sigma a_k^2 < +\infty$. The vector operations on these sequences are performed term-by-term, so that $\mathbf{a} + \mathbf{b}$ is the sequence $(a_1 + b_1, a_2 + a_2, a_3 + b_3, \ldots)$. Inner products are defined by the formula

(28) $(\mathbf{a}, \mathbf{b}) = \sum_{k=1}^{\infty} a_k b_k = a_1 b_1 + a_2 b_2 + a_3 b_3 + \cdots$

LEMMA 1. *The space (ℓ_2) is a Hilbert space.*

Proof. Our problem is to prove completeness. To this end, let $\{\mathbf{a}^n\}$ be any Cauchy sequence of square-summable sequences. That is, let

$$\lim_{m,n \to \infty} \|\mathbf{a}^m - \mathbf{a}^n\|^2 = \lim_{m,n \to \infty} \left\{ \sum_{k=1}^{\infty} (a_k^m - a_k^n)^2 \right\} = 0$$

For each fixed k, the sequence of real numbers a_k^n ($n = 1, 2, \ldots$) (the kth components of $\mathbf{a}^n$) is a Cauchy sequence and, therefore, converges to some real number a_k. Let $\mathbf{a} = \{a_k\}$ ($k = 1, 2, 3, \ldots$). We must prove that the sequence $\mathbf{a}$ is square-summable and that $\mathbf{a}^n \to \mathbf{a}$ in the Euclidean vector space.

Since $|\,\|\mathbf{a}^n\| - \|\mathbf{a}^m\|\,| \leq \|\mathbf{a}^n - \mathbf{a}^m\|$ by the triangle inequality, it follows that the sequence $\|\mathbf{a}^n\|$ is bounded. Let $\sqrt{M}$ be an upper bound. Then, for all integers n, we have $\Sigma_{k=1}^{N} (a_k^n)^2 \leq M$. Letting $n \to \infty$ in this finite sum, we get $\Sigma_{k=1}^{N} (a_k)^2 \leq M$. Since N is arbitrary, it follows that $\|\mathbf{a}\|^2 = \Sigma_{k=1}^{\infty} (a_k)^2 \leq M$.

---

† Courant and John, p. 94 ff.; Widder, p. 277.

Hence, $\mathbf{a}$ is a square-summable sequence. Moreover, given $\epsilon > 0$, n can be found so large that $\|\mathbf{a}^n - \mathbf{a}^m\|^2 < \epsilon$ for all $m > n$. Therefore, for every integer N we have $\Sigma_{k=1}^{N} (a_k^n - a_k^m)^2 < \epsilon$. Letting $m \to \infty$, we obtain $\Sigma_{k=1}^{N} (a_k^n - a_k)^2 \leq \epsilon$. Since N is arbitrary, this implies $\Sigma_{k=1}^{\infty} (a_k^n - a_k)^2 \leq \epsilon$, q.e.d.

The property of Hilbert space that is most useful for establishing the completeness of eigenfunctions is the following.

THEOREM 12.   *An orthogonal sequence $\{\phi_k\}$ of vectors of a Hilbert space is complete if and only if there is no nonzero vector f orthogonal to all the ϕ_k.*

Proof.   Let f be a vector, and let $\{\phi_k\}$ be a sequence of orthogonal vectors in a Hilbert space $\mathcal{H}$. Let $g_n = \Sigma_{k=1}^{n} c_k \phi_k$ be the nth least-square approximation to f by a linear combination of $\phi_1, \ldots, \phi_n$. Then, as before, we have $c_k = (f, \phi_k)/(\phi_k, \phi_k)$ and, for $m > n$,

$$\|g_m - g_n\|^2 = \sum_{n+1}^{m} \left[\frac{(f, \phi_k)^2}{(\phi_k, \phi_k)} \right] \leq \sum_{n+1}^{\infty} \left[\frac{(f, \phi_k)^2}{(\phi_k, \phi_k)} \right]$$

By the Bessel inequality (24), the series $\Sigma_{k=1}^{\infty} [(f, \phi_k)^2/(\phi_k, \phi_k)]$ of positive numbers is convergent; hence, the last sum in the preceding display tends to zero as $n \to \infty$. That is, the sequence $\{g_n\}$ is a *Cauchy sequence*.

It follows that $\mathcal{H}$, being complete, contains a vector g to which the g_n converge; let $h = f - g = \lim_{m \to \infty} (f - g_m)$. Then, since $(f - g_m, \phi_k) = 0$ for all $m \geq k$, we have in the limit, as $m \to \infty$, $(h, \phi_k) = 0$ for all k. By Theorem 9 (Parseval's equality), $h = 0$ for all f if and only if $\{\phi_k\}$ is complete. This completes the proof.

Remark.   A complete orthonormal sequence $\{\phi_k\}$ in the space (ℓ_2) is obtained by choosing $\phi_k = \mathbf{e}^k$, the kth unit vector whose kth component is 1 and whose other components are all 0.

The Euclidean vector space $\mathcal{C}[a, b]$ of all continuous functions on a finite interval $a \leq x \leq b$ is not complete; that is, it is not a Hilbert space. The following is an example of a mean-square Cauchy sequence of continuous functions that does not converge to any continuous function. In $-1 \leq x \leq 1$, let $f_n(x) = 0$ for $-1 \leq x \leq 0$, $f_n(x) = nx$ for $0 \leq x \leq 1/n$, and $f_n(x) = 1$ for $1/n \leq x \leq 1$. The limit function $f_\infty(x)$ equals 0 for $-1 \leq x \leq 0$ and 1 for $0 < x \leq 1$.

Though the Euclidean vector space $\mathcal{C}[a, b]$ is not complete, it can be embedded in the (complete) Hilbert space (ℓ_2), as follows. First, make the change of independent variable $t = \pi(x - a)/(b - a)$, in order to map the interval $[a, b]$ on $[0, \pi]$. Then for each $f(x) \in \mathcal{C}[a, b]$, expand $\tilde{f}(t) = f(a + (b - a)t/\pi)$ into the *cosine series* $\tilde{f}(t) = \Sigma c_k \cos kt$, $k = 1, 2, \ldots$. The vector $\mathbf{f} = (c_0, c_1, c_2, \ldots)$ defines an element of the space (ℓ_2) by the Bessel inequality.

The correspondence $f \to \mathbf{f}$ maps the space $\mathcal{C}[a, b]$ into the space (ℓ_2); moreover, it preserves vector operations: $f + g \to \mathbf{f} + \mathbf{g}$ and $\lambda f \to \lambda \mathbf{f}$. By Parseval's generalized equality (23′), which is applicable since the cosine functions are complete (Theorem 3, Corollary 3), we know that

$$(\mathbf{f}, \mathbf{g}) = \left[\frac{\pi}{(b-a)} \right] \int_a^b f(x)g(x) \, dx = \int_0^\pi \tilde{f}(t)\tilde{g}(t) \, dt$$

Hence, the correspondence $f \to \mathbf{f}$ also preserves *inner products* (up to a constant normalizing factor). Therefore, it also preserves lengths $(\mathbf{f}, \mathbf{f})^{1/2}$. In particular, $\mathbf{f} = \mathbf{0}$ implies that $\int_0^\pi f^2(t) \, dt = 0$. Hence, since f is continuous, it also implies that $f(t) = 0$. In conclusion, we have proved the following result.

LEMMA 2. *The Euclidean vector space $\mathcal{C}[a, b]$ $(-\infty < a < b < +\infty)$ can be embedded in the Hilbert space (ℓ_2), with preservation of vector operations and inner products.*

More generally, let ϕ_n be a complete sequence of orthonormal vectors in an arbitrary Euclidean space E. Then the mapping $f \to \mathbf{c} = \{c_k\}$, where $c_k = (f, \phi_k)$ for f in E, defines an embedding of E as a subspace of (ℓ_2). In the same way as before we obtain the following lemma.

LEMMA 3. *Every Euclidean vector space with a complete sequence of orthonormal vectors can be embedded in (ℓ_2), with preservation of vector operations and inner products.*

In view of this lemma, it suffices to prove Theorem 10 under the assumption that the Euclidean vector space E is complete, that is, that E is a Hilbert space. We shall make this assumption from now on.

*11 PROOF OF COMPLETENESS

We are now ready to prove Theorem 10. But to bring out more clearly the idea of the proof, we first treat a special case.

We define a sequence of orthonormal vectors of a Hilbert space to be an *orthonormal basis* if and only if it is complete.

LEMMA. *Let $\{\phi_k\}$ be an orthonormal basis in the Hilbert space $\mathcal{H}$. Let $\{\psi_k\}$ be an orthonormal sequence in $\mathcal{H}$ satisfying the condition*

$$(29) \qquad \sum_{k=1}^\infty \|\psi_k - \phi_k\|^2 < 1$$

Then the sequence $\{\psi_k\}$ is also an orthonormal basis in $\mathcal{H}$.

Proof. If the sequence ψ_k were not a basis, then we could find a nonzero vector h orthogonal to every ψ_k by Theorem 12. The inner product of this vector with ϕ_k would be given by

$$(h, \phi_k) = (h, \psi_k) + (h, \phi_k - \psi_k) = (h, \phi_k - \psi_k)$$

Squaring and using the Schwarz inequality, we would have

(29') $$(h, \phi_k)^2 = (h, \phi_k - \psi_k)^2 \leq \|h\|^2 \|\phi_k - \psi_k\|^2$$

Summing with respect to k, we would obtain

$$\|h\|^2 = \sum_{k=1}^{\infty} (h, \phi_k)^2 \leq \|h\|^2 \sum_{k=1}^{\infty} \|\phi_k - \psi_k\|^2 < \|h\|^2$$

in evident violation of Parseval's equality (Theorem 9), since the ϕ_k are an orthonormal basis.

COROLLARY 1.   *Replace condition (29) by the weaker condition*

(30) $$\sum_{k=N+1}^{\infty} \|\psi_k - \phi_k\|^2 < 1$$

for some integer N. Then every element h of $\mathcal{H}$ orthogonal to $\phi_1, \ldots, \phi_N$ and to ψ_{N+1}, $\psi_{N+2}, \ldots$ must vanish.

Proof.   Any such element h satisfies the inequality (29') for $k > N$. Indeed, summing over all k, we have, since $(h, \phi_k) = 0$ for $k = 1, 2, \ldots, N$,

$$\|h\|^2 = \sum_{k=1}^{\infty} (h, \phi_k)^2 = \sum_{f=N+1}^{\infty} (h, \phi_k)^2 \leq \|h\|^2 \sum_{k=N+1}^{\infty} \|\phi_k - \psi_k\|^2 < \|h\|^2$$

again contradicting Parseval's equality.

COROLLARY 2.   *If ϕ_k is an orthonormal basis and ψ_k an orthonormal sequence satisfying (30), then every element of $\mathcal{H}$ orthogonal to $\psi_{N+1}, \psi_{N+2}, \ldots$ and to the elements*

(31) $$\eta_n = \phi_n - \sum_{k=N+1}^{\infty} (\phi_n, \psi_k)\psi_k, \qquad n = 1, 2, \ldots, N$$

must vanish.

Proof.   For any such element h, we have

$$(h, \phi_n) = (h, \eta_n) + \sum_{k=N+1}^{\infty} (\phi_n, \psi_k)(h, \psi_k) = 0$$

for $n = 1, 2, \ldots, N$. Thus h also satisfies the conditions of Corollary 2 and therefore vanishes.

The proof of Theorem 10 can now be completed as follows. Choose an integer N so that

$$\sum_{k=N+1}^{\infty} \|\psi_k - \phi_k\|^2 < 1$$

By Corollary 2, any element h of $\mathcal{H}$ which is orthogonal to the elements ψ_{N+1}, $\psi_{N+2}, \ldots$ and to the elements η_n $(n = 1, 2, \ldots, N)$ defined by formula (31) must vanish. Denote by S the set of all elements of $\mathcal{H}$ orthogonal to $\psi_{N+1}, \psi_{N+2}, \ldots$. Evidently S is a vector space containing $\eta_1, \ldots, \eta_N$. By virtue of the previous remark, the vector space S contains only the linear combinations of these elements. In other words, S is a finite-dimensional vector space whose dimension is at most N.

But the elements $\psi_1, \psi_2, \ldots, \psi_N$ also belong to the vector space S, and they are linearly independent (they are an orthonormal sequence!). Therefore, the elements $\psi_1, \psi_2, \ldots, \psi_N$ are a *basis* for the vector space S.† Hence, the elements $\eta_1, \eta_2, \ldots, \eta_N$ are linear combinations of $\psi_1, \psi_2 \ldots, \psi_N$. It follows that any element of $\mathcal{H}$ that is orthogonal to all the ψ_k must vanish, because such an element is also orthogonal to $\eta_1, \ldots, \eta_N$ and to $\psi_{N+1}, \psi_{N+2}, \ldots$. We conclude that the sequence $\{\psi_k\}$ is complete, proving Theorem 10.

Theorem 10 implies Theorem 11, as was already shown in §9.

It is natural to ask whether the sums $\Sigma c_k \phi_k$ having square-summable coefficient sequences can also be interpreted as functions. This question can also be answered in the affirmative, by use of the Lebesgue integral. Given any complete family $\{\phi_k\}$ of orthonormal functions on $[a, b]$, the partial sums $\Sigma_{k=1}^n c_k \phi_k$ with *square-summable* coefficient sequences $\{c_k\}$ converge in the mean square of a function $f(x)$, which is *square-integrable* for the *Lebesgue integral*. Conversely, if $f(x)$ is a Lebesgue square-integrable and if $c_k = (f, \phi_k)$, the partial sums $f_n = \Sigma_{k=1}^n c_k \phi_k$ converge to $f(x)$ in the mean square. That is, the metric completion of the space $\mathcal{C}[a, b]$ is precisely the Hilbert space $\mathcal{L}_2[a, b]$ of all functions on $[a, b]$ whose squares are Lebesgue integrable. It is this space that is really appropriate for the theory of expansions in eigenfunctions.

EXERCISES E

1. Show that in a finite-dimensional Euclidean vector space E, a vector subspace S is dense if and only if $S = E$.

2. Show that an orthonormal sequence ϕ_k $(k = 1, 2, 3, \ldots)$ in a Euclidean vector space is complete if Parseval's equality (23) holds for all f in a dense subset.

3. (Vitali). Show that an orthonormal sequence $\phi_k(x)$ $(k = 1, 2, \ldots)$ of continuous functions on $a \leq x \leq b$ is complete if and only if

$$\sum_{k=1}^{\infty} \left(\int_a^x \phi_k(t)\, dt \right)^2 = x - a, \qquad a \leq x \leq b$$

---

† As shown in Apostol, Vol. 2, p. 12.

[HINT: Show that linear combinations of the functions $f(x) = |x - c|$ form a dense subset, of $\mathcal{C}[a, b]$, and apply Ex. 2.]

*4. (Dalzell). Show that a sequence of continuous orthonormal square-integrable functions $\phi_k(x)$, $a \le x \le b$, is complete if and only if

$$\sum_{k=1}^{\infty} \int_a^b \left(\int_a^x \phi_k(t) \, dt \right)^2 dx = \frac{(b - a)^2}{2}$$

[HINT: Let $q(x) = x - a - \sum_{k=1}^{\infty} (\int_a^x \phi_k(t) \, dt)^2$, and show that $q(x) \equiv 0$ by establishing $q(x) \ge 0$ and $\int_a^b q(x) \, dx = 0$, q continuous, applying Ex. 3.]

5. Assuming the equality $\sum_{k=1}^{\infty} 1/k^2 = \pi^2/6$, infer from Ex. 4 that the trigonometric functions $\cos kx$, $\sin kx$, $k = 0, 1, 2, \ldots$, are complete on $(-\pi, \pi)$.

6. (Moment problem). Let $f(x)$ be continuous, $a \le x \le b$, and let

$$m_n = \int_a^b x^n f(x) \, dx$$

Show that if all the moments m_n vanish, $f(x) \equiv 0$. [HINT: Use the Weierstrass Approximation Theorem.]

*7. Prove the completeness of the eigenfunctions of any regular S-L system with boundary conditions $u(a) = u(b) = 0$.

*8. Let $G(f_1, f_2, \ldots, f_n) = \det (a_{ij})$, where $a_{ij} = (f_i, f_j)$. Show that the minimum of $\|f - \sum c_k f_k\|$ is equal to $G(f, f_1, f_2, \ldots, f_n)/G(f_1, f_2, \ldots, f_n)$. [HINT: Interpret the determinants as volumes.]

9. Show by a counterexample that Theorem 10 does not remain valid if the ψ_n are allowed to be nonorthogonal unit vectors.

10. Consider the orthogonal sequence ϕ_k, and let $\psi_k = \phi_k$, $k \ge 2$, $\psi_1 = 0$. Then $\sum \|\phi_k - \psi_k\|^2 < \infty$, but $\{\psi_k\}$ is not complete. Which hypothesis of Theorem 10 is violated?

11. Show that the first-order complex DE $iu' + [q(x) + \lambda]u = 0$, $a \le x \le b$, with the boundary condition $u(a) = u(b)$ and $i = \sqrt{-1}$, has a complete sequence of eigenfunctions for any real continuous $q(x)$.

12. Show that linear combinations of the trigonometric functions $\cos kx$, $\sin kx$ are dense in $-\pi \le x \le \pi$, relative to any continuous positive weight function.

APPENDIX A
LINEAR SYSTEMS

1 MATRIX NORM

As in Ch. 6, §4, any system of linear DEs can be reduced to a system of n first-order DEs and so written in matrix notation as

$$(1) \qquad\qquad \mathbf{x}'(t) = A(t)\mathbf{x}$$

where $\mathbf{x}(t)$ is a column vector of length n and $A(t)$ an $n \times n$ matrix, both depending on t. In Ch. 6, we proved the existence and uniqueness of solutions of (1) in any interval of continuity of $A(t)$, for any initial $\mathbf{x}(0) = \mathbf{c}$. In this appendix, we derive some properties of solutions of linear systems (1) whose proofs depend on deeper properties of matrices.

As before, $\mathbf{x}'(t)$ is the limit

$$\lim_{\Delta t \to 0} \frac{1}{\Delta t}^{-1}[\mathbf{x}(t + \Delta t) - \mathbf{x}(t)]$$

where the limit is taken separately in each component $\Delta x_k/\Delta t$. Here we extend this definition, calling the *norm* of any matrix A the real number

$$(2) \qquad\qquad \|A\| = \sup_{\mathbf{x} \neq 0} \frac{|A\mathbf{x}|}{|\mathbf{x}|}$$

where

$$|\mathbf{x}| = (\mathbf{x}^T\mathbf{x})^{1/2} = \left(\sum x_k^2\right)^{1/2} \qquad \text{and} \qquad |A\mathbf{x}| = |\mathbf{x}^T A^T A \mathbf{x}|^{1/2}$$

similarly.† Then $A(t) \to A(t_0)$ means equivalently:

(i) $a_{jk}(t) \to a_{jk}(t_0)$ for all $j, k = 1, \ldots, n$.

(ii) $\|A(t) - A(t_0)\|$ tends to zero as $t \to t_0$.

---

† We use standard notation here: A^T signifies the transpose of A, and $\mathbf{x}^T\mathbf{x}$ the inner product of $\mathbf{x}$ with itself; see Birkhoff-MacLane, Chs. 8–10 for the facts assumed in this Appendix.

The equation $\|I\| = 1$, the triangle inequality $\|A + B\| \leq \|A\| + \|B\|$, the multiplicative inequality $\|AB\| \leq \|A\| \cdot \|B\|$, and $\|tA\| = |t| \cdot \|A\|$ all follow quite directly from (2). From these relations, in turn, one can prove the convergence of the *exponential series*

$$(3) \qquad \exp tA = I + tA + (t^2/2!)A^2 + (t^3/3!)A^3 + \cdots$$

for any matrix A and any scalar t. It suffices to copy the usual proof for real or complex exponential series, replacing absolute values by norms throughout.

2   CONSTANT-COEFFICIENT SYSTEMS

We now introduce the alternative notation e^{tA} for $\exp tA$ as defined above, and prove the differentiability relation

$$(4) \qquad \frac{d(e^{tA})}{dt} = Ae^{tA} \qquad \text{for any constant matrix } A$$

Because of the commutativity of all terms, we have, as in the scalar case, $e^{tA}e^{uA} = e^{(t+u)A}$ and, hence,

$$e^{(t+\Delta t)A} - e^{tA} = (e^{\Delta tA} - I)e^{tA} = \left(\sum_{k=1}^{\infty} \frac{\Delta t^k A^k}{k!} \right) e^{tA}$$

Dividing the series in parenthesis through by Δt, we get the identity matrix I plus a series of matrices whose norms are bounded by $\Delta t^{k-1}\alpha^k/(k!)$, $k = 2$, $3, \ldots$, where $\alpha = \|A\|$. Hence, the norm of the sum tends to zero with Δt, proving (4).

From (4), there follows a very beautiful result.

THEOREM 1.   *The constant-coefficient first-order linear system* $\mathbf{x}'(t) = A\mathbf{x}$ *has the general solution* $\mathbf{x}(t) = e^{tA}\mathbf{x}(0)$.

Proof.   Differentiating $e^{tA}\mathbf{x}(0) = \mathbf{y}(t)$, we get $Ae^{tA}\mathbf{x}(0) = A\mathbf{y}$ by (4), where $\mathbf{x}(0) = \mathbf{y}(0)$ trivially.

THEOREM 2.   *If* $\mathbf{v}$ *is a (column) eigenvector of the matrix* A *with eigenvalue* $\lambda = \mu + iv$, *then the vector-valued function* $\mathbf{w}(t) = e^{\lambda t}\mathbf{v}$ *is a solution of* $w'(t) = A\mathbf{w}$.

Proof.   Since $A\mathbf{v} = \lambda \mathbf{v}$, we have

$$\mathbf{w}'(t) = \lambda e^{\lambda t}\mathbf{v} = e^{\lambda t}(\lambda \mathbf{v}) = e^{\lambda t}(A\mathbf{v}) = Ae^{\lambda t}\mathbf{v} = A\mathbf{w}$$

where the successive steps are easily justified.

Complex Solutions. The preceding results hold for complex as well as for real solution vectors, coefficient-matrices, and independent variables, provided that we define the *norm* of a complex vector $\mathbf{z} = (z_1, \ldots, z_n)^T$ as the square root of the *Hermitian* inner product

$$\mathbf{z}^H\mathbf{z} = x_1^2 + y_1^2 + \cdots + x_n^2 + y_n^2 = \sum_{k=1}^{n} z_k z_k^*$$

of $\mathbf{z}$ with its conjugate transpose $\mathbf{z}^H = (z_1^*, \ldots, z_n^*)$. Hence, we can apply them to complex eigenvalues and eigenvectors of real matrices.

Example 1. Consider the linear system

$$\begin{aligned} dz_1/dt &= az_1 - bz_2 \\ dz_2/dt &= bz_1 + az_2 \end{aligned} \qquad \text{or} \qquad \frac{d}{dt}\begin{pmatrix} z_1 \\ z_2 \end{pmatrix} = \begin{pmatrix} a & -b \\ b & a \end{pmatrix}\begin{pmatrix} z_1 \\ z_2 \end{pmatrix}$$

where the variables t, z_1, z_2 may be real or complex, but the coefficients a and b are real. The characteristic polynomial of the coefficient-matrix $A = \begin{pmatrix} a & -b \\ b & a \end{pmatrix}$ in (5) is $(-a)^2 + b^2$; hence the eigenvalues of A are $= a \pm ib$ (cf. Ex. B1, Ch. 5). The corresponding column eigenvectors are $\begin{pmatrix} 1 \\ -i \end{pmatrix}$ and $\begin{pmatrix} 1 \\ i \end{pmatrix}$. Hence the vector-valued functions

$$e^{ct}\begin{pmatrix} 1 \\ -i \end{pmatrix} \qquad \text{and} \qquad e^{c^*t}\begin{pmatrix} 1 \\ i \end{pmatrix}, \qquad c = a + ib, \qquad c^* = a - ib$$

are a basis of complex solutions of the system (5).

In general, every matrix A with distinct eigenvalues has a basis of real or complex (column) eigenvectors $\mathbf{v}_j$, with corresponding eigenvalues λ_j $(j = 1, \ldots, n)$. The vector-valued functions $\boldsymbol{\phi}_j(t) = e^{\lambda_j t}\mathbf{v}_j$ then form a basis of solutions of the constant-coefficient linear system $\mathbf{z}'(t) = A\mathbf{z}$ in the entire complex t-plane.

The Secular Equation. We now generalize the concept of the secular equation from second-order linear systems with constant coefficients to nth-order systems. Namely, if $p_A(\lambda)$ is the characteristic polynomial of the matrix A, we define the *secular equation* of the constant-coefficient system $\mathbf{x}'(t) = A\mathbf{x}$ to be the DE $p_A(D)u = 0$, where D denotes d/dt, and $p_A(D)$ is the scalar differential operator obtained from the characteristic polynomial $|A - \lambda I| = p_A(\lambda)$ of A by substituting D for λ. We first prove Theorem 3 of Ch. 5.

THEOREM 3. *Any component $x_i(t)$ of any vector solution $\mathbf{x}'(t) = A\mathbf{x}$ satisfies the secular equation $p_A(D) = 0$.*

Proof.   Rewrite the given system as $D\mathbf{x} = A\mathbf{x}$, and let $C(D) = \|C_{jk}(D)\|$ denote the matrix of *cofactors* of the entries $a_{jk} - D\delta_{jk}$ of $A - DI$. Premultiplying the equation $(A - D)\mathbf{x} = \mathbf{0}$ by the matrix $C(D)^T$, we get the vector DE

$$(6) \qquad\qquad p_A(D)I\mathbf{x} = C(D)^T(A - D)\mathbf{x} = \mathbf{0}$$

But (6) asserts precisely that for each $j = 1, \ldots, n$, $p_A(D)x_j = 0$, q.e.d.

Example 2.   When $n = 2$, the identity $C(D)^T(A - D) = p_A(D)I$ reduces to the easily verified matrix identity

$$\begin{pmatrix} a_{11} - D & a_{12} \\ a_{21} & a_{22} - D \end{pmatrix} \begin{pmatrix} a_{22} - D & -a_{12} \\ -a_{21} & a_{11} - D \end{pmatrix} = \begin{pmatrix} p_A(D) & 0 \\ 0 & p_A(D) \end{pmatrix}$$

Stability.   Theorem 3 shows that, if every eigenvalue of A has a negative real part, then by Theorem 5′ of Ch. 5 the constant-coefficient systems $\mathbf{x}'(t) = A\mathbf{x}$ is *strictly stable*. A more explicit proof can be given of this result and its converse by allowing vectors and matrices to have *complex* components and replacing $\mathbf{x}^T\mathbf{x}$ and A^TA by the corresponding Hermitian formulas $\mathbf{x}^H\mathbf{x}$ and A^HA, where the superscript H signifies conjugate transpose.† This permits us to make a change of basis in the vector space of $\mathbf{x}(t)$ that replaces A by its *Jordan normal form* $J = PAP^{-1}$ (P complex nonsingular).

In this Jordan normal form, A is the direct sum of square diagonal blocks J_i having some eigenvalue λ_i on the diagonal and 1's just above the diagonal. The given system $\mathbf{x}'(t) = A\mathbf{x}$ then has for its typical DE

$$\frac{dx_i}{dt} = \lambda_i x_i \qquad \text{or} \qquad \frac{dx_i}{dt} = \lambda_i x_i + x_{i+1}$$

depending on whether x_i is the last variable in its block or not.

Now consider, for example, a typical elementary Jordan block submatrix on the diagonal of J, say the 4×4 matrix

$$J_i = \begin{bmatrix} \lambda_i & 1 & 0 & 0 \\ 0 & \lambda_i & 1 & 0 \\ 0 & 0 & \lambda_i & 1 \\ 0 & 0 & 0 & \lambda_i \end{bmatrix}$$

---

† A^H is also often denoted A^*, and called the "adjoint" of A.

We can exponentiate this block explicitly

$$e^{tJ_i} = e^{\lambda_i t} \begin{bmatrix} 1 & t & t^2/2! & t^3/3! \\ 0 & 1 & t & t^2/2! \\ 0 & 0 & 1 & t \\ 0 & 0 & 0 & 1 \end{bmatrix}$$

Since negative exponentials die out faster than any power, it follows that $e^{tJ} \to 0$ as $t \to \infty$ if and only if every eigenvalue of A has a negative real part. Finally, since $e^{tA} = P^{-1}e^{tJ}P$, it follows that the same is true of e^{tA}. (The matrices A and $J = PAP^{-1}$ have the same eigenvalues.)

3 THE MATRIZANT

We next consider the general case of $A(t)$ variable but continuous in (1). If $\mathbf{x}_1(t), \ldots, \mathbf{x}_m(t)$ is any list of m (column) vector solutions of (1), the $n \times m$ matrix $X(t)$ composed of these columns will, trivially, satisfy the matrix DE $X'(t) = A(t)X$. In particular, we can construct in this way an $n \times n$ matrix solution of

(7) $\qquad M'(t) = A(t)M, \qquad M(0) = I \qquad$ (identity matrix)

The matrix function $M(t)$ so defined is called the *matrizant* of the system (1). By (4), the matrizant of the constant-coefficient system $\mathbf{x}'(t) = A\mathbf{x}$ is just e^{tA}.

Since the multiplication of matrices is associative, the vector-valued function $M(t)\mathbf{c} = \mathbf{x}(t)$ will satisfy $\mathbf{x}(0) = M(0)\mathbf{c} = I\mathbf{c} = \mathbf{c}$ for any given $\mathbf{c}$. Since

$$\mathbf{x}'(t) = M'(t)\mathbf{c} = A(t)M(t)\mathbf{c} = A(t)\mathbf{x}$$

it will also satisfy (1). By the uniqueness theorem proved in Ch. 6, it follows that *every* solution of (1) has the form $M(t)\mathbf{c}$.

The determinant det $M(t)$ of the matrizant also has a remarkable property. We know by (7) that

$$M(t + \Delta t) = M(t) + \Delta M(t) = M(t) + A(t)\,\Delta t M(t) + o(\Delta t)$$
$$= [I - A(t)\,\Delta t]M(t) + o(\Delta t).$$

By the multiplication rule for determinants, therefore, we have

$$\det M(t + \Delta t) = \det [I + A(t)\,\Delta t] \det M(t) + o(\Delta t)$$

Expanding in minors, on the other hand, we obtain

$$\det [I + A(t)\,\Delta t] = 1 + \Delta t \operatorname{Tr} A(t) + O(\Delta t^2),$$

where $\operatorname{Tr} A = \Sigma\, a_{kk}$ is the *trace* of A. Combining,

$$\det M(t + \Delta t) = (1 + \Delta t \operatorname{Tr} A) \det M(t) + o(\Delta t)$$

or

$$\frac{\det M(t + \Delta t) - \det M(t)}{\Delta t} = \operatorname{Tr} A \det M(t) + o(1)$$

Now, passing to the limit as $\Delta t \to 0$, we get [using the alternative notation $|M(t)|$ for $\det M(t)$]:

$$\frac{d\,|M(t)|}{dt} = \operatorname{Tr} A\,|M(t)|$$

Integrating this first-order DE by the methods of Ch. 1, we obtain our final result.

THEOREM 4.   *The determinant of the matrizant of* (1) *satisfies*

$$(8) \qquad \det M(t) = \exp\left\{ \int_0^t [\operatorname{Tr} A(t)]\, dt \right\}$$

COROLLARY.   *If* $\operatorname{Tr} A \equiv 0$, *the matrizant of* (1) *has a determinant identically* 1 (*is "unimodular"*).

This is the case, for example, with systems (1) that come from nth-order DEs having the normal form

$$u^{(n)}(t) + a_2(t)u^{(n-2)}(t) + \cdots + a_n(t)u = 0$$

Next, we define the *adjoint* of the system (1) to be the first-order linear system

$$(9) \qquad \mathbf{f}'(t) = -A^T(t)\mathbf{f} \qquad \left(\text{that is, } \frac{df_i}{dt} = -\sum_j a_{ji} f_j \right)$$

We consider the solution curves of (9) as lying in the dual space of linear functionals on the space of $\mathbf{x}(t)$, given by the linear forms

$$(10) \qquad \mathbf{f}(\mathbf{x}) = \Sigma f_j\, x_j = (\mathbf{f},\, \mathbf{x}) = \mathbf{f}^T\mathbf{x} \qquad \text{(inner product)}$$

THEOREM 5.   *In order that* $[\mathbf{f}(t),\, \mathbf{x}(t)]$ *be constant for every solution* $\mathbf{x}(t)$ *of the vector DE* (1), *it is necessary and sufficient that* (9) *hold.*

Proof. By straightforward differentiation, we have

$$
\frac{d}{dt}\left[\sum_j f_j(t)x_j(t)\right] = \sum_j f_j(t)x_j'(t) + \sum_j f_j'(t)x_j(t)
$$

$$
= \sum_j f_j(t)a_{jk}(t)x_k(t) + \sum_k f_k'(t)x_k(t)
$$

Since there exists a solution of (1) with $\mathbf{x}(t_0) = \mathbf{c}$ for any t_0 and $\mathbf{c}$, the condition that $(\mathbf{f}(t), \mathbf{x}(t))$ be constant for all solutions of (1), and hence that the derivative above vanish identically, is precisely that

(11) $$0 = \sum_j f_j(t)a_{jk}(t) + f_k'(t) \qquad \text{for all} \quad k$$

which is obviously equivalent to (9), as claimed.

4 FLOQUET THEOREM; CANONICAL BASES

We next consider *periodic* linear systems, that is, systems of the form (1) with

(12) $$A(t + T) = A(t) \qquad \text{for some fixed period } T$$

A very special example is furnished by the Mathieu equation

$$u'' + (\lambda - 16d\cos 2x)u = 0$$

which is equivalent in the phase plane to

$$u' = v, \qquad v' = (-\lambda + 16d\cos 2x)u = 0$$

Equations of Hill's type,

$$u'' + p(t)u = 0, \qquad p(t + \pi) = p(t)$$

provide a more general class of examples.

In any case, let $M(T)$ be the matrizant of the first-order periodic linear system satisfying (12), and let the eigenvectors of $M(T)$ be $\mathbf{v}_1, \ldots, \mathbf{v}_r$ (hopefully, $r = n$), with $M(T)\mathbf{v}_i = \lambda_i\mathbf{v}_i$. Then, if $\mathbf{x}_i(t)$ is the solution of (1) and (12) with the initial-value vector $\mathbf{v}_i$, then by definition we have $\mathbf{x}_i(T) = \lambda_i\mathbf{x}_i(0) = \lambda_i\mathbf{v}_i$; moreover, $\lambda_i \neq 0$ since $\mathbf{x}(t) = \mathbf{0}$ is the only solution with $\mathbf{x}(T) = \mathbf{0}$ (Uniqueness Theorem). Since $A(t + T) = A(t)$, $\mathbf{x}_i(t + T)$ is, therefore, the solution of the same initial value problem as $\mathbf{x}_i(t)$ but with the initial value vector multiplied by

λ_i. We conclude the quasiperiodic relation

$$(13) \qquad\qquad \mathbf{x}_i(t + T) = \lambda_i \mathbf{x}_i(t)$$

This is essentially Floquet's Theorem. Since $\lambda_i \neq 0$, $\lambda_i = e^{\alpha_i T}$ for some real or complex constant α_i, we can define $\mathbf{y}_i(t) = e^{-\alpha_i t}\mathbf{x}_i(t)$ with the assurance that, by (13), $\mathbf{y}_i(t + T) = \mathbf{y}_i(t)$. We have proved the following theorem.

THEOREM 6. *If the matrizant $M(T)$ of (1) has r independent eigenvectors and (12) holds, then (1) has r linearly independent solutions satisfying*

$$(14) \qquad \mathbf{x}_i(t) = e^{\alpha_i t}\mathbf{y}_i(t), \qquad \text{where} \qquad \mathbf{y}_i(t + T) = \mathbf{y}_{i,}(t)$$

so that each $\mathbf{x}_i(t)$ is an exponential (scalar) function $e^{\alpha_i t}$ times a periodic function.

As in Ch. 5, §10, any second-order equation of Hill's type (including the Mathieu equation) leaves invariant a periodic, time-dependent "energy" function $u'^2 + \int p(t)dt$. It follows that, for such equations, all the α_r in (14) must vanish if

$$\oint p(t)dt = \int_0^T p(t)dt = 0$$

In particular, referring back to Ch. 10 (Example 4 in §1, Ex. A5, and §8), we can conclude that there exist infinite sequences of even and odd Mathieu functions, and that the functions in each sequence have periods 2π and π alternately.

Finally, let $t = a$ be an isolated singular point of the matrix $A(t)$, considered as a function of the *complex* independent variable t. By this we mean that all $a_{jk}(t)$ are analytic in some punctured disk $0 < |t - a| < \rho$. Much as in Ch. 9, we can consider the effect of making a simple loop $t = a + e^{i\theta}\tau$ around $t = a$ on some basis of solutions of the linear system (1); we do *not* assume periodicity. If the basis of solutions has for initial values $\mathbf{x}_i(t_0) = \mathbf{e}_i$, where $\mathbf{e}_i = (\delta_{i,1} \cdots \delta_{i,n})$ is the ith unit vector, the vectors resulting from going once around the loop are the column vectors of the "matrizant" of (1) for the given loop. In any event, we can write

$$(15) \qquad\qquad \mathbf{x}(c + e^{2\pi i}\tau) = M\mathbf{x}(c + \tau)$$

where M is a suitable nonsingular *circuit matrix*. An argument like that used to prove Theorem 5 (and Theorem 4 of Ch. 9) gives the following result.

THEOREM 7. *If the circuit matrix of $A(t)$ in (1) for the isolated singular point $t = a$ has r independent eigenvectors, then (1) has r solutions*

$$(16) \qquad\qquad \mathbf{x}(t) = (t - a)^{\alpha_i}\mathbf{f}_i(t),$$

where the $\mathbf{f}_i(t)$ are holomorphic in $0 < |t - a| < \rho$ for some $\rho > 0$.

EXERCISES

1. Compute the matrix $B = e^{tA}$ for the following A:

(a) $\begin{pmatrix} 0 & 1 \\ -1 & 0 \end{pmatrix}$ (b) $\begin{pmatrix} 0 & 1 \\ 1 & 0 \end{pmatrix}$ (c) $\begin{pmatrix} 1 & 1 \\ 0 & 1 \end{pmatrix}$ (d) $\begin{pmatrix} a & 0 \\ 0 & d \end{pmatrix}$

2. Apply the Ex. 1 to the following 3×3 matrices:

(a) $\begin{pmatrix} 0 & 0 & 1 \\ 1 & 0 & 0 \\ 0 & 1 & 0 \end{pmatrix}$ (b) $\begin{pmatrix} 0 & 1 & 0 \\ 0 & 0 & 1 \\ 0 & 0 & 0 \end{pmatrix}$ (c) $\begin{pmatrix} \lambda & 1 & 0 \\ 0 & \lambda & 1 \\ 0 & 0 & \lambda \end{pmatrix}$

3. For each A in Ex. 1, discuss the stability of the DE $\mathbf{x}'(t) = A\mathbf{x}$.

4. Do the same for each matrix in Ex. 2.

5. Compute the matrizant of $\mathbf{x}'(t) = A\mathbf{x}$ for the following matrices A:

(a) $\begin{pmatrix} 0 & t \\ -t & 0 \end{pmatrix}$ (b) $\begin{pmatrix} t & t^2 \\ 0 & t \end{pmatrix}$ (c) $\begin{pmatrix} 0 & 0 & t^2 \\ t^2 & 0 & 0 \\ 0 & t^2 & 0 \end{pmatrix}$

6. (a) For the system $x' = y$, $y' = z$, $z' = x$, find column vectors $\boldsymbol{\phi}_j$ ($j = 1, 2, 3$) such that $\mathbf{x}_1(t) = \boldsymbol{\phi}_1$, $\mathbf{x}_2(t) = e^{\omega t}\boldsymbol{\phi}_2$, and $\mathbf{x}_3(t) = e^{\omega^2 t}\boldsymbol{\phi}_3$ are a basis of solutions.
   (b) Discuss the behavior of solutions for large t.

7. Find a basis of solutions for the system

$$\frac{dx_1}{dt} = -2x_1 + x_2 \qquad\qquad \frac{dx_n}{dt} = x_{n-1} - 2x_n$$

$$\frac{dx_j}{dt} = x_{j-1} - 2x_j + x_{j+1} \qquad (j = 2, \ldots, n - 1)$$

[HINT: Try the initial conditions $c_j = \sin rj\pi$, $r = 1, \ldots, n - 1$.]

8. Show that the matrizant $M(T)$ of any DE of Hill's type has determinant $|M(T)| = 1$.
   [HINT: Show that, in the phase plane, $TrA(t) \equiv 0$.]

BIFURCATION
THEORY

1   WHAT IS BIFURCATION?

In recent decades, the qualitative theory of ordinary differential equations has been revolutionized by a series of new concepts, loosely characterized by such words as "bifurcation," "control," "strange attractor," "chaos," and "fractal." Several fascinating books are now available which describe one or more of these ideas in some detail and depth.† The purpose of this appendix is to give readers some idea of the variety and richness of the phenomena which they try to analyze, by introducing them to the nature and typical examples of "bifurcation."

In general, the qualitative behavior of solutions of differential equations and systems of differential equations of given form often depends on the *parameters* involved. We studied this dependence for the DE $\ddot{x} + p\dot{x} + qx = 0$ in Chapter 2, §2, and for the system

$$(1) \qquad \frac{d}{dt}\begin{pmatrix} x \\ y \end{pmatrix} = \begin{pmatrix} a & b \\ c & d \end{pmatrix}\begin{pmatrix} x \\ y \end{pmatrix}$$

in Chapter 5, §5. The qualitative dependence on the paramater λ of solutions of Sturm-Liouville equations

$$(2) \qquad \frac{d}{dx}\left[p(\mathrm{x})\,\frac{dy}{dx} \right] + [\lambda\rho(x) - q(x)]u = 0$$

provides another classic illustration, to which Chapters 10 and 11 were devoted. This appendix will introduce a typical concept associated with this parameter dependence: that of *bifurcation.*

A striking example of bifurcation is provided by the van der Pol equation

$$(3) \qquad \ddot{y} - \mu(1 - y^2)\dot{y} + y = 0$$

---

† See the books by Hale and Chow-Hale, Guckenheimer-Holmes, and Thompson-Stewart listed in the bibliography.

When μ is positive, as was shown in Chapter 5, §12, the equilibrium solution $y \equiv 0$ is unstable, while all other solutions tend to stable periodic oscillations $y = f(t - \tau)$ associated with the same limit cycle in the phase plane, as $t \uparrow +\infty$.

When μ is negative, the behavior of solutions is totally different. This is evident since the substitution $t \mapsto -t$ carries (3) into $\ddot{y} + \mu(1 - y^2)y + y = 0$, thus effectively *reversing* the sign of μ and the sense of trajectories in the phase plane. It follows that, when $\mu < 0$, the equilibrium solution $y \equiv 0$ of (3) is stable; the *periodic* solutions $y = f(t - \tau)$ associated with the limit cycle are unstable; and all solutions having initial values $(y_o, \dot{y}_o)$ located outside this limit cycle spiral out to infinity as $t \uparrow +\infty$.

To describe this qualitative change, the value $\mu = 0$ where it occurs is called a *bifurcation point* of the parameter μ in (3). More precisely, it is called a *Hopf bifurcation* (contrast with Example 2 in §4 below).

An analogous, even simpler bifurcation occurs when $\mu = 0$ for the first-order DE

(4) $$\dot{x} + \mu x = C$$

in the stable case $\mu < 0$, all solutions approach the same equilibrium solution $x = C/\mu$; when $\mu < 0$, they all diverge from it.

By fixing one parameter (e.g., q), one can also apply the concept of bifurcation to DEs depending on two parameters, such as

(5) $$\ddot{x} + p\dot{x} + qx = 0$$

see the exercises at the end of this appendix.

A more novel example is provided by the so-called Brusselator equations of chemical kinetics:

(6) $$\dot{x} = A - (B + 1)x + x^2y \qquad \dot{y} = Bx - x^2y$$

Setting $A = 1$, the phase portrait of (6) has a stable focal point for $B < 2 = 1 + A^2$, and an unstable focal point for $B > 2$; hence $B = 2$ is a bifurcation point for B in (6), if $A = 1$.

*2 POINCARÉ INDEX THEOREM

Let $(X(x,y), Y(x,y))$ be a plane vector field, and let γ be a simple closed curve which does not go through any critical point of the vector field. By the Jordan curve theorem,† the interior of γ is a well-defined, simply connected region; we will assume the vector field to be of class $\mathcal{C}^2$ in γ and its interior.

---

† It is notoriously difficult to prove the Jordan curve theorem rigorously.

DEFINITION. Let $\psi(x,y) = \arctan[Y(x,y)/X(x,y)]$ be the angle made by the vector $(X(x,y),Y(x,y))$ with the horizontal. Then the (Poincaré) *index* of γ for the given vector field is

(7)
$$I(\gamma) = \frac{1}{2\pi} \oint_{\gamma} d\psi - 1$$

It is understood that γ is traversed counterclockwise.

Note that although the arctangent function is only defined up to an integral multiple of π, the integral in (7) is independent of which branch of this function is chosen, as well as of the initial point O from which the integral is computed. Further, one can prove that $I(\gamma)$ is an additive function of domains, in the following sense.

LEMMA 1.   *Let γ' and γ'' enclose domains D' and D'', whose union is a simply connected domain D with boundary γ, as in Figure B.1. Then*

(8)
$$I(\gamma) = I(\gamma') + I(\gamma'')$$

The proof depends on the fact that γ consists of $\gamma' \cup \gamma''$ with their common segment $\gamma' \cap \gamma''$ deleted; since this segment is traversed in opposite directions by γ' and γ'', the contributions from it cancel.

LEMMA 2.   *If a sufficiently small curve γ contains* no *critical point, then $I(\gamma) = 0$. If it contains a single critical point (x_j, y_j), and*

(9)
$$X(x,y) = a_j(x - x_j) + b_j(y - y_j) + O(r^2)$$
$$Y(x,y) = c_j(x - x_j) + d_j(y - y_j) + O(r^2)$$

where $r^2 = (x - x_j)^2 + (y - y_j)^2$, then

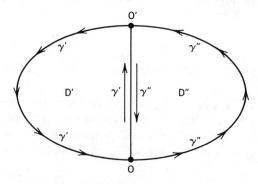

Figure B.1   Additivity of Poincaré index.

(9') $$I(\gamma) = \begin{cases} 1 & if \quad a_j d_j > b_j c_j \\ -1 & if \quad a_j d_j < b_j c_j \\ indeterminate & if \quad a_j d_j = b_j c_j \end{cases}$$

In other words, $I(\gamma)$ is the *sign* of the determinant $|a_j| = a_j d_j - b_j c_j$ of the matrix of the linearization of the DE $\dot{x} = X(x,y), \dot{y} = Y(x,y)$ unless this critical point (x_j, y_j) is *degenerate*, in the sense that $|A_j| = 0$. Getting down to cases, we see that focal, nodal, and vortex points have index $+1$, while saddle points have index -1.

The proof of Lemma 2, in the case of nondegenerate critical points, is straightforward but tedious. One first observes that, since the vector field $\mathbf{X}(\mathbf{x}) \in \mathcal{C}^2$, there is some neighborhood of (x_j, y_j) in which not only is there no other singular point, but the *direction* of $\mathbf{X}(\mathbf{x})$ differs by less than $\pi/4$ radians from that of the *linearized* vector field

$$\mathbf{X}_j(x,y) = (a_j(x - x_j) + b(y - y_j), c_j(x - x_j) + d(y - y_j))$$

One then takes up individually each of the non-degenerate cases of Figure 5.5; see the exercises at the end of this appendix.

The reason why the degenerate case $|A_j| = 0$ has been excluded in Lemma 2 is easily explained by examples. First, the degenerate field $Z(z) = z^n (n > 1)$ has index $I(\gamma) = n$ for any contour γ containing the origin. Thus $(x^2 - y^2, 2xy)$ has index 2, $(x^3 - 3xy^2, 3x^2y - y^3)$ has index 3, and so on. And again, the vector field $(x, x^4 \sin(1/x))$ has infinitely many critical points where the angle ψ is undefined, in any neighborhood of the origin. However, using Lemmas 1 and 2, we can prove the following Poincaré index theorem:

THEOREM 1. *Let γ be any simple closed curve not containing any degenerate critical point of the plane vector field $\mathbf{X}(\mathbf{x})$. Then γ contains only a finite number of critical points $\mathbf{x}_j = (x_j, y_j)$, and*

(10) $$I(\gamma) = \sum_j I_j = \sum_j \text{sgn} \begin{vmatrix} \partial X/\partial x & \partial Y/\partial y \\ \partial X/\partial y & \partial Y/\partial y \end{vmatrix}_j$$

is the sum of the (Poincaré) indices of these critical points.

3 HAMILTONIAN SYSTEMS

Classical dynamics (including celestial mechanics) is primarily concerned with systems of first-order DEs having a very special form. These are so-called *Ham-*

iltonian systems of DEs of the form

(11)
$$\dot{q_i} = \frac{\partial H}{\partial p_i}, \quad \dot{p_i} = -\frac{\partial H}{\partial q_i},$$

where $i = 1, \ldots, n$ and $H(\mathbf{q}; \mathbf{p})$ is a given real-valued *Hamiltonian* function.

Most of the energy-conserving autonomous systems arising in classical dynamics are "Hamiltonian," with $H = T + V$ the sum of the kinetic energy T and the potential energy $V(\mathbf{q})$. The q_i are *position* coordinates and the p_i the corresponding *momenta*. Thus, for the particle of mass m discussed in Chapter 5, §9, $q = x$, $p = m\dot{x}$, and $H = p^2/2m + V(q)$. Hence, $\dot{x} = \dot{q} = \partial H/\partial p = p/m$ (which checks), and $m\ddot{x} = \dot{p} = -\partial H/\partial q = -V'(x) = F(x)$, where $F(x) = -q(x)$ is the *force*.

Likewise, for the pendulum discussed in Chapter 5, §3, letting the (generalized) position coordinate be $\theta = q$, and $p = \ell\dot{\theta}$, we have the energy function or Hamiltonian

(12)
$$H = g\ell(1 - \cos\theta) + \ell^2\dot{\theta}^2/2$$
$$= 2g\ell \sin^2\frac{q}{2} + p^2/2 = V + T$$

formulas from which (11) can easily be checked.

Many other Hamiltonian systems having no clear connection with physics are also of current interest. An ingenious one-parameter family of such systems is the following.†

Example 1. Consider the one-parameter family of plane autonomous systems

(13)
$$\dot{x} = -\mu y + xy, \quad \dot{y} = \mu x + \tfrac{1}{2}(x^2 + y^2)$$

For any fixed μ, since $\dot{x} = 0$ when $(x - \mu)y = 0$, the critical points of (13) lie where $\dot{y} = 0$ on the lines $x = \mu$ and on $y = 0$. When $x = \mu$, $\dot{y} = 0$ where $x = 0$ or $x = -2\mu$; when $x = \mu$, $\dot{y} = \mu^2 + \tfrac{1}{2}(\mu^2 + y^2) = 0$ when $y = \pm\sqrt{3}\mu$.

Hence, if $\mu \neq 0$, the system (13) has four bifurcation points: one located at the origin, and the other three at the vertices $(-2\mu, 0)$ and $(\mu, \pm\sqrt{3}\mu)$ of an equilateral triangle. (When $\mu = 0$, there is only one critical point. This is at the origin, and is degenerate.)

One easily verifies that the system (13) *is* Hamiltonian because, for

(14)
$$V = -\frac{\mu}{2}(x^2 + y^2) + \frac{1}{2}\left(xy^2 - \frac{x^3}{3}\right)$$

---

† This example is taken from Sec. 1.8 of Guckenheimer-Holmes.

the DE (13) can be rewritten as

(14')
$$\dot{x} = \frac{\partial V}{\partial y}(x,y;\mu), \quad \dot{y} = -\frac{\partial V}{\partial x}(x,y;\mu)$$

which is of the form (12) in another notation.

We next generalize Theorem 1 of Chapter 5 from plane Hamiltonian systems to Hamiltonian systems in general.

THEOREM 2.   *The flow defined by any Hamiltonian system (11) is volume-conserving; moreover, the solution curves (orbits or trajectories) lie on level surfaces* $H(q;p) = C$.

The first statement follows from a basic theorem of vector analysis, which states that a velocity field $\mathbf{X}(\mathbf{x})$ defines a volume conserving flow $\mathbf{x}'(t) = \mathbf{X}(\mathbf{x})$ if and only if its divergence is zero.† For (11), evidently

(15)
$$\text{div } \mathbf{X} = \sum_{i=1}^{n} \frac{\partial x_i}{\partial x_i} = \sum_{i=1}^{n} \frac{\partial^2 H}{\partial q_i \partial p_i} + \sum_{i=1}^{n} \frac{\partial}{\partial p_i}\left(-\frac{\partial H}{\partial q_i}\right) = 0$$

The second statement follows since

$$\frac{d}{dt}[H(x(t))] = \sum_{i=1}^{2n} \frac{\partial H}{\partial x_i}\dot{x}_i = \sum_{i=1}^{n} \frac{\partial H}{\partial q_i}\dot{q}_i + \sum_{i=1}^{n} \frac{\partial H}{\partial p_i}\dot{p}_i$$
$$= \sum_{i=1}^{n} \frac{\partial H}{\partial q_i}\frac{\partial H}{\partial p_i} + \sum_{i=1}^{n} \frac{\partial H}{\partial p_i}\left(-\frac{\partial H}{\partial q_i}\right) = 0.$$

Critical Points.   By (11), the critical points of the Hamiltonian $H(\mathbf{p},\mathbf{q})$, where its gradient vanishes, are the points where the velocity vector $(\mathbf{q}'(t),\mathbf{p}'(t))$ is $\mathbf{0}$. This shows that the critical points of the function H are the stagnation points of the associated flow.

In the plane case $(n = 1)$, we can say more. Near any critical point (ξ,η), an expansion in Taylor series gives

(16) $2V(x,y) = a_{11}(x - \xi)^2 + 2a_{12}(x - \xi)(y - \nu) + a_{22}(y - \eta)^2 + \cdots$

where $a_{11} = V_{xx}$, $a_{12} = V_{xy}$, $a_{22} = V_{yy}$. Hence

(17)
$$\frac{d}{dt}\begin{pmatrix} x - \xi \\ y - \eta \end{pmatrix} = \begin{bmatrix} a_{11} & a_{22} \\ -a_{11} & -a_{22} \end{bmatrix}\begin{pmatrix} x - \xi \\ y - \eta \end{pmatrix} + \cdots$$

---

† Courant and John, vol. II, p. 602. The fact that the Hamiltonian flows are volume-conserving is called *Liouville's Theorem*.

The eigenvalues of the coefficient matrix of the linearization (17) at (x_j, y_j) are the roots of the quadratic equation

$$(18) \qquad \lambda^2 = a_{12}^2 - a_{11}a_{22} = - \det \begin{bmatrix} a_{11} & a_{12} \\ a_{12} & a_{22} \end{bmatrix}$$

THEOREM 3. *A nondegenerate critical point of a planar Hamiltonian system is a vortex point at maxima and minima of H, where $a_{11}a_{22} > q_{12}^2$, and a* saddle point *where H has a saddle point (where $a_{11}a_{22} < a_{12}^2$).*

Where $a_{11}a_{22} = a_{12}^2$, the local behavior of trajectories is indeterminate. Note that since the surface $z = H(x,y)$ is horizontal at critical points, $\det \begin{bmatrix} a_{11} & a_{12} \\ a_{12} & a_{22} \end{bmatrix}$ is precisely its *Gaussian curvature.*

4 HAMILTONIAN BIFURCATIONS

We now consider the bifurcations of one-parameter families of Hamiltonian systems, with Hamiltonians

$$(19) \qquad \dot{x} = \frac{\partial V}{\partial y}(x,y;\mu), \ \dot{y} = - \frac{\partial V}{\partial x}(x,y;\mu)$$

The fact (Theorem 1) that for any μ, the orbits of (19) are the level curves of the Hamiltonian $V(x,y;\mu)$, makes it easy to see how bifurcations arise. One can think of the orbits as the "shores" of the "lakes" obtained by deforming a flexible bowl $z = V(x,y;\mu)$ as the "time" μ varies.†

In Example 1, the critical points (singular points of the phase portraits or critical points $\nabla V = \mathbf{0}$ of V) move around as μ varies, changing their nature only at the "bifurcation value" $\mu = 0$. The next example is more typical of bifurcation, as that word is generally used.

Example 2. Consider the one-parameter family of systems defined by the Hamiltonians

$$(20) \qquad V = y - \mu e^{-r^2}, \ r^2 = x^2 + y^2$$

The corresponding Hamiltonian flows satisfy

$$(21) \qquad \dot{x} = 1 + 2\mu y e^{-r^2}, \ \dot{y} = - 2\mu x e^{-r^2}$$

---

† Of course, this "time" parameter μ is unrelated to the variable t associated with the "time" derivatives $\dot{x} = dx/dt$ and $\dot{y} = dy/dt$.

In terms of the simile in the preceding paragraph, the sloping plane $z = y$ becomes deformed by a deepening circular depression centered at the origin. When this depression becomes deep enough, the resulting dimple can hold water; at the same time, the retaining ridge on the downside has a saddle point. Thus this *bifurcation* gives rise to a *pair* of singularities: a vortex point and a related *saddle point*.

In detail, clearly $\dot{y} = 0$ for $r > 0$ in Example 2 if and only if $x = 0$—that is, on the y-axis. Moreover, at points $(0,\eta)$ on the y-axis, $\dot{x} = 0$ if and only if $e^{\eta^2} = 2\mu\eta$. Bifurcation takes place when the line $\xi = -2\mu\eta$ in the (η,ξ)-plane touches the graph of $\xi = e^{\eta^2}$. For larger slopes $\mu > \mu_o$, there are two points of intersection and the surface $z = V = y - \mu e^{-r^2}$ has corresponding critical points: a vortex point and a saddle point, as already stated.

Bifurcations involving the simultaneous appearance of such a paired vortex point and saddle-point are called *saddle-point* (or "fold") bifurcations, to distinguish them from *Hopf* bifurcations (see §1), in which a stable equilibrium point is replaced by an unstable one enclosed in a surrounding limit cycle.

The Poincaré Index.   Theorem 3 explains why the critical points arising from the preceding "bifurcation" are neither nodal points nor focal points. The fact that they are paired and of opposite indices is, however, a simple consequence of the Poincaré index theorem (Theorem 1), as we shall now explain.

For a general one-parameter family of plane autonomous systems,

$$(22) \qquad \dot{x} = X(x,y;\mu) \qquad \dot{y} = Y(x,y;\mu)$$

consider the variation with μ of the Poincaré index of a fixed simple closed curve γ, as defined by (7). Unless the curve passes through a critical point of the system (22) for some value of μ, the angle $\psi(\gamma,\mu)$ between the vector $(X(x,y;\mu),Y(x,y;\mu))$ and the horizontal will vary continuously with μ. Hence, the contour integral

$$I(\gamma,\mu) = \oint_{\gamma} d\psi(\gamma,\mu)$$

will also vary continuously, and *cannot jump* by an integral multiple of 2π. We conclude

THEOREM 4.   *Inside any simple closed curve not passing through a critical point, new nondegenerate critical points appearing or disappearing at bifurcation values μ must arise or be destroyed in adjacent pairs having indices of opposite signs.*

In particular, they cannot appear or disappear singly.

5   POINCARÉ MAPS

The concept of a Poincaré map arises in three different contexts: (i) limit cycles (Ch. 5, §13) and other periodic orbits of *autonomous systems*, (ii) *periodically forced* systems, of which linear constant-coefficient systems like

$$(26) \qquad\qquad \ddot{x} + p\dot{x} + qx = A \cos \omega t$$

yield familiar examples (in the phase plane), and (iii) systems with periodic coefficient-functions, like the phase plane representation

$$\dot{u} = v, \qquad \dot{v} = -p(t)u \qquad\qquad (27)$$

of the Hill equation mentioned in Appendix A.

Example 3. Consider plane autonomous systems of the special form

$$(28) \qquad \dot{x} = f(r)x - y, \quad \dot{y} = x + f(r)y, \quad r = \sqrt{x^2 + y^2}$$

In polar coordinates, the system (28) simplifies to

$$(28') \qquad\qquad \dot{\theta} = 1, \qquad \dot{r} = g(r) = rf(r)$$

and has a periodic orbit of radius $r = a$ and period 2π whenever $g(a) = 0$. This orbit is evidently *stable* (a "limit cycle") if $g'(a) < 0$, and *unstable* if $g'(a) > 0$.

The concept of a Poincaré map provides another way of visualizing perturbation approximations to such periodic orbits. Let us call a small interval $a - \epsilon < r < a + \epsilon$ of the r-axis $\theta = 0$ a *Poincaré section* of the periodic orbit $r = a$. The *Poincaré map* of this "section" maps each point $(a + \eta, 0)$ in it onto the point where the orbit (solution curve) passing through that point next crosses the section. If $f \in \mathcal{E}^1$, then $d\eta/d\theta = d\eta/dt = \eta f'(a) + O(\eta)$, and so the Poincaré map multiplies η by approximately $\exp(2\pi f'(a))$. It is thus locally a *contraction* if $f'(a) < 0$, but an *expansion* if $f'(a) > 0$; the periodic orbit is correspondingly stable or unstable.

More generally, by a *Poincaré section* of a periodic orbit at a point P is meant any smooth curve through P not tangent to the orbit, and the Poincaré map of this section is then defined (locally) as in the preceding paragraph. Using the theory of perturbations (Ch. 6, §12), one can show that if $f'(a) \neq 0$, the *linearization* $\eta \rightarrow \exp(2\pi f'(a))\eta$ is the same for all Poincaré sections—and that the Poincaré maps of the "sections" crossing a given periodic orbit are all "equivalent" under diffeomorphism (Ch. 5, §6).

Bifurcation. Now consider a one-parameter family $\dot{x} = f(r,\mu)x - y, \dot{y} = x + f(r, \mu)y$ of DEs of the form (28), with

$$(29) \qquad\qquad g(r,\mu) = rf(r, \mu) = -r^2 + \mu \sin^2 r$$

When $\mu = 0$, the origin is a stable equilibrium point toward which all orbits tend. But when μ exceeds 1, the origin is unstable, and a *limit cycle* occurs whenever $(\sin r)/r = 1/\mu$. The nearest limit cycle is locally stable; the next is unstable, the third is stable, and so on. Looking more closely at the intersections of the curve $s = (\sin r)/r$ with $s = 1/\mu$, we see that new periodic orbits arise in pairs as μ increases, one stable and the next unstable. The analogy with bifurcations of equilibrium points is obvious!

6 PERIODICALLY FORCED SYSTEMS

The Poincaré maps of sinusoidally forced (first-order) *linear constant-coefficient* systems are easily determined algebraically, by expressing in matrix notation the ideas introduced in Ch. 3, §7. Let

$$(30) \qquad\qquad \mathbf{x}'(t) = A\mathbf{x} + \mathbf{c}e^{\lambda t}, \quad \lambda = ik \ (k \text{ real})$$

be such a system. Then, unless $|\lambda I - A| = 0$ (i.e., unless λ is an eigenvalue of A, the case of *resonance*), the periodic solution is

$$(30') \qquad\qquad \mathbf{x} = (\lambda I - A)^{-1}\,\mathbf{c}e^{\lambda t} = \mathbf{C}e^{\lambda t}$$

where $\mathbf{C}$ is the solution of $(\lambda I - A)\mathbf{C} = \mathbf{c}$. If the eigenvalues λ_j of A are distinct, then the *section* $|\mathbf{y} - \mathbf{C}| < \epsilon$ surrounding the point $\mathbf{x} = \mathbf{C}$ through which the periodic orbit passes when $t = 0$ is transformed during the period $T = 2\pi/k$ into a neighborhood $|\mathbf{z} - \mathbf{C}| < \epsilon$ of $\mathbf{x}(T) = \mathbf{C}$, by the formula

$$(31) \qquad\qquad \mathbf{y} = \mathbf{C} + \Sigma a_j\phi_j, \qquad \mathbf{z} = \mathbf{C} + \Sigma e^{\lambda_j T}a_j\phi_j$$

Just as in Ch. 4, §7, this periodic orbit is *strictly stable* if an only if every eigenvalue λ_j of A has a negative real part, so that $|e^{\lambda_j T}| < 1$.

Periodic Linear Systems. More generally, the periodic, homogeneous linear systems

$$(32) \qquad\qquad \mathbf{x}'(t) = A(t)\mathbf{x}, \qquad A(t + T) = A(t)$$

considered in Appendix A, §4, also give rise to Poincaré maps. Indeed, every such system has the trivial *equilibrium* periodic solution $\mathbf{x}(t) \equiv \mathbf{0}$, and the linear transformation $\mathbf{x}(t) = M(T)\,\mathbf{x} \equiv \mathbf{x}(t + T)$ defined by the matrizant (Floquet matrix) referred to in Theorem 6 of Appendix A is the Poincaré map associated with this equilibrium solution. Clearly, the equilibrium solution ("equilibrium") is strictly *stable* if and only if all the eigenvalues λ_j of this Floquet matrix have magnitudes $|\lambda_j| < 1$.

Example 5. Consider for example the Mathieu equation of Ch. 10, §1, Example 4:†

$$(33) \qquad\qquad u'' + (\mu + 16\,d\cos 2x)u \equiv 0$$

---

† We have written μ in place of λ, to avoid confusion with the eigenvalues of $(33'')$.

The initial conditions

(33')
$$\phi(0) = 1, \qquad \phi'(0) = 0$$
$$\psi(0) = 0, \qquad \psi'(0) = 1$$

determine a basis of solutions of (33), the first of which is an even function and the second odd. Moreover, the associated Floquet matrix has determinant one (see Ex. 8 of Appendix A). Hence its eigenvalues are the roots of a real characteristic equation of the form

(33'')
$$\lambda^2 - 2B\lambda + 1 = 0$$

with roots $\lambda_j = B \pm \sqrt{B^2 - 1}$. The parameter $B = B(\mu,d)$ in (33'') must be computed.

LEMMA   *The Mathieu equation* (33) *is stable or unstable according as* $|B| < 1$ *or* $|B| > 1$.

For, since $\lambda_1\lambda_2 = 1$, they are complex conjugate and on the unit circle if $|B| < 1$, but real and one exceeding one if $|B| > 1$. On the other hand, the values $B = \pm 1$ yield the eigenvalues $\lambda_j = \pm 1$ associated with the *Mathieu functions* of periods π and 2π. There follows:

THEOREM 5.   *The bifurcation associated with transition between stability and instability occurs at values* μ *and* d *associated with the periodic Mathieu function solutions of* (33).

Duffing's Equation.   More interesting, and much harder to understand theoretically, are the Poincaré maps of periodically forced *nonlinear* systems. A good introduction to these is provided by Duffing's equation

(34)
$$\ddot{x} + c\dot{x} - x + x^3 = A \cos \omega t.$$

In the *unforced* case $A = 0$, (34) evidently has three equilibrium solutions: $x(t) \equiv 0$ and $x(t) = \pm 1$. Of these, $x = 0$ is unstable, while $x = \pm 1$ are stable. The trajectories in the phase plane are easily drawn in the Hamiltonian case $A = c = 0$: they are the level curves of the energy function $H = x^4 - 2x^2 + 2v^2$, with "separatrices" $v = \pm x\sqrt{1 - (x^2/2)}$ through the origin. These separatrices separate the periodic orbits corresponding to local oscillations about one stable equilibrium point from the oscillations about all three, whose amplitudes exceed $2\sqrt{2}$. The tangents to all trajectories in the phase plane are horizontal where they cross the vertical lines $x = 0$ and $x = \pm 1$.

When $A = 0$ but $c > 0$, most trajectories in the (x,v)-plane spiral clockwise from "infinity" into the two stable *focal points* at $(\pm 1,0)$. There are, however, two special trajectories which originate (at $t = -\infty$) in the unstable *saddle point* at $(0,0)$, each of which spirals into one of the focal points at $(\pm 1,0)$. Likewise, there are two special *separating* trajectories which spiral in from "infinity" but come to rest at $(0,0)$. These separate the trajectories which spiral into $(1,0)$ from those which spiral into $(-1,0)$.

In the *forced* case $A \neq 0$, a rich variety of qualitatively different kinds of behavior can arise, depending on the choice of the three coefficients c, A, and ω in (34). These can be explored most efficiently by using modern computers to compute and display the sequences of points $(x(nT),\dot{x}(nT))$ in the phase plane arising by iterating the Poincaré map, applied to selected initial states $(x_o,\dot{x}_o)$ for selected choices of c, A, and ω.

Of particular interest here are the fixed points of the Poincaré map, that is, points such that $(x(T),\dot{x}(T)) = (x(0),\dot{x}(0))$, on periodic orbits of period T. For example, with small A, there are three such fixed points for (34): two stable fixed points (on stable periodic orbits) near $(\pm 1,0)$, and an unstable fixed point near the origin. A family of trajectories asymptotic to the unstable fixed point forms a manifold separating trajectories attracted to the two stable periodic orbits.

EXERCISES

1. Explain why, for fixed $q > 0$, the value $p = 0$ is a "bifurcation value" of p in (5).

2. (a) Show that for given A and B, $(A, B/A)$ is the only equilibrium point of (6).
   (b) Derive the system (6′) of variational equations for perturbations of the equilibrium solution.

*3. Compute numerically the stable limit cycle of (6) for $A = 1$, $B = 3$.

4. (a) Show that the system $\dot{x} = 2y$, $\dot{y} = 3 - 3x$ leaves invariant the energy (Hamiltonian) function $V = x^3 - 3x + y^2$.
   (b) Show that this system has a saddle point at $(1,0)$, a vortex point at $(-1,0)$, and no other critical points.
   (c) Show that the cubic curve $x^3 - 3x + y^2 = -2$ through the saddle point is a "separatrix."
   (d) Sketch the phase portrait of the system.

5. (a) Show that $\mu = 0$ is a bifurcation value for the one-parameter family of Hamiltonian systems with Hamiltonians $H = y^2 + 3\mu x^2 = 3x$.
   (b) Describe the qualitative change that takes place in the phase portrait when μ changes sign.

   [HINT: The case $\mu = 1$ is treated in Ex. 4.]

6. (a) For the Hamiltonian $H = \mu y - e^{-r^2/2}$, $r^2 = x^2 + y^2$, write down the Hamiltonian DEs.
   (b) Locate the critical points of the system, if any.
   (c) Show that the bifurcation values are $\mu = \pm\sqrt{e}$, and describe the changes in the phase portrait that take place when μ crosses these values.

7. As $\epsilon \to 0$, the DE $\epsilon u'' + u = 1$ is said to constitute a singular perturbation of $u = 1$.
   (a) Solve this DE explicitly for the endpoint conditions $u(0) = u(1) = 0$, when $\epsilon > 0$ and when $\epsilon < 0$.
   (b) Contrast the increasingly irregular behavior of solutions as $\epsilon \downarrow 0$ with the smooth "boundary layer" behavior as $\epsilon \uparrow 0$.

BIBLIOGRAPHY

GENERAL REFERENCES

Abramowitz, Milton, and I. A. Stegun, *Handbook of Mathematical Functions.* New York: Dover, 1965.

Ahlfors, L. V., *Complex Analysis,* 2nd ed. New York: McGraw-Hill, 1966.

Apostol, Tom M., *Calculus,* 2 vols., 2nd ed. New York: Blaisdell, 1967, 1969.

Birkhoff, Garrett, and S. MacLane, *A Survey of Modern Algebra,* 4th ed. New York: Macmillan, 1977.

Carrier, George F., Max Krook, and C. E. Pearson, *Functions of a Complex Variable.* New York: McGraw-Hill, 1966.

Courant, Richard, and D. Hilbert, *Methods of Mathematical Physics, Vol. 1.* New York: Wiley-Interscience, 1953.

Courant, Richard, and F. John, *Introduction to Calculus and Analysis,* 2 vols. New York: Wiley, 1965, 1972.

Dwight, H. B., *Tables of Integrals and Other Mathematical Data,* rev. ed. New York: Macmillan, 1947.

Fletcher, A., J. C. P. Miller, and L. Rosenhead, *Index of Mathematical Tables, Vol. I and II,* 2nd ed. Reading, Mass.: Addison-Wesley, 1962.

Hildebrand, Francis B., *Introduction to Numerical Analysis,* 2nd ed. New York: McGraw-Hill, 1974.

Hille, Einar, *Analytic Function Theory, Vol. 1.* Waltham, Mass.: Blaisdell, 1959; *Vol. 2,* 1962.

Kaplan, Wilfred, *Advanced Calculus.* Reading, Mass.: Addison-Wesley, 1952.

Picard, E., *Traité d'Analyse,* 3 vols., 2nd ed. Paris: Gauthier-Villars, 1922–1928.

Rudin, Walter, *Principles of Mathematical Analysis,* 2nd ed. New York: McGraw-Hill, 1964.

Stiefel, E. L., *An Introduction to Numerical Mathematics,* translated by W. C. Rheinboldt. New York: Academic Press, 1963.

Taylor, A. E., *Advanced Calculus.* Waltham, Mass.: Blaisdell, 1955.

Widder, David V., *Advanced Calculus,* 2nd ed. Englewood Cliffs, N.J.: Prentice-Hall, 1961.

WORKS ON ORDINARY DIFFERENTIAL EQUATIONS

Arnold V. I., *Ordinary Differential Equations*, translated by R. A. Silverman. Cambridge: M.I.T. Press, 1973.

Bellman, Richard, and Kenneth L. Cooke, *Differential-Difference Equations*. New York: Academic Press, 1953.

Cesari, L., *Asymptotic Behavior and Stability Problems in Ordinary Differential Equations*, 2nd ed. New York: Academic Press, 1963.

Chow, S.-N., and J. R. Hale, *Methods in Bifurcation Theory*. New York: Springer, 1982.

Coddington, Earl A., and N. Levinson, *Theory of Ordinary Differential Equations*. New York: McGraw-Hill, 1955.

Cronin, Jane S., *Differential Equations: Introduction and Qualitative Theory*. Dekker, 1981.

Davis, Philip J., and Philip Rabinowitz, *Numerical Integration*. Blaisdell, 1966.

Gear, C. William, *Numerical Initial Value Problems in Ordinary Differential Equations*. Prentice-Hall, 1971.

Guckenheimer, John, and Philip Holmes, *Nonlinear Oscillations, Dynamical Systems, and Bifurcations of Vector Fields*. New York: Springer, 1983.

Hale, Jack K., *Ordinary Differential Equations*. New York: Wiley, 1969.

Hartman, Philip, *Ordinary Differential Equations*. New York: Wiley, 1964.

Henrici, Peter, *Discrete Variable Methods in Ordinary Differential Equations*. New York: Wiley, 1961.

Henrici, Peter, *Error Propagation for Difference Methods*. New York: Wiley, 1963.

Hille, Einar, *Lectures on Ordinary Differential Equations*, Reading, Mass.: Addison-Wesley, 1969.

Hirsch, M. W., and S. Smale, *Differential Equations, Dynamical Systems, and Linear Algebra*. New York: Academic Press, 1974.

Hurewicz, Withold, *Lectures on Ordinary Differential Equations*. Cambridge: M.I.T. Press, 1958.

Ince, E. L., *Ordinary Differential Equations*, 4th ed. New York: Dover, 1953.

Jordan, D. W., and P. Smith, *Nonlinear Ordinary Differential Equations*. Oxford: Oxford University Press, 1977.

Kamke, E., *Differentialgleichungen: Lösungsmethoden und Lösungen*. Leipzig: Akademische Verlag, 1943; Chelsea reprint, 1971.

Kaplan, Wilfred, *Ordinary Differential Equations*. Reading, Mass.: Addison-Wesley, 1958.

LaSalle, J. P., and S. Lefschetz, *Stability by Liapounov's Direct Method, with Applications*. New York: Academic Press, 1961.

Lefschetz, Solomon, *Differential Equations, Geometric Theory*, 2nd ed. New York: Wiley-Interscience, 1963.

Liapounoff, A., *Problème Générale de la Stabilité de Mouvement*, reprinted by Princeton University Press, 1949.

Magnus, Wilhelm, and S. Winkler, *Hill's Equation*. New York: Wiley-Interscience, 1966.

McLachlan, N. W., *Ordinary Non-Linear Differential Equations in Engineering and Physical Sciences*, 2nd ed. New York: Oxford University Press, 1956.

Nemytskii, V. V., and V. V. Stepanov, *Qualitative Theory of Differential Equations.* Princeton, N.J.: Princeton University Press, 1960.

Petrowski, I. G., *Ordinary Differential Equations.* New York: Prentice-Hall, 1966.

Picard, E., *Lecons sur Quelques Problèmes aux Limites de la Théorie des Equations Différentielles.* Paris: Gauthier-Villars, 1930.

Pliss, V. A., *Nonlocal Problems of the Theory of Oscillations.* New York: Academic Press, 1966. (Russian ed., 1964.)

Poincaré, H., *Les Méthodes Nouvelles de la Mécanique Céleste, Vol. I–III.* New York: Dover, 1957.

Reid, W. T., *Ordinary Differential Equations.* New York: Wiley, 1971.

Sansone, G., and R. Conti, *Non-Linear Differential Equations.* New York: Pergamon, 1952.

Simmons, G. F., *Differential Equations with Applications and Historical Notes.* New York: McGraw-Hill, 1972.

Stoker, J. J., *Nonlinear Vibrations in Mechanical and Electrical Systems.* New York: Wiley-Interscience, 1950.

Thompson, J. M. T., and H. B. Stewart, *Nonlinear Dynamics and Chaos.* New York: Wiley, 1986.

Tricomi, F. G., *Equazioni Differenziali,* 3rd ed. Turin: Einaudi, 1961.

Wasow, W., *Asymptotic Expansions for Ordinary Differential Equations.* New York: Wiley-Interscience, 1966.

INDEX